Published By

Oxford Book Company

Regd. Off. "Bony Residency", Gate No. 2
Opp. Tilak Public School, Vishweswariya Nagar,
Gopalpura Road, Jaipur - 302018, India
Ph: 0141 - 2761280, Telefax: 2761381
email - oxfordbookjpr@gmail.com
website - oxfordbookcompany.com

Delhi Office: 102, 1st Floor, Satyam House,
4327/3, Ansari Road, Darya Ganj,
New Delhi - 110002, India
Ph: 011-45652440

Distributed by

KEYA INTERNATIONAL PRESS LLP
E-mail: keyainternationalpressllp@gmail.com

Published in India, 2025

ISBN : 9789380179216

Typeset by Shivangi Computers, Jaipur

Printed and bound in India by Replika Press Pvt. Ltd.

INTRODUCTORY MICROBIOLOGY

Uma Shankar Singh
Kiran Kapoor

Oxford Book Company
Jaipur ▪ New Delhi

Introductory Microbiology

Contents

1

Introduction to Microbiology

Microbiology is the study of microorganisms, which are unicellular or cell-cluster microscopic organisms. This includes eukaryote such as fungi and protists, and prokaryotes, which are bacteria and archaea. Viruses, though not strictly classed as living organisms, are also studied. In short; microbiology refers to the study of life and organisms that are too small to be seen with the naked eye. Microbiology is a broad term which includes virology, mycology, parasitology, bacteriology and other branches. A microbiologist is a specialist in microbiology.

Microbiology is researched actively, and the field is advancing continually. Although microbes were first observed over three hundred years ago, the field of microbiology can be said to be in its infancy relative to older biological disciplines such as zoology and botany. Microbiology could be defined as the study of organisms too small to be seen with the naked eye. Figure 1 shows the relative size of microbes compared to other living things. However, the recent discovery of bacteria of near 1 mm in size has made this definition somewhat inaccurate and in the grand tradition of science, a new definition is in order.

Though microbes are small, they nevertheless span a large range of sizes from the smallest bacterial cells at ~0.15 μm to giant bacteria larger than 700 μm. The viruses depicted at the far left of the scale are even smaller. We will consider microbiology to be the study of organisms that can exist as single cells, contain a nucleic acid genome for at least some part of their life cycle, and are capable of replicating that genome. This broad description encompasses an understandably large group of organisms including fungi, algae, protozoa and bacteria.

Many different organisms fall under the definition of microorganisms. Shown here are: A, the bacterium Escherichia coli; B, a photosynthetic cyanobacterium; C, a fungus; D, Ebola virus; E, the protozoan malaria parasite. Microbiology also involves a collection of techniques to study and manipulate these small creatures. Because of their size, special instruments and methods had to be developed to allow the performance of interpretable experiments on microorganisms.

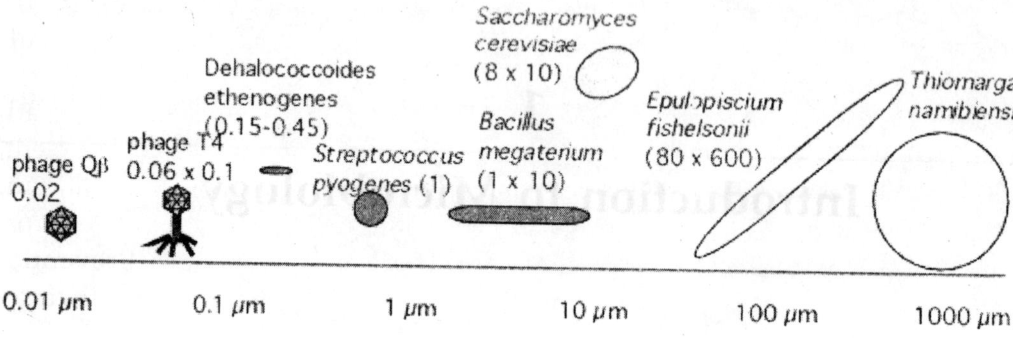

Figure 1. The Relative size of Microbes

These methods are not restricted to microbes alone, but have also found utility in working with populations of cells from higher organisms. With apologies to other small organisms, this book will mostly focus on bacteria and their impact on the rest of the biosphere. This can be weakly justified by the fact that bacteria have a major impact on the world around us and, because of their perceived importance, more research and knowledge has been accumulated about them.

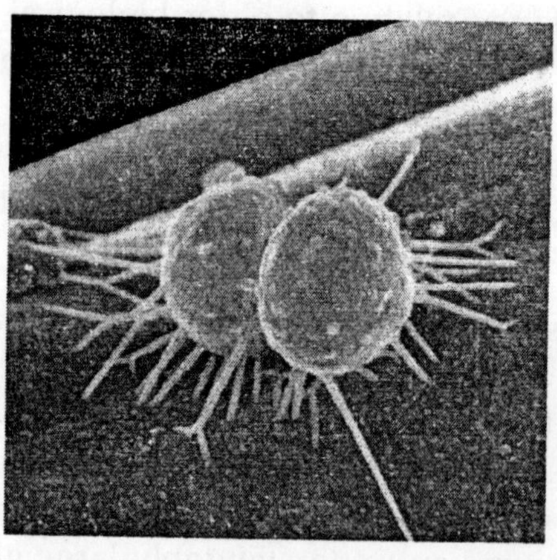

Figure 2. Photosynthetic cyanobacteria

Microorganisms are everywhere, but why are they worth learning about? The short answer is that they affect your life in many different ways. Before we begin our study of these creatures, we will first take a tour of some of their important habitats and point out why your existence depends upon them. We will then briefly explore the history of microbiology.

MICROBES HAVE A LARGE IMPACT ON HUMAN HEALTH

If you ask the average person how microbes (or germs) impact their lives, they would immediately think of disease. This is not a silly view, as Figure 3 shows a number of important pathogens. While death from infectious disease in the U.S. has been greatly diminished, infection rates in developing nations remain unacceptably high. "Ancient" diseases continue to be a problem where nutrition and sanitation are poor, and emerging diseases such as Acquired Immunodeficiency Syndrome (AIDS) are even more dangerous for such populations. The Centers for Disease Control and Prevention estimate that about 9% of adults between the ages of 18-49 in Sub-Saharan Africa are infected with HIV.

Many of the new diseases are viral in nature, making them notoriously difficult to treat and they have no known cure. Influenza and pneumonia are leading killers of the elderly. Even the common cold causes illness and misery for almost everyone and drains the productivity of all nations. Disease due to food-borne pathogens also remain a problem, largely because of consumption of improperly processed or stored foods. Understanding the sources of contamination and developing ways to limit the growth of pathogens in food is the job of food microbiologists.

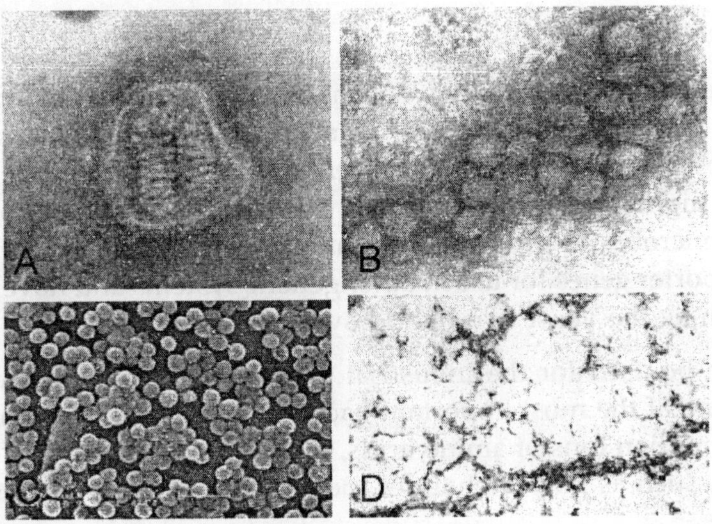

Figure 3. Some Important Pathogens

New infections continually appear. Having an available food source to grow on (humans) inevitably results in a microorganism that will take advantage. Some of these feeders will interfere with our own well being, causing disease. Surprisingly, many diseases that were previously thought to have only behavioral or genetic components have been found to involve microorganisms.

The clearest case is that of ulcers, which was long thought to be caused by stress and poor diet. However the causative agent is actually a bacterium, *Helicobacter pylori,* and many ulcers can be cured with appropriate antibiotics. Work on other non-infectious diseases such as heart disease, stroke and some autoimmune diseases also suggest a microbial component that triggers the illness.

Finally, some pathogenic microbes that had been "controlled" through the use of antibiotics are beginning to develop drug resistance and therefore reemerge as serious threats in the industrialised world as well as developing nations. Tuberculosis is an illness that was on the decline until the middle 80's. It has recently become more of a problem, partly due to drug resistance and partly due to a higher population of immunosuppressed individuals from the AIDS epidemic.

Staphylococcus aureus strains are emerging that are resistant to many of the antibiotics that were previously effective against them. These staph infections are of great concern in hospital settings around the world. Understanding both familiar killers and new pathogens will require an understanding of their biology, and thus an understanding of the field of microbiology.

BENEFITS OF MICROBIOLOGY

Whilst there are undoubtedly some who fear all microbes due to the association of some microbes with many human illnesses, many microbes are also responsible for many beneficial processes such as industrial fermentation, antibiotic production and as vehicles for cloning in higher organisms such as plants. Scientists have also exploited their knowledge of microbes to produce biotechnologically important enzymes such as Taq polymerase, reporter genes for use in other genetic systems and novel molecular biology techniques such as the yeast two-hybrid system.

Bacteria can be used for the industrial production of amino acids. *Corynebacterium glutamicum* is one of the most important bacterial species with an annual production of more than two million tons of amino acids, mainly L-glutamate and L-lysine. A variety of biopolymers, such as polysaccharides, polyesters, and polyamides, are produced by microorganisms.

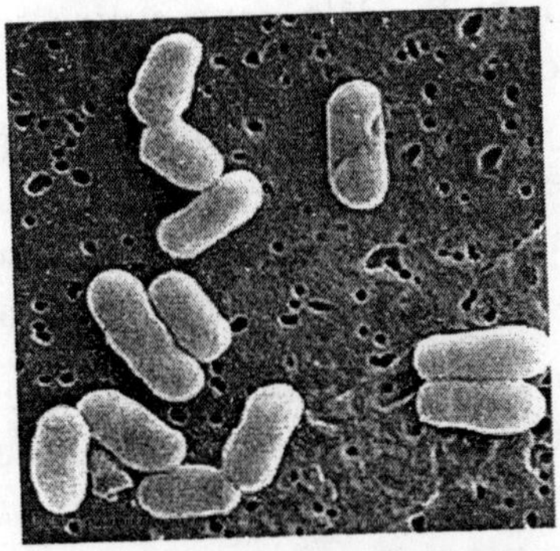

Figure 4. Corynebacterium glutamicum

Microorganisms are used for the biotechnological production of biopolymers with tailored properties suitable for high-value medical application such as tissue engineering and drug delivery. Microorganisms are used for the biosynthesis of xanthan, alginate, cellulose, cyanophycin, poly(gamma-glutamic acid), levan, hyaluronic acid, organic acids, oligosaccharides and polysaccharide, and polyhydroxyalkanoates. Microorganisms are beneficial for microbial biodegradation or bioremediation of domestic, agricultural and industrial wastes and subsurface pollution in soils, sediments and marine environments.

The ability of each microorganism to degrade toxic waste depends on the nature of each contaminant. Since most sites are typically comprised of multiple pollutant types, the most effective approach to microbial biodegradation is to use a mixture of bacterial species and strains, each specific to the biodegradation of one or more types of contaminants. There are also various claims concerning the contributions to human and animal health by consuming probiotics and/or prebiotics.

Recent research has suggested that microorganisms could be useful in the treatment of cancer. Various strains of non-pathogenic clostridia can infiltrate and replicate within solid tumors. Clostridial vectors can be safely administered and their potential to deliver therapeutic proteins has been demonstrated in a variety of preclinical models.

Microbes are often Helpful

From the beginning of microbiology, significant resources have been spent to understand and fight disease-causing microorganisms. You may be surprised to learn that only a small fraction of microbes are involved in disease; many other microbes actually enhance our well being. The harmless microbes that live in our intestines and on our skin actually help us fight off disease. They actively antagonise other bacteria and take up space, preventing potential pathogens from gaining a foothold on our bodies.

The microbial community in humans not only protects us from disease, but also provides needed vitamins, such as B_{12}. We have entire communities of microorganisms in our digestive systems that contribute to our overall health. In fact, like all other large organisms, humans are actually consortia of different organisms — there are more non-human cells in and on our bodies than there are human cells! Human health and nutrition also depends on healthy farm animals. Cows, sheep, horses and other ruminant animals utilise their microbial associates to degrade plant material into useful nutrients.

Commercial crops are also central to human prosperity, and much of agriculture depends upon the activities of microbes. For example, an entire group of plants, the legumes, forms a cooperative relationship with certain bacteria. These bacteria convert nitrogen gas to ammonia for the plant, an important nutrient that is often limiting in the environment. Microbes also serve as small factories, producing valuable products such as cheese, yogurt, beer, wine, organic acids and many other items. In conclusion, while it is less apparent to us, the positive role of microbes in human health is at least as important as the negative impact of pathogens.

Microbes have Profound Effects on the Environment

Whether measured by the number of organisms or by total mass, the vast majority of life on this planet is microscopic. These teaming multitudes profoundly influence the make-up and character of the environment in which we live. Presently, we know very little about the microbes that live in the world around us because less than 2 % of them can be grown in the laboratory. Understanding which microbes are in each ecological niche and what they are doing there is critical for our understanding of the world.

Microbes are the major actors in the synthesis and degradation of all sorts of important molecules in environments. Cyanobacteria and algae in the oceans are responsible for the majority of photosynthesis on Earth. They are the ultimate source of food for most ocean creatures (including whales) and replenish the world's oxygen supply. Cyanobacteria also use carbon dioxide to synthesize all of their biological molecules and thus remove it from the atmosphere. Since carbon dioxide is a major greenhouse gas, its removal by cyanobacteria affects the global carbon dioxide balance and may be an important mitigating factor in global warming.

In all habitats, microorganisms make nutrients available for the future growth of other living things by degrading dead organisms. Microbes are also essential in treating the large volume of sewage and wastewater produced by metropolitan areas, recycling it into clean water that can be safely discharged into the environment. Less helpfully (from the view of most humans), termites contain microorganisms in their guts that assist in the digestion of wood, allowing the termites to extract nutrients from what would otherwise be indigestible.

Understanding of these systems helps us to manage them responsibly and as we learn more we will become ever more effective stewards. Energy is essential for our industrial society and microbes are important players in its production. A significant portion of natural gas comes from the past action of methanogens (methane-producing bacteria). Numerous bacteria are also capable of rapidly degrading oil in the presence of air and special precautions have to be taken during the drilling, transport and storage of oil to minimise their impact.

In the future, microbes may find utility in the direct production of energy. For example, many landfills and sewage treatment plants capture the methane produced by methanogens to power turbines that produce electricity. Excess grain, crop waste and animal waste can be used as nutrients for microbes that ferment this biomass into methanol or ethanol. These biofuels are presently added to gasoline and thus decreasing pollution. They may one day power fuel cells in our cars, causing little pollution and having water as their only emission.

Finally, we are increasingly taking advantage of the versatile appetite of bacteria to clean up environments that we have contaminated with crude oil, polychlorinated biphenyls (PCBs) and many other industrial wastes. This process is termed bioremediation and is a cheap and increasingly effective way of cleaning up pollution.

Microbes Helps us to Understand the World around us

Microorganisms used in research have many useful properties. They grow on simple, cheap medium and often give rise to large populations in a matter of 24 hours. It is easy to isolate their genomic material, manipulate it in the test tube and then place it back into the microbe. Due to their large populations it is possible to identify rare events and then, with the use of powerful selective techniques, isolate interesting bacterial cells and study them. These advantages have made it possible to test hypotheses rapidly. Using microbes scientists have expanded our knowledge about life.

Microorganisms have been indispensable instruments for unlocking the secrets of life. The molecular basis of heredity and how this is expressed as proteins was described through work on microorganisms. Due to the similarity of life at the molecular level, this understanding has helped us to learn about all organisms, including ourselves. Some prokaryotes are capable of growing under unimaginably harsh conditions and define the extreme limits of where life can exist.

Some species have been found growing at near 100 °C in hot springs and well above that temperature near deep-sea ocean vents. The ability of microbes to live under such extreme conditions is forcing scientists to rethink the requirements necessary to support life. Many now believe it is entirely possible that Jupiter's moon Europa may harbor living communities in waters deep below its icy crust. What may the rest of the universe hold?

Until recently, while we could study specific types of bacteria, we lacked a cohesive classification system, so that we could not readily predict the properties of one species based on the known properties of others. Visual appearance, which is the basis for classification of large organisms, simply does not work with many microbes because there are few distinguishing characteristics for comparison between species even under the microscope.

However, analysis of their genetic material in the past 20 years has allowed such classification and spawned a revolution in our thinking about the evolution of bacteria and all other species. The emergence of a new system organising life on Earth into three domains is attributable to this pioneering work with microorganisms. The fruits of basic research on microbes have been used by scientists to understand microbial activity and therefore to shape our modern world. Human proteins, especially hormones like insulin and human growth factor, are now produced in bacteria using genetic engineering. Our understanding of the immune system was developed using microbes as tools.

Microorganisms also play a role in treating disease and keeping people healthy. Many of the drugs available to treat infectious disease originate from bacteria and fungi. Lastly, microbes have informed us about our world through the tools they provide for molecular biology. Enzymes purified from bacterial strains are useful as tools to perform many types of analyses. Such analyses allow us to determine the complete genome sequence of almost any organism and manipulate that DNA in useful ways. We now know the entire sequence of the human genome, with the exception of regions of repetitive DNA, and this will hopefully lead to medical practices and treatments that improve health. We also know the entire genome sequences of hundreds of microbes, including those of many important pathogens.

Analysis of these data will eventually lead to an understanding of the function of critical enzymes in these microbes and the development of tailor-made drugs to stop them. The tools of molecular biology will also affect agriculture. For example, we now know the complete genome sequence of the plant *Arabidopsis*. This opens a new avenue to a better understanding of all plants and hopefully improvements in important crops.

Microbes have a profound impact on every facet of human life and everything around us. Pathogens harm us, yet other microbes protect us. Some microbes are pivotal in the growth of food crops, but others can kill the plants or spoil the produce. Bacteria and fungi eliminate the wastes produced in the environment, but also degrade things we would rather preserve. Clearly they affect many things we find important as humans. In the remainder of this how scientists came to be interested in microbes and follow a few important developments in the history of microbiology.

HISTORY OF MICROBIOLOGY

The history of microbiology, like all human history, is not a catalog of linear progress, but is more of an interweaving of the careers of bright individuals and their insights. Each new discovery relied on previous ones and in turn spawned further inquiry. A web of interdependent concepts evolved over time through the work of scientists in many related disciplines and nations. Often the research of one individual impacted the efforts of another studying a completely different problem. Keep this in mind as you look at this history.

The science of microbiology had its most significant early impact on human health, uncovering the cause of the major killers of the day, and then methods to treat them. As microbiology matured, scientists began to look at what non-pathogenic microbes were doing in the environment and we will look a bit at the history of general microbiology.

Finally, this will end with an examination of the events that lead to the understanding of life at the molecular level and the profound impact this has had on microbiology and on society in general. For years the existence of microorganisms was suspected, but could not be proven, since bacteria were too small to be seen with the naked eye. It took the microscope to expose their tiny world and that instrument has been linked to microbiology ever since. In 1664, Robert Hooke devised a compound microscope and used it to observe fleas, sponges, bird feathers, plants and molds, among other items.

Several years later Anton van Leeuwenhoek, a fabric merchant and amateur scientist, became very adept at grinding glass lenses to make telescopes and microscopes. They also produced clearer images than the compound microscopes of the time. Human societies had neither the technical prowess nor the inclination to develop the science of microorganisms. It was not until the rise of the industrial revolution that governments and people dedicated the financial and physical resources to understand these small inhabitants of our world.

With the development of better microscopes in the 19th century, scientists returned to an examination of microorganisms. After finishing his education, Ferdinand Julius Cohn was able to convince his father to lay down the large sum necessary to purchase a microscope for him, one better than that available at the University in Breslau, then part of Germany. He used it to carefully examine the world of the microbe and made many observations of eukaryotic microorganisms and bacteria. Cohn's work with microscopes popularised their use in microbiology. This and his other work inspired many other scientists to examine microbes. Cohn's encouragement of Robert Koch, a German physician by training, began the field of medical microbiology.

Development of Microbiological Techniques

A major contribution to bacterial techniques was the development of methods using solid medium for the cultivation of bacteria. Koch was convinced that microbes caused some diseases. However, to test this idea, he needed to isolate the causative agent. Almost all samples from diseased animals or any natural surface contained many different microbes and it was impossible to tell which one was the problem. A method was needed to separate these different bacteria.

The most common method of isolation was to continually dilute a sample in liquid broth in hopes that at high enough dilution, only one type of microbe would be found. A problem with this method is that only the most populous microbe would be isolated, but that might not be one causing the disease. There were other technical problems as well with such a liquid-based system, so a solid medium would seem to provide distinct advantages. Koch had tried gelatin for these experiments with unsatisfactory results. Building on the work of Brefeld and Schroeter, Koch used potato slices as a solid medium and observed that a boiled potato left in the open air would develop tiny circular raised spots.

Examination of these spots revealed they were made up of microorganisms and each spot had just one type of microbe in it. He realised that these colonies were pure cultures of bacteria and probably arose from a single species of microbe from the air that landed on the potato. By boiling a potato, slicing it with a hot knife and keeping it in a sterile container with a lid, Koch could keep the potato sterile. But if a sample from a disease animal was smeared across the potato, colonies arose, each being pure isolates from the animal. By then testing these isolates in animals, Koch was able to isolate the cause of anthrax, *Bacillus anthracis*.

Potatoes failed to support the growth of many microorganisms and Koch and his laboratory were constantly frustrated by the lack of a good solid medium. Walter Hesse joined Koch's laboratory to do studies on air quality, showing a remarkable attention to detail and patience in his work. His wife, Angelina Fannie Hesse along with raising their three sons, also would assist her husband with his research in the laboratory. Walter was attempting to do his air quality experiments using medium containing gelatin as the solidifying agent.

In the summertime, temperatures would often rise above the melting point of gelatin. In addition, microbes would often grow in the cultures that were capable of degrading gelatin and in both cases this would cause liquefaction of the medium, ruining the experiments. One day while eating lunch, the frustrated scientist asked Lina why her jellies and puddings stayed solid even in the hot summer temperatures. She told him about agar-agar, a heat resistant gelling agent that she had learned about while growing up in New York from a Dutch neighbor who had emigrated from Java.

Development of the new agent by Angelina and Walter led to a resounding success. Few microbes are able to degrade agar and it melts at 100 °C yet remains molten at temperatures above 45 °C. This allows the mixing of the agar with heat-sensitive nutrients and microbes. After solidification, it does not melt until a temperature of 100 °C is again attained, facilitating the easy cultivation of pathogens. It can also be stored for long periods of time, allowing the cultivation of slow-growing microorganisms.

Any type of broth can be mixed with agar, giving great flexibility in the kinds of medium that can be made. Thus, many more types of microbes could be cultivated. Koch's laboratory also developed methods of pure culture maintenance and aseptic technique. Aseptic technique involves the manipulation of pure cultures in a manner that prevents their contamination by outside microorganisms.

Equally important, aseptic technique prevents their spread into the environment. Remember that Koch was studying some of the most devastating microbial pathogens of the period, and their release could potentially cause disease in the scientists working on them. These procedures were also absolutely critical because they allowed careful study of pure microorganisms, making it possible to identify the role of each microbe in a given situation.

Another problem in the cultivation of microbes was solved by Julius Petri while working in Koch's laboratory. Solid medium was poured on glass plates and allowed to spread and harden. Once cooled it allowed a solid surface for streaking.

However, creation of these plates required great care since exposure to the air often lead to contamination. In addition, to prevent contamination of plates during incubation, a cumbersome bell jar was used. If one wanted to view samples, the plate had to be removed from the jar, further exposing it to unwanted microbes in the air. In 1887 Petri developed shallow glass dishes, with one having a slightly larger diameter than the other.

Medium is poured into the smaller dish and the larger one serves as a cover. This simple device solved all of the above problems and took on the name of its inventor, the petri plate. These same techniques are essential in studying all microorganisms. Collectively the above techniques have been used to isolate and identify thousands of different microorganisms. As a testament to the significance of their achievement, these techniques are practiced with remarkably little change in every laboratory that works with microorganisms today.

Spontaneous Generation Hypothesis

Spontaneous generation is the hypothesis that some vital force contained in or given to organic matter can create living organisms from inanimate objects. Spontaneous generation was a widely held belief throughout the middle ages and into the latter half of the 19th century. In fact, some people still believe in it today.

The idea was attractive because it meshed nicely with the prevailing religious views of how God created the universe. There was a strong bias to legitimise the idea because this vital force was considered a strong proof of God's presence in the world. Many recipes and experiments were offered in proof. To create mice, a recipe called for dirty underwear and wheat grain to be mixed in a bucket and left open outside.

In 21 days or less, you would have mice. The real cause may seem obvious from a modern perspective, but to the proponents of this idea, the mice spontaneously arose from the wheat kernels. Another often-used example was the generation of maggots from meat that was left in the open. The failing here was revealed by Francesco Redi in 1668 with a classic experiment. Redi suspected that flies landing on the meat laid eggs that eventually grew into maggots.

The concept were revived in 1745 by the experiments of John Needham. It was known at the time that heat was lethal to living organisms. Needham theorised that if he took chicken broth and heated it, all living things in it would die. After heating some broth, he let a flask cool and sit at a constant temperature. The development of a thick turbid solution of microorganisms in the flask was strong proof to Needham of the existence of spontaneous generation.

Lazzaro Spallanzani later repeated the experiments of Needham, but removed air from the flask, suspecting that the air was providing a source of contamination. No growth occurred in Spallanzani's flasks and he took this as evidence that Needham was wrong. Proponents of spontaneous generation discounted the experiment by asserting that air was required for the vital force to work.

It was not until almost 100 years later that the great French chemist Louis Pasteur, put the debate to rest. He first showed that the air is full of microorganisms by passing air through gun cotton filters. The filter trapped tiny particles floating in the air. By dissolving the cotton with a mixture of ether and alcohol, the particles were released and then settled to the bottom of the liquid. Inspection of this material revealed numerous microbes that resembled the types of bacteria often found in putrefying media. Pasteur realised that if these bacteria were present in the air then they would likely land on and contaminate any material exposed to it.

Pasteur then entered a contest sponsored by The French Academy of Sciences to disprove the theory of spontaneous generation. Similar to Spallanzani's experiments, Pasteur experiment, pictured in Figure 5, used heat to kill the microbes, but left the end of the flask open to the air. In a simple, but brilliant modification, the neck of the flask was heated to melting and drawn out into a long S-shaped curve, preventing the dust particles and their load of microbes from ever reaching the flask. After prolonged incubation the flasks remained free of life and ended the debate for most scientists.

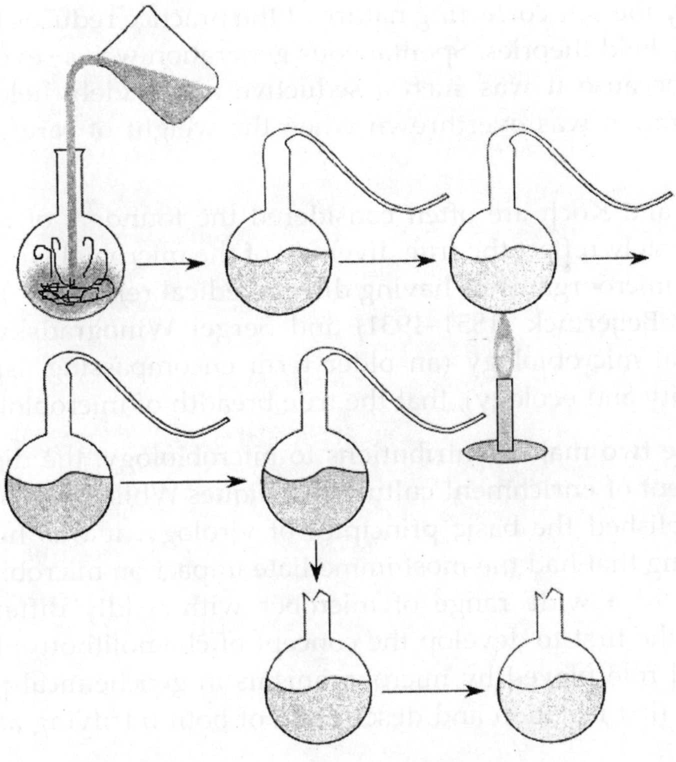

Figure 5. The Swan Neck Flask Experiment

Pasteur filled a flask with medium, heated it to kill all life, and then drew out the neck of the flask into a long S shape. This prevented microorganisms in the air from easily entering the flask, yet allowed some air interchange. If the swan neck was broken, microbes readily entered the flask and grew.A final footnote on the topic was added when John Tyndall showed the existence of heat-resistant spores in many materials. Boiling does not kill these spores and their presence in chicken broth, as well as many other materials, explains the results of Needham's experiments. While this matter may seem silly from a modern perspective, remember that the scientists of the time had little knowledge of microorganisms. Koch would not isolate microbes until 1881. The proponents of spontaneous generation were neither sloppy experimenters nor stupid. They did careful experiments and interpreted them with their own biases.

Detractors of the theory of spontaneous generation were just as guilty of bias, but in the opposite direction. In fact, it is somewhat surprising that Pasteur and Spallanzoni did not get growth in their cultures, since the sterilisation conditions they used would often not kill endospores. Luck certainly played a role. It is important keep in mind that the discipline of science is performed by humans with all the fallibility and bias inherent in the species. Only the self-correcting nature of the practice reduces the impact of these biases on generally held theories. Spontaneous generation was a severe test of scientific experimentation, because it was such a seductive and widely held belief. Yet, even spontaneous generation was overthrown when the weight of careful experimentation argued against it.

While Pasteur and Koch are often considered the founders of microbiology, their work did not accurately reflect the true diversity of the microbial world because of their exclusive focus on microorganisms having direct medical relevance. It was not until the work of Martinus Beijerinck (1851–1931) and Sergei Winogradsky (1856–1953), the founders of general microbiology (an older term encompassing aspects of microbial physiology, diversity and ecology), that the true breadth of microbiology was revealed.

Beijerinck made two major contributions to microbiology: the discovery of viruses and the development of enrichment culture techniques.While his work on the Tobacco Mosaic Virus established the basic principles of virology, it was his development of enrichment culturing that had the most immediate impact on microbiology by allowing for the cultivation of a wide range of microbes with wildly different physiologies. Winogradsky was the first to develop the concept of chemolithotrophy and to thereby reveal the essential role played by microorganisms in geochemical processes. He was responsible for the first isolation and description of both nitrifying and nitrogen-fixing bacteria.

REFERENCES

Madigan M, Martinko J (eds). 2006. *Brock Biology of Microorganisms* (11th ed.). Prentice Hall.

Ryan KJ, Ray CG (eds). 2004. *Sherris Medical Microbiology* (4th ed.). McGraw Hill.

Bordenave G. 2003. "Louis Pasteur (1822-1895)". *Microbes Infect.* 5 (6): 553–60. PMID 12758285.

Burkovski A (editor). 2008. *Corynebacteria: Genomics and Molecular Biology.* Caister Academic Press.

2

Microbial Cell Structure

Microorganisms typically face the world as single cells rather than as the multi-cellular assemblies of higher organisms. Each single cell must therefore contain all the structures necessary for managing its internal state and dealing with the outside environment. Not surprisingly, this evolutionary process results in the use of rather similar structures and processes to solve similar need in different microorganisms. However, prokaryotes have been on this Earth for a long period of time and this has allowed them to differentiate into a dizzying number of different species. Eukaryotic microbes are not quite so diverse, but they still display a remarkable range of properties.

The diversity of the microbial organisms also means that this survey of structures is not exhaustive. No one cell contains all the structures that we describe here, but we will explore the more common structures that have been observed by scientists in the past 150 years as show in Figure 1.

A distinction made between the two types of prokaryotes: the Archaea and their cousins, the Bacteria. We will initially focus on the Bacteria, since that is what we know the most about. Many of the structures we will examine are found in both the Bacteria and the Archaea. Finally, we will talk about the features that are distinctive among the microbial eukaryotes.

Electron microscopes have been important, of course, but so have genetics, molecular biology and biochemistry. Microscopes help scientists to visualise where these structures are located and how they are arranged spatially in the microbe.

Bacterial genetics and molecular biology identify and analyse the genes necessary for the synthesis and regulation of these structures. Biochemistry permits the detailed examination of each part separately, with implications for its role in the living bacterium. The powerful combination of these disciplines has provided a deep understanding of how a bacterium is put together, but there is still much to learn.

The microbial structure is separated into two parts. In the first part, the chemical nature of the types of molecules and polymers that are important in carrying out the business of biology. In the second part, we examine the functional units in the cell, describing how the various chemical structures in the cell interact to carry out important cellular functions.

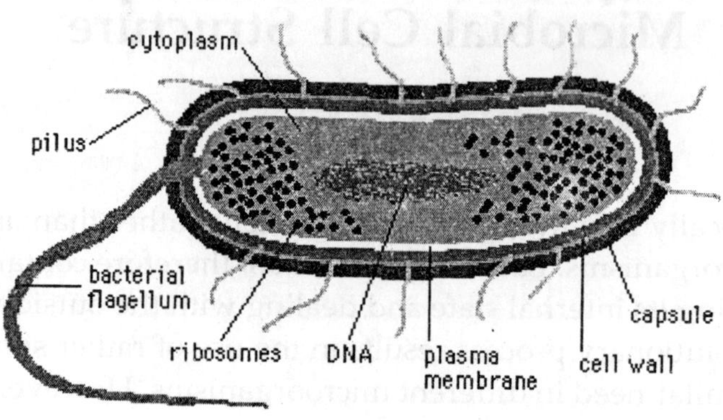

Figure 1. The Generalised Bacterium

First and foremost, it is important to point out that there are some universal structures that all living cells contain. They are the basic building blocks of life: DNA, RNA, protein and cellular membranes. Most, but not all, bacteria also possess cell walls. Beyond these essentials, the frequency of the rest of the structures we mention here ranges in the bacterial world from quite common to very rare.

NUCLEIC ACIDS

Deoxyribonucleic acid (DNA) and ribonucleic acid (RNA) are involved in information storage and processing. DNA serves as the cell's hereditary information, while RNA is involved in converting that information into functional products, such as proteins. RNA and DNA are long polymers of only 4 nucleotides: adenine, guanine, cytosine and thymine.

The nucleotides for adenosine (A), guanine (G), cytosine (C) and thymine (T) as found in DNA are shown. The first three are also found in RNA, but when incorporated into that polymer, the associated sugar has two hydroxyls, as shown in the model of uracil (U). Uracil is the RNA equivalent of the DNA nucleotide thymine. The bases of the four nucleotides are different, but there is also a pattern. Adenine (A) and guanine

(G) are purines, and therefore have a distinctive two-ring structure; they differ in the chemical groups attached to the rings. Likewise, cytosine (C), thymine (T) and uracil (U) are all pyrimidines and share a single-ringed structure, but also differ in their attached groups. Not surprisingly, as these extra chemical groups distinguish the different purines and pyrimidines structurally, they are also responsible for their important functional differences as well.

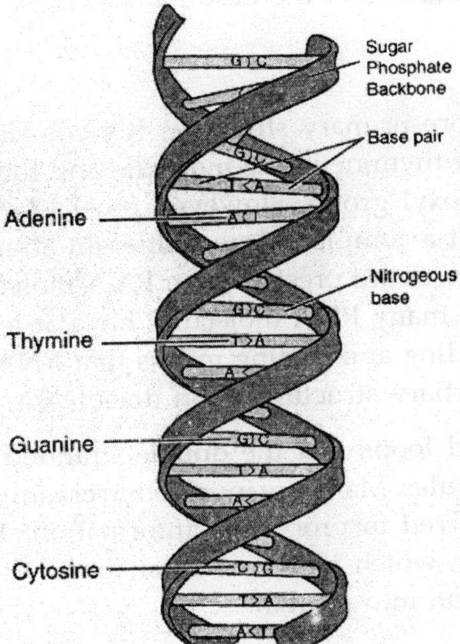

Figure 2. Structure of DNA

The bases in a nucleic acid polymer are capable of forming hydrogen bonds with neighboring bases on a second strand of nucleic acid, a process termed base pairing. However, there are rules to this association. Adenine is capable of forming two hydrogen bonds with thymine and cytosine can base pair with guanine, forming three hydrogen bonds. Note how the two nucleic acid strands spiral around each other in a regular repeating pattern. There are 10 bases per turn of the double helix. Bases pair with one another in the center of the helix cylinder and form what some have likened to a spiral staircase.

Secondary and Tertiary DNA Structures

DNA almost always exists in cells as a double-stranded structure of complementing strands. It happens that this double-stranded form is rather stable, and resists tight twists and turns. It is often assumed that the stability is due to the hydrogen bonding between the bases, but this is not the case. Finally, the larger organisation of the DNA strands with respect to each other, termed the tertiary structure, is also fairly similar in all DNA molecules. One implication is that proteins that want to distinguish between different DNA molecules must do so by reading different primary structure sequences by interacting with the outer surface of the base pairs.

Structures of RNA

In composition and therefore primary structure, RNA is similar to DNA, except that uracil (U) takes the place of thymine in the molecule and the ribose unit on each sugar contains a additional hydroxyl group. However, most RNA in cells exists as single-stranded molecules and not a complex of two different strands as with DNA. Now if complementary base sequences are present in an RNA molecule, it can fold back upon itself and base pair, so that many RNA molecules have at least some double-stranded regions. However, this bending and folding means that RNA molecules typically have much more complicated tertiary structures than does DNA.

Both the single-stranded loops and the double-stranded stems are critical for the function of most RNA molecules. Many are involved in creating physical structures, such as ribosomes, that are involved in processing information. The other general class of RNAs are messenger RNAs, which represent a version of the DNA primary structure that is suitable for translation into protein.

PROTEINS

Proteins and peptides are essential to the cell and serve two major functions. Many proteins are enzymes that catalyse almost all biological reactions in a living organism. Other proteins perform a structural role for the cell - either in the cell wall, the cell membrane or in the cytoplasm. In this section, we will look at the basic structural elements shared by all proteins.

Structure of Amino Acids

Proteins are polymers of amino acids. Amino acids, with rare exception, contain an a carbon that is connected to an amino (NH_3) group, a carboxyl group (COOH), and a variable side group (R).

Primary Structure of Proteins

As with nucleic acids, primary structure refers to the ordered sequence of the different amino acids in a protein. The carboxyl group and the amino group of amino acids are reactive. The peptide bond between an alanine residue and a valine residue is identified by the arrow. Peptide bonds can form between any two of the 20 amino acids and link the carboxyl group of one amino acid and the amine group of the next.

There are 20 common amino acids found in proteins and these amino acids can be roughly classified into 3 groups: polar, non-polar and charged. Polar and charged amino acids are hydrophilic and are often found on the surface of a protein, interacting with the surrounding water. In contrast, non-polar (or hydrophobic) amino acids avoid water. While this categorisation is adequate for most purposes, you should recognise that it is a bit simplistic. For example, arginine does have a charged hydrophilic group at one end, but the $-CH_2-$ backbone that makes up most of the amino acid is actually quite hydrophobic.

Secondary Structure

Peptides and proteins are formed when a ribosome and the rest of the translation machinery link amino acids together in polymers that range from 10 to 10,000 residues in length. During and after protein synthesis, the residues of the primary sequence dictate how the protein folds. The simplest aspect of protein folding is termed its secondary structure, which refers to the geometry of the local polypeptide chain with respect to their immediate neighbors. How a protein folds is dictated by the primary sequence of amino acids, but predicting the overall structure from the primary sequence remains one of the most important unsolved problems in biology. Nevertheless, it is clear that the major determinants of this final structure are hydrophobic interactions. During protein folding, hydrophobic amino acids must be hidden from the water interface by being buried in the interior of the protein. This burying defines the protein core which then influences the immediate structure around it and greatly affects the protein's overall structure.

Common Secondary Structures

Peptide bonds between adjacent amino acids can rotate and twist to allow a large number of interactions, but two local organisation schemes, the a helix and the ß sheet, are found in many proteins. Their prevalence is certainly because they happen to form particularly energetically favorable structures. Formation of these structures is driven by favorable hydrogen bonding and hydrophobic interactions between nearby amino acids in the protein. The a helix resembles a ribbon of adjacent amino acids wrapped around a tube to form a staircase-like structure.

Tertiary Structure

The relationships mentioned above do not fully define the structure of proteins. For example, we need a way to describe larger organisations of a helices and ß sheets, as well as other parts of proteins that do not fit these two patterns. This larger organisation is termed tertiary structure. The most important stabilising force in proteins is burying hydrophobic residues from the surrounding water, but there are other chemical features that are important in creating and stabilising tertiary structure. These include hydrogen bonds, ionic interactions, and sulfhydryl bonds.

Ionic interactions are attractions between groups of opposite charge. There are amino acids with negatively charged side groups (aspartate, glutamate) and amino acids with positively charged side groups (typically lysine and arginine). If opposite charges are brought close enough together, there will be an attractive force that can contribute to protein structure. It also follows that like charges repel each other and this can also dictate protein structure. Sulfhydryl linkages are covalent bonds between cysteine groups. Cysteine is a unique amino acid in that it has a sulfur group at the end of its variable side group that is available for binding to other groups. Often in proteins, nearby sulfhydryl groups on cysteines form a covalent bond and these are often crucial for the stabilisation of the mature protein or for it to perform its function.

Quaternary Structure

Many proteins are actually complexes of several polypeptides. The arrangement of more than a single polypeptide into one protein is termed the quaternary structure of that protein. Protein complexes might contain two or more copies of the same protein or they may consist of any number of different polypeptides in various ratios. Such complexes are certainly not random, but reflect precise interactions among the protein subunits based on the same sorts of interactions (e.g. hydrophobic, hydrogen bonds, etc) described above.

LIPIDS

Lipids are molecules with two personalities. One part of the molecule wants to associate with water and the other does not. Molecules with these properties are termed amphipathic. Hydrophhobic, long-chain fatty acids attach to two hydroxyl groups on the glycerol. To the third hydroxyl group, a polar, and therefore hydrophilic, group is attached. Many bacteria contain phospholipids in which this third group contains a phosphate connected to a carbon molecule. The amphipathic nature of lipids is important in their function in the cell.

SMALL MOLECULES

There are also a number of important small molecules that shuttle protons, electrons or small carbon moieties around the cell. These small entities typically do their job in association with proteins to which they can be either loosely or tightly bound. All of life on this planet seems to have settled on a surprisingly small set of molecules to perform these tasks. Almost certainly this is because the use of these molecules evolved early and has been maintained through evolution.

Proton and Electron carriers

Most amino acids are not particularly good at either donating or accepting electrons and when they do, it is under a limited range of conditions.

Carbon Carriers

There are also small molecules in the cell that serve as carriers of important carbon compounds. Essentially, these carriers have the right chemical properties that make it relatively easy for enzymes to add or remove a particular carbon unit. Tetrahydrofolate and cobalamin (vitamin B_{12}) are often involved in adding or removing one-carbon units during the synthesis of various structures in the cell. Coenzyme A is necessary for the transfer of small 2 to 4 carbon units (acetyl, propyl) from one enzyme to another. It finds utility in both the synthesis and breakdown of organic molecules. The beauty of using a small set of carriers is that it allows the easy movement of carbon from one pathway to another.

Important Minerals

Many types of minerals are important for the proper functioning of enzymes. For example, magnesium ions are essential for ATP-binding by many enzymes. Zinc is important in the proper folding of some enzymes and iron, in the form of iron-sulfur centers and hemes, is critical in many electron transport proteins. Minerals also help bind structures in the cell together. For example, magnesium and calcium are necessary for the stabilisation of membranes. Potassium ions in the cell shield the large amount of negative charge on the DNA allowing it to pack more tightly together.

CELL ORGANISATION

Now that you have had an introduction to the chemicals that make up the typical cell, we will now look at how this chemistry combines to form major functional units. These can be thought of as the organisations that carry out the major business of the cell: growth, replication, feeding and movement. We will first start with membranes because so many things interact with them. Next internal structures in the cytoplasm will be described and finally structures outside of the membrane.

First, however, we should describe an important evolutionary hypothesis that will make sense of much of the following details. As you will see, there are a number of curious similarities and differences in the details of cellular structure among bacteria, archaea and eukaryotes. In general much of the machinery in a eukaryotic nucleus and in the cytoplasm looks rather a lot like what is present in the archaea. However, the organelles of eukaryotes, such as the mitochondria and chloroplasts, have properties that are much more similar to those of bacteria. How is this possible?

One clue comes from observing organisms in nature. It is very common to find cooperative relationships between different species and this is also true in the microbial world. In some instances these relationships involved close physical contact between their participants, sometimes with one participant engulfing the other.

In 1968 Dr. Lynn Margulis extended this observation and proposed that some of the organelles found in eukaryotes, specifically mitochondria and chloroplasts, were originally endosymbionts of their host. Originally these two microbes probably were able to live independently, but over time, the endosymbiont lost functionality that its host was already providing and then became dependent. Over the years ample evidence has accumulated to support this exceptional insight.

Membranes

General properties

The cytoplasmic membrane immediately surrounds the inside of the cell and is perhaps the most conserved structure in living cells. Membranes are thin structures, measuring about 8 nm thick and every living thing on this planet has some type of membrane They are the major barrier separating the inside of the cell from the outside and allow cells to selectively interact with their environment. Membranes are highly organised and asymmetric. This asymmetry comes from the fact that the membrane that faces the environment performs very different functions than does the side that faces the cytoplasm. Membranes are also dynamic, constantly adapting to changing environmental conditions.

Physical structure

Membranes are composed of lipids and proteins. The molecules have a similar structure to sterols found in eukaryotic membranes and serve to help stabilise the membrane. Proteins are more numerous in bacterial membranes than in eukaryotic membranes. This is because bacteria in general only contain a single membrane in contact with the cytoplasm and this has to carry out all the functions of the cell. In eukaryotes these functions are divided amongst the cytoplasmic membrane and the other organelles.

Membranes as selective barrier

The concentration of solutes, sugars, and most ions is generally much higher within the cell than outside. A fundamental principle of nature is that different concentrations of a given solute tend to equilibrate across the boundary due to diffusion. However, the cell boundary is the membrane and its hydrophobic core prevents this diffusion for polar molecules.

Compounds such as amino acids, organic acids and inorganic salts must therefore be specifically transported across the membrane by proteins and once inside these molecules cannot escape. The cell can therefore control the nature and amount of these compounds that enter or leave the cell. Though hydrophilic, water is not very polar and can flow freely across the membrane, as can some small non-polar molecules. This creates a serious problem.

The inside of the cell is full of many types of solutes: proteins, nucleic acids, other small molecules and ions. In comparison the outside environment, in most cases, is very dilute. Because of this there is a higher concentration of water outside the cell than inside the cell.

Nature hates imbalances such as this and in an effort to correct the problem; water tends to flow into the cell, by a process called osmosis. Osmosis causes a high pressure against the cell membrane. This pressure would rapidly cause lysis of most cells and one of the major purposes of the peptidoglycan of the cell wall is to prevent the cell membrane from bursting. For molecules that are soluble in both the lipid membrane and the surrounding aqueous environment, the law of simple diffusion directs transport.

The membrane is not a barrier for such molecules. These types of molecules are uncommon since solubility in both a hydrophobic and a hydrophilic environment is unusual. There is no transport protein for such compounds, so there is no specificity of control or energy cost. The cell cannot create a concentration gradient of these molecules. One important example is water. Water can pass freely into and out of cells.

There are three basic types of transport systems

— Facilitated Diffusion

— Group Translocation

— Active Transport

Many of the proteins in the membrane function to help carry out selective transport, particularly of polar compounds. These proteins typically span the entire membrane, making contact with the outside environment and the cytoplasm. They often require the expenditure of energy to help compounds move across the membrane, though cells can also use concentration gradients of these compounds to generate energy, as described below

Facilitated diffusion

This process involves a protein that binds the molecule to be transported and physically moves that compound through the membrane. Binding of the molecule to the protein causes a conformational change in the protein so that the molecule now faces the opposite side from where it was. However, these small molecules are readily moved in and out of the cell, so a gradient cannot be formed nor is energy required. One example of a protein involved in facilitated diffusion is the glycerol facilitator protein. In E. coli this enzyme binds glycerol and a few other polyalcohols and allows their diffusion into the cell. Once inside, the glycerol is immediately phosphorylated, preventing its diffusion back outside the cell.

An animation of the migration of solutes in and out of the cell as facilitated by a protein. Notice that this mechanism does not lead to a solute concentration inside the cell that is higher than outside. Rather, it leads to an equilibrium of that solute across the gradient.

Group translocation

In this process, a protein specifically binds the target molecule and transports it inside the cell while simultaneously modifying it chemically. Most group translocations require energy and tend to be unidirectional, unlike facilitated diffusion. The substrates of catabolic pathways, such as sugars, are sometimes transported by group translocation. This is an efficient way to both bring substrate into the cell and begin the breakdown process. An animation of group translocation. The glucose olecule that is being transported into the cell is modified by the addition of a phosphate from phosphoenolpyruvate to form glucose-6-phosphate.

Active transport

In active transport, energy is expended to transport the small molecules, but they are not chemically altered. The process is efficient enough to cause the internal concentration in the cell to reach many times its external concentration. Active transport proteins are molecular pumps that expend energy to pump their substrates against a concentration gradient. This energy comes in two forms: ATP and ion gradients.

In ATP-based active transport, ATP hydrolysis is coupled to the movement of the small molecule across the membrane. One large group of proteins involved in this type of transport is the ATP binding cassette (ABC) transporters. ABC transporters have been found in all living species with 80 identified in the E. coli genome and 48 in the human genome. The mechanism of ABC transporters is exemplified by the maltose binding protein of *E. coli.*

Three separate types of transport are shown. In the first, an antiporter moves two different small molecules across the cell membrane, but in the opposite directions. In the second, a symporter moves two or more different molecules to move into the cell simultaneously.

Typically, the desired molecule is being concentrated against a gradient and that transport is driven by the transport of the other molecule, which is moving with a gradient. Finally, a uniporter binds and transports the target molecule only. Energy is required for these processes and the cell can accumulate molecules inside the cell using this mechanism. Ion gradient active transport uses the energy of one chemical gradient, that of the specific ion, to drive the creation of a different gradient, the uptake of the small molecule.

The ion gradient that supports the work is at a higher concentration on one side of the membrane than the other. The transport protein both binds its small molecule to transport and provides a gateway for this ion to fall down its concentration gradient. When the ion moves through its gateway, it causes a conformational change in the protein and this change is used to transport the target small molecule into the cell. Active transport proteins may be highly specific for only one molecule or may be able to carry a class of chemically related molecules.

Translation Involves Messenger RNA

Translation is the process of converting the instructions coded in the DNA into the proteins that actually carry out the work. The macromolecules that perform this task consist of mRNA, transfer RNA (tRNA) and the ribosome, which is made of ribosomal RNA (rRNA) and ribosomal proteins. In brief, this process consists of making an mRNA copy of a region in the DNA that gives directions for the synthesis of a protein or proteins.

The mRNA is then bound by a ribosome that translates the mRNA into a amino acid sequence. The amino acids necessary for the protein are carried to the ribosome by tRNA that actually read the information in the mRNA and add the appropriate amino acid to the nascent protein chain.

We will now examine the structure of the molecules involved in translation, starting with mRNA, where the primary structure is simple - merely unmodified A, G, C and

U bases. In almost all prokaryotic mRNAs there is not a great deal of secondary and tertiary structure, since they are typically being translated by ribosomes and the translating ribosome certainly removes any structure as it moves along the mRNA. What structure there is tends to be in the regions that are not translated, notably the 5' and 3' ends of the mRNA. One of the roles of structure, especially at the 3' end, is to stabilise the mRNA. Now it happens that most prokaryotic mRNAs are not very stable in the cell because they are rapidly degraded by RNases. However, different types of RNA structures can impede the progress of nucleases, particularly the type that degrades from the 3' end of the mRNA.

As a complication, however, there are some RNA structures that actually serve as a specific target for other types of RNases and thus lead to destabilisation of the mRNA. This is biologically important because a more stable mRNA is translated by more ribosomes and therefore leads to more protein product. As a final structural feature, most if not all prokaryotic mRNAs have short stretches of adenosine residues added to the 3' end after transcription.

The presence of these also tends to lead to mRNA degradation. This is a bit surprising because the presence of long adenosine stretches on the 3' ends of eukaryotic mRNAs actually tends to stabilise those mRNAs. This appears to be a case where evolution has taken a single feature, addition of adenosines to mRNAs, and changed its functional importance through evolution. In eukaryotes, the fact that mRNA is transcribed in the nucleus and must be exported to the cytoplasm for translation changes some details. One of these is that eukaryotic mRNAs are rather more stable than in prokaryotes.

Ribosomal RNA

In contrast to the case with mRNAs, the other RNAs involved in translation, tRNA and rRNA, have very distinct structures. Each rRNA folds into a known secondary structure and has a complex tertiary structure containing many short helical regions and long range base pair interactions. These structures are also maintained by interactions between the RNAs and protein.

The composition of ribosomes is 62 % RNA and 38 % protein by weight. Two complexes of RNA and protein make up the ribosome, the 30S subunit and the 50S subunit. The 30S subunit is composed of 21 proteins and a single-stranded rRNA molecule of about 1,500 nucleotides, termed the 16S rRNA.

The 50S subunit contains 31 proteins and two RNA species, a 5S rRNA of 150 nucleotides and a 23S rRNA of about 2,900 nucleotides. Several of the nucleotides on the 16S and 23S rRNAs have been modified by methylation and these modifications are probably critical to the function of the rRNAs since they always happen in regions conserved through evolution. The ribosome-associated proteins are positively charged, with a high proportion of lysine and arginine residues. This facilitates complex formation between the acidic RNA and these basic proteins.

Transfer RNA

Transfer RNA (tRNA) is the ferry that transports the amino acids to the ribosome. There are one or more different tRNA molecules for each of the 20 amino acids. Each consists of 70 to 80 nucleotides of single-stranded RNA that is extensively base-paired to form four short helical domains. Now the tertiary structures of tRNAs are all rather similar, so the critical features that make each appropriate to a specific amino acid are largely found in the primary structure itself. Many bases in tRNA molecules are chemically modified by enzymes to help the molecule carry out its function.

The aminoacyl-tRNA synthetases are the enzymes that add the amino acid to the tRNAs. There is a single synthetase for each amino acid and it binds each of its appropriate tRNA molecules and charges it with its appropriate amino acid. The synthetases avoid both the wrong tRNAs as well as the wrong amino acids. However, the process is somewhat trickier than it first appears.

First, one might expect that the synthetase might recognise the proper tRNAs by examining the anticodon loop, the part recognised by the ribosome to match the tRNA to the mRNA. After all, the anticodon certainly does define the amino acid in translation. This is not the case, however, perhaps because the anticodon is very far away from the end of the tRNA that is charged with the amino acid. In any event, it is clear that most of the basis for proper synthetase-tRNA recognition lies elsewhere in the tRNA. This then raises another problem: when one looks at the different tRNAs that all carry a given amino acid and are therefore all recognised by a single synthetase, no obvious pattern emerges. In other words, the important features that allow the alanine synthetase to recognise only alanine tRNAs are not completely clear, though great strides in understanding this process have been made.

One final thought before we move on. Think about the chicken-and-egg conundrum that translation brings up. This whole process has the express purpose of synthesizing proteins, yet it involves proteins at every step. How could these proteins have evolved to serve this function when they are needed for their own synthesis? In other words, how could you synthesize any protein until a complete set of translation proteins had evolved? Part of the answer is that primordial translation was probably much simpler, though less accurate and efficient.

Perhaps only very few and somewhat general protein functions were actually required. Alternatively, perhaps early translation used no proteins at all: Some scientists believe that early life employed RNA molecules that were capable of both carrying out necessary enzymatic functions and storing hereditary information. They posit that it was only later that proteins came along and started to assist in their own synthesis. However, this hypothesis still does not explain how one simultaneously evolved functional proteins and a process for creating them.

INCLUSIONS AND OTHER INTERNAL STRUCTURES

We said before that there were few structures in the prokaryotic cytoplasm that are visible by microscopy. In general, they serve specific purposes in the cell and are often found only in certain cell types or under certain growth conditions.

Inclusions

Inclusions are dense aggregates of specific chemical compounds in the cell. Typically, the aggregated chemical serves as a reservoir of either energy-rich compounds or building blocks for the cell. Forming polymers costs energy and it may seem wiser for the cell to keep the excess monomers around for when they are needed. The benefit of polymerisation is that it decreases the osmotic pressure on the cell, a serious problem as described later. Inclusions often accumulate under laboratory conditions when a cell is grown in the presence of excess nutrients. However, the role of some inclusions is unclear. Growth on rich medium causes their creation, but subsequent starvation in the test tube does not always result in the use of these reserves.

Poly-ß-hydroxyalkanoate

One of the more common storage inclusions involves poly-ß-hydroxyalkanoate (PHA). It is a long polymer of repeating hydrophobic units that can have various carbon chains attached to it. The function of PHA in bacteria is as a carbon and energy storage product. Just as we store fat, some bacteria store PHA. Some PHA polymers have plastic-like qualities and there is some interest in exploiting them as a form of biodegradable plastic.

Glycogen

Glycogen is another common carbon and energy storage product. Humans also synthesize and utilise glycogen, which is a polymer of repeating glucose units.

Phosphate Granules and Sulfur Globules

Given the opportunity, many organisms accumulate granules containing long chains of phosphate, since this is often a limiting nutrient in the environment. These polyphosphate polymers, also called volutin, form visible granules in some microbes. These granules are readily stained by many basic dyes such as toluidine blue and turn reddish violet in color. These inclusions are often referred to as metachromatic granules because they become visible by "metachromasy" (a color change). Polyphosphate is found in all known cells (eukaryotes, bacteria and archaea) and appears to serve many important roles.

1. It serves as a phosphate reservoir
2. It is an alternative substrate in place of ATP when phosphorylating sugars during catabolism.
3. It is a chelator for divalent cations
4. It can be a buffer under alkaline stress
5. It is an important factor for DNA uptake.
6. Finally, phosphate polymers are important regulators in response to stress

Oxidation of the compounds is linked either to energy metabolism or photosynthesis. Oxidation of sulfide initially yields elemental sulfur, which accumulates in globules inside or outside the cell. If the sulfide is exhausted the sulfur may be further oxidised to sulfate.

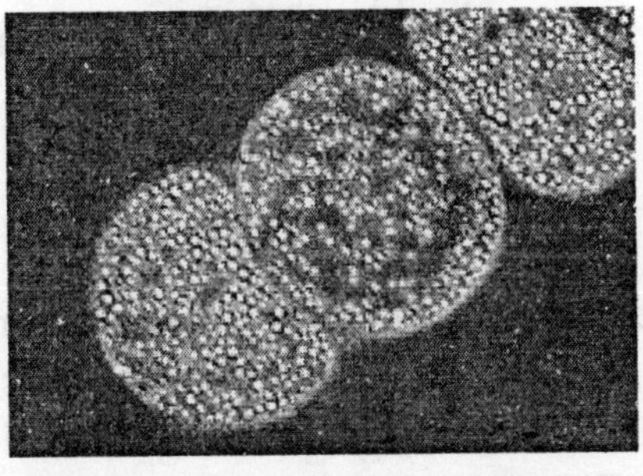

Figure 3. Sulfur Globules

Gas Vesicles

Figure 4 shows an example of gas vesicles, also known as gas vacuoles, that are found in cyanobacteria. Cyanobacteria are photosynthetic and live in lakes and oceans. In these environments, the cyanobacteria use gas vesicles to control their position in the water column to obtain the optimum amount of light and nutrients.

Gas vesicles are often aggregates of hollow cylindrical structures composed of rigid proteins. They are impermeable to water, but permeable to gas. The amount of gas in the vesicle is under the control of the microorganism. Release of gas from the cell causes the bacteria to fall in the water column, while filling the vesicle with gas causes the cells to rise.

Magnetosomes

magnetosomes are intracellular crystals of iron magnetite (Fe_3O_4) that impart a permanent magnetic dipole to prokaryotic cells that have them. They allow these microbes to orient themselves in a magnetic field. This process does not appear to involve any special machinery beside the magnetosome, Each microbe can be thought of as having a tiny magnet that is responding to the magnetic field in the environment.

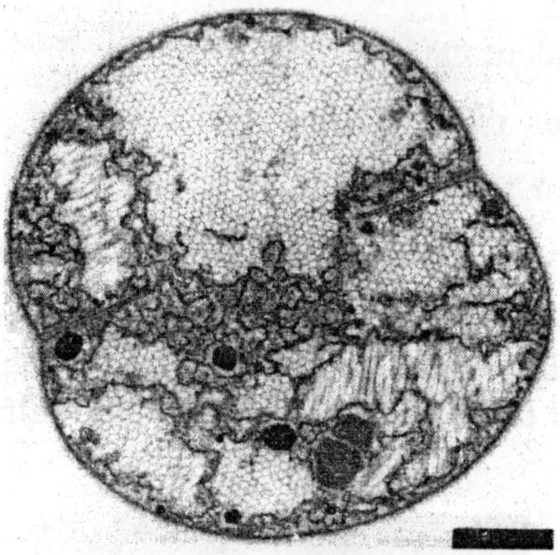

Figure 4. Gas Vesicles

These magnetosomes allow the microbes to follow the magnetic field of the earth. Some species of magnetotatic bacteria have the following behavior. In the northern hemisphere magnetotatic bacteria swim north along the magnetic field, while in the southern hemisphere they swim south. Because of the inclination of the earth's magnetic field, this causes the microbes to swim downward.

Many microbes containing magnetosomes are aquatic organisms that do not grow well in the presence of atmospheric concentrations of oxygen and they move away from the oxygen higher up in the water column by detecting the magnetic field and swimming downward. A special membrane surrounds magnetosomes that confines the magnetite to a defined area. The membrane likely plays a role in precipitating the iron as Fe_3O_4 in the developing magnetosome. Magnetosomes can be square, rectangular or even spike-shaped. Magnetosomes are primarily found in aquatic bacteria and in some unicellular algae (eukaryotes).

Cytoplasm and outer Membranes in Gram-negative Bacteria

The periplasm is found in gram-negative bacteria and is the space in between the cytoplasmic and outer membranes. The periplasm is filled with water and proteins and is therefore somewhat reminiscent of the cytoplasm. However, pools of small molecules in the periplasm are not like those in the cytoplasm because the membrane prevents the free exchange between these two compartments. Also, the proteins found in the periplasm are distinct from those in the cytoplasm and are specifically guided to this site during translation through specific signal sequences typically near their N-termini.

The peptidoglycan shell that provides the strength to prokaryotic membranes is also found in the periplasmic space of gram-negative bacteria, while in gram-positive bacteria it provides the outside border to the periplasm.

Cell Wall and Microbe Survival

The cell wall is essential to the survival of most microorganisms. Many microbes live in environments in relatively dilute environments and the cell wall's most important function is to prevent the cell from bursting due to the osmotic stress placed upon. The cell wall also determines the shape of the cell. Any cell that has lost its cell wall, either artificially or naturally, becomes roughly spherical and lyses due to osmotic pressure, unless placed in certain concentrated solutions. Finally, the cell wall helps to support any structure that penetrates from the cell out into the environment.

The structure and synthesis of prokaryotic cell walls is unique and many compounds found in the bacterial cell wall are found nowhere else in nature. It is true that plants also make cell walls, but they are chemically and structurally different. There are two basic types of bacterial cell wall structures that have been studied in detail: gram-positive and gram-negative. These two classes of bacterial cells look very different following staining with the Gram stain and this has been a standard basis for starting to identify different bacterial species.

When the Gram stain was developed by Hans Christian Gram in 1884 the molecular basis of the stain was unknown. In fact very little was understood about bacteria in general. He just determined empirically that when bacterial smears were run through a four-step staining procedure using two different dyes, some cells retained the first dye and stained purple, while other only retained the second dye and stained pink.

The gram-positive wall is much thicker than is the gram-negative wall and its external appearance is smoother. gram-positive and gram-negative cells do share one thing in common that is unique to bacteria - peptidoglycan. We will talk about the structure of this and then move on to examine the various structures found in each cell wall type. Peptidoglycan is a thick rigid layer composed of an overlapping lattice of two sugars, N-acetyl glucosamine (NAG) and N-acetyl muramic acid (NAM). The exact molecular makeup of these cross-bridges is species-specific. NAM is only found in the cell walls of bacteria and nowhere else. Attached to NAM is a side chain generally composed of four amino acids. In the best-studied bacterial cell walls (E. coli) the cross-bridge is most commonly composed of L-alanine, D-alanine, D-glutamic acid and diaminopimelic acid (DPA).

The NAM, NAG and amino acid side chain form a single peptidoglycan unit that can link with other units via covalent bonds to form a repeating polymer. The polymer is further strengthened by covalent bonds between cross-bridges and the degree of cross-linking determines the degree of rigidity. In the *E. coli,* the penultimate D-alanine of one unit is linked to DPA of the next cross-bridge.

In some gram-positive microbes there is a peptide composed of various amino acids that serves as a link between the cross-bridges. In *Staphylococcus aureus* strains, five glycines make up the linker between peptidoglycan monomers. The sequence of these linkers varies considerably between species. The completed peptidoglycan layer forms a strong mesh that can be thought of as a chain link fence. The complete cell wall contains one or more layers of peptidoglycan one atop the other, providing much of the strength of the cell wall.

While both gram-negative and gram-positive bacteria have peptidoglycan, its physical arrangement in the cell wall is different. In gram-positive cells the peptidoglycan is a heavily cross-linked woven structure that encircles the cell in many layers. It is very thick with peptidoglycan accounting for 50% of weight of cell and 90% of the weight of the cell wall.

Electron micrographs show the peptidoglycan to be 20-80 nm thick. In gram-negative bacteria the peptidoglycan is much thinner with only 15-20% of the cell wall being peptidoglycan and it is only intermittently cross-linked. In both cases peptidoglycan is not a barrier to solutes, as the openings in the mesh are large enough for most molecules including proteins to pass through.

There are numerous antibacterial agents that target the bacterial cell wall because mammals do not synthesize walls and therefore are not susceptible to the toxic effects of these agents. Penicillin inhibits the linking of the amino acid side chains of peptidoglycan units, which therefore weakens the stability of the wall eventually, causing the cells to rupture. Humans and other animals even synthesize an enzyme that specifically attacks bacterial cell walls. The enzyme lysozyme is found in many body fluids and hydrolyses the NAM-NAG bond in the cell wall. It serves as a critical part of the mammalian defense against bacterial invasion.

The Gram-positive Cell Wall

Another structure in the gram-positive cell wall is teichoic acid. It is a phosphodiester polymer of glycerol or ribitol joined by phosphate groups. Amino acids such as D-alanine are attached. Teichoic acid is covalently linked to muramic acid and stitches various layers of the peptidoglycan mesh together. Teichoic acid stabilises the cell wall and makes it stronger.

Gram-negative Cell Structure

Gram-negative cell walls have a more complicated structure than do those of gram-positive organisms. Outside the cytoplasmic membrane is the periplasm, which contains the thin layer of peptidoglycan. The peptidoglycan in gram-negative cells contains less cross-linking than in gram-positive cells with no peptide linker. Covalently bound to the peptidoglycan is Braun's lipoprotein, which has a hydrophobic anchor in the outer membrane that helps to strongly bind the peptidoglycan to the outer membrane.

The Outer Membrane

The outer membrane of gram-negative bacteria is another lipid bilayer similar to the cytoplasmic membrane, and contains lipids, proteins, and also lipopolysaccharides (LPS). The membrane has distinctive sides, with the side that faces the outside containing all the LPS. LPS is composed of two parts: Lipid A and the polysaccharide chain that reaches out into the environment. Attached to Lipid A is a conserved core polysaccharide that contains KDO, heptose, glucose and glucosamine sugars. The rest of the polysaccharide consists of repeating sugar units and this is called the O-antigen. The O-antigen varies among bacterial species and even among various isolates of the same species. Many bacterial pathogens vary the make-up of the O-antigen in an effort to avoid recognition by the host's immune system.

LPS confers a negative charge and also repels hydrophobic compounds including certain drugs and disinfectants that would otherwise kill the cell. Some gram-negative species live in the gut of mammals and LPS repels fat-solubilising molecules such as bile that the gal bladder secretes. This repulsion enables these bacteria to survive in this environment. The O-antigen and other molecules on the outer membrane are also used by certain viruses that infect bacteria, as a means to identify the correct hosts for infection.

LPS is medically important because when LPS is released from bacterial cells it is toxic to mammals and is therefore called endotoxin. It creates a wide spectrum of physiological reactions including the induction of a fever (endotoxins are said to be pyrogenic), changes in white blood cell counts, leakage from blood vessels, tumor necrosis and lowered blood pressure leading to vascular collapse and eventually shock. At high enough concentrations the LPS endotoxin is lethal. Finally, the outer membrane keeps the enzymes in the periplasm from floating away from the cell.

There are fewer total proteins and fewer unique types of proteins in the outer membrane than in the cytoplasmic membrane. Porins are particularly important components because of their role in the permeability of the outer membrane to small molecules. Porins are proteins that form pores in the outer membrane large enough to allow passage of most small hydrophilic molecules. All known porins have a similar structure, with the protein containing a central channel that allows the passage of molecules. This allows migration of these molecules into the periplasmic space for possible transport across the cytoplasmic membrane.

Some porins in the outer membrane are general, doing simple discrimination on size and charge, but having little substrate specificity. Examples include OmpF that is selective for positively charged molecules and PhoE that is permeable to negatively charged molecules. Other porins are more specific. The best studied is LamB, which recognises the sugar polymer maltooligosaccharide and transports it through the outer membrane. Very large or hydrophobic molecules cannot penetrate the outer membrane, so the outer membrane serves as a permeability barrier to at least some molecules.

REFERENCES

Kitani, Osumu and Carl W. Hall. 1989. *Biomass Handbook*. Taylor & Francis US. p. 256.

Langford, Roland E. 2004. *Introduction to Weapons of Mass Destruction: Radiological, Chemical, and Biological*. Wiley-IEEE. p. 140

Moses, Vivian et al. 1999. *Biotechnology: The Science and the Business*. CRC Press. p. 563

Schmidt A, Ragazzi E, Coppellotti O, Roghi G. 2006. "A microworld in Triassic amber". *Nature* 444 (7121): 835.

Soni, S.K. 2007. *Microbes: A Source of Energy for 21st Century*. New India Publishing.

3

Bacteria: Structure and Functions

Bacteria are single-celled microorganisms that lack a nuclear membrane, are metabolically active and divide by binary fission. Medically they are a major cause of disease. Superficially, bacteria appear to be relatively simple forms of life; in fact, they are sophisticated and highly adaptable. Many bacteria multiply at rapid rates, and different species can utilise an enormous variety of hydrocarbon substrates, including phenol, rubber, and petroleum. These organisms exist widely in both parasitic and free-living forms. Because they are ubiquitous and have a remarkable capacity to adapt to changing environments by selection of spontaneous mutants, the importance of bacteria in every field of medicine cannot be overstated. The discipline of bacteriology evolved from the need of physicians to test and apply the germ theory of disease and from economic concerns relating to the spoilage of foods and wine.

The initial advances in pathogenic bacteriology were derived from the identification and characterisation of bacteria associated with specific diseases. During this period, great emphasis was placed on applying Koch's postulates to test proposed cause-and-effect relationships between bacteria and specific diseases. Today, most bacterial diseases of humans and their etiologic agents have been identified, although important variants continue to evolve and sometimes emerge, e.g., Legionnaire's Disease, tuberculosis and toxic shock syndrome. Major advances in bacteriology over the last century resulted in the development of many effective vaccines as well as of other vaccines that are less effective or have side effects.

Another major advance was the discovery of antibiotics. These antimicrobial substances have not eradicated bacterial diseases, but they are powerful therapeutic tools. Their efficacy is reduced by the emergence of antibiotic resistant bacteria. In reality, improvements in sanitation and water purification have a greater effect on the incidence of bacterial infections in a community than does the availability of antibiotics or bacterial vaccines. Nevertheless, many and serious bacterial diseases remain. Most diseases now known to have a bacteriologic etiology have been recognised for hundreds of years. Some were described as contagious in the writings of the ancient Chinese, centuries prior to the first descriptions of bacteria by Anton van Leeuwenhoek in 1677. There remain a few diseases that are thought by some investigators to be caused by bacteria but for which no pathogen has been identified. Occasionally, a previously unrecognised diseases is associated with a new group of bacteria. An example is Legionnaire's disease, an acute respiratory infection caused by the previously unrecognised genus, Legionella. Also, a newly recognised pathogen, Helicobacter, plays an important role in peptic disease.

Another important example, in understanding the etiologies of venereal diseases, was the association of at least 50 percent of the cases of urethritis in male patients with Ureaplasma urealyticum or Chlamydia trachomatis. Recombinant bacteria produced by genetic engineering are enormously useful in bacteriologic research and are being employed to manufacture scarce biomolecules needed for research and patient care. The antibiotic resistance genes, while a problem to the physician, paradoxically are indispensable markers in performing genetic engineering. Genetic probes and the polymerase chain reaction (PCR) are useful in the rapid identification of microbial pathogens in patient specimens.

Genetic manipulation of pathogenic bacteria continues to be indispensable in defining virulence mechanisms. As more protective protein antigens are identified, cloned, and sequenced, recombinant bacterial vaccines will be constructed that should be much better than the ones presently available. In this regard, a recombinant-based and safer pertussis vaccine is already available in some European countries. Also, direct DNA vaccines hold considerable promise. In developed countries, 90 percent of documented infections in hospitalised patients are caused by bacteria. These cases probably reflect only a small percentage of the actual number of bacterial infections occurring in the general population, and usually represent the most severe cases. In developing countries, a variety of bacterial infections often exert a devastating effect on the health of the inhabitants. Malnutrition, parasitic infections, and poor sanitation are a few of the factors contributing to the increased susceptibility of these individuals to bacterial pathogens.

The World Health Organisation has estimated that each year, 3 million people die of tuberculosis, 0.5 million die of pertussis, and 25,000 die of typhoid. Diarrheal diseases, many of which are bacterial, are the second leading cause of death in the world, killing 5 million people annually. Many bacterial diseases can be viewed as a failure of the bacterium to adapt, since a well-adapted parasite ideally thrives in its host without causing significant damage. Relatively nonvirulent microorganisms can cause disease under special conditions - for example, if they are present in unusually large numbers, if the host's defenses are impaired, or if anaerobic conditions exist. Pathogenic bacteria constitute only a small proportion of bacterial species; many nonpathogenic bacteria are beneficial to humans and participate in essential processes such as nitrogen fixation, waste breakdown, food production, drug preparation, and environmental bioremediation. This textbook emphasizes bacteria that have direct medical relevance. In recent years, medical scientists have concentrated on the study of pathogenic mechanisms and host defenses. Understanding host-parasite relationships involving specific pathogens requires familiarity with the fundamental characteristics of the bacterium, the host, and their interactions. Therefore, the first presents with the basic concepts of the immune response, bacterial structure, taxonomy, metabolism, and genetics.

STRUCTURE OF BACTERIA

All bacteria, both pathogenic and saprophytic, are unicellular organisms that reproduce by binary fission. Most bacteria are capable of independent metabolic existence and growth, but species of Chlamydia and Rickettsia are obligately intracellular organisms. Bacterial cells are extremely small and are most conveniently measured in microns (10-6 m). They range in size from large cells such as Bacillus anthracis (1.0 to 1.3 µm X 3 to 10 µm) to very small cells such as Pasteurella tularensis (0.2 X 0.2 to 0.7 µm) Mycoplasmas (atypical pneumonia group) are even smaller, measuring 0.1 to 0.2 µm in diameter. Bacteria therefore have a surface-to-volume ratio that is very high: about 100,000.

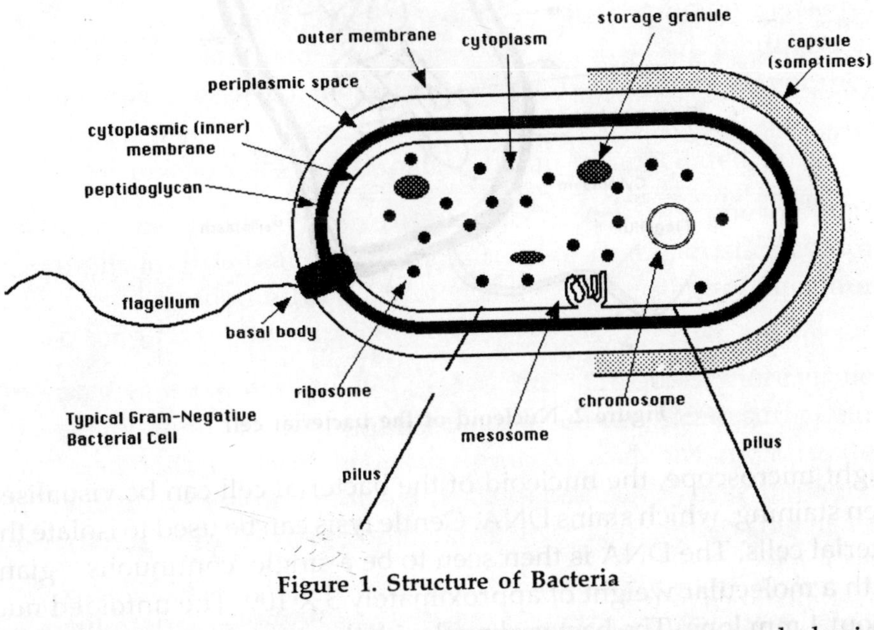

Figure 1. Structure of Bacteria

Bacteria have characteristic shapes. The common microscopic morphologies are cocci ; rods, such as Bacillus and Clostridium species; long, filamentous branched cells, such as Actinomyces species; and comma-shaped and spiral cells, such as Vibrio cholerae and

Treponema pallidum, respectively. The arrangement of cells is also typical of various species or groups of bacteria. Some rods or cocci characteristically grow in chains; some, such as Staphylococcus aureus, form grapelike clusters of spherical cells; some round cocci form cubic packets. Bacterial cells of other species grow separately. The microscopic appearance is therefore valuable in classification and diagnosis. The higher resolving power of the electron microscope not only magnifies the typical shape of a bacterial cell but also clearly resolves its prokaryotic organisation.

The Nucleoid

Prokaryotic and eukaryotic cells were initially distinguished on the basis of structure: the prokaryotic nucleoidthe equivalent of the eukaryotic nucleusis structurally simpler than the true eukaryotic nucleus, which has a complex mitotic apparatus and surrounding nuclear membrane. The bacterial nucleoid, which contains the DNA fibrils, lacks a limiting membrane.

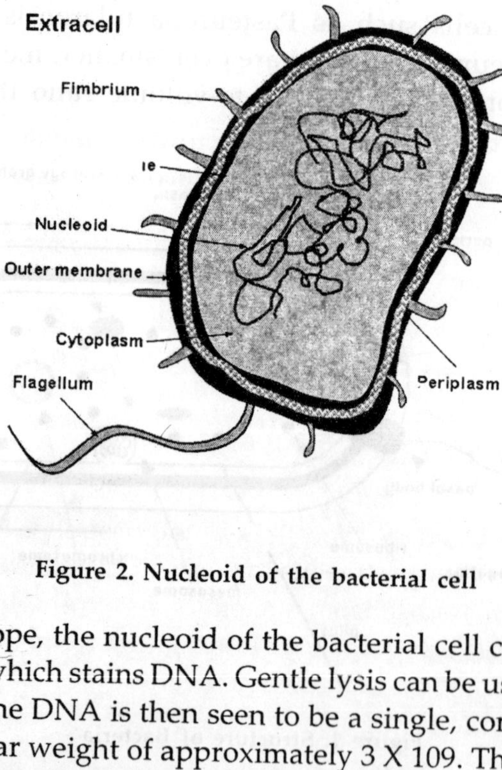

Figure 2. Nucleoid of the bacterial cell

Under the light microscope, the nucleoid of the bacterial cell can be visualised with the aid of Feulgen staining, which stains DNA. Gentle lysis can be used to isolate the nucleoid of most bacterial cells. The DNA is then seen to be a single, continuous, "giant" circular molecule with a molecular weight of approximately 3 X 109. The unfolded nuclear DNA would be about 1 mm long. The bacterial nucleoid, then, is a structure containing a single chromosome. The number of copies of this chromosome in a cell depends on the stage of the cell cycle. Although the mechanism of segregation of the two sister chromosomes following replication is not fully understood, all of the models proposed require that the chromosome be permanently attached to the cell membrane throughout the various stages of the cell cycle. Bacterial chromatin does not contain basic histone proteins, but low-molecular-weight polyamines and magnesium ions may fulfill a function similar to that of eukaryotic histones. Despite the differences between prokaryotic and eukaryotic DNA, prokaryotic DNA from cells infected with bacteriophage g, when visualised by electron microscopy, has a beaded, condensed appearance not unlike that of eukaryotic chromatin.

Surface Appendages

Two types of surface appendage can be recognised on certain bacterial species: the flagella, which are organs of locomotion, and pili (Latin hairs), which are also known as fimbriae (Latin fringes). Flagella occur on both Gram-positive and Gram-negative bacteria, and their presence can be useful in identification. For example, they are found on many species of bacilli but rarely on cocci. In contrast, pili occur almost exclusively on Gram-negative bacteria and are found on only a few Gram-positive organisms (e.g., Corynebacterium renale).

Flagella

Structurally, bacterial flagella are long (3 to 12 μm), filamentous surface appendages about 12 to 30 nm in diameter. The protein subunits of a flagellum are assembled to form a cylindrical structure with a hollow core. A flagellum consists of three parts: (1) the long filament, which lies external to the cell surface; (2) the hook structure at the end of the filament; and (3) the basal body, to which the hook is anchored and which imparts motion to the flagellum. The basal body traverses the outer wall and membrane structures. It consists of a rod and one or two pairs of discs. The thrust that propels the bacterial cell is provided by counterclockwise rotation of the basal body, which causes the helically twisted filament to whirl. The movement of the basal body is driven by a proton motive force rather than by ATP directly.

The ability of bacteria to swim by ns of the propeller-like action of the flagella provides them with the mechanical means to perform chemotaxis. Response to chemical stimuli involves a sophisticated sensory system of receptors that are located in the cell surface and/or periplasm and that transmit information to methyl-accepting chemotaxis proteins that control the flagellar motor. Genetic studies have revealed the existence of mutants with altered biochemical pathways for flagellar motility and chemotaxis.

Chemically, flagella are constructed of a class of proteins called flagellins. The hook and basal-body structures consist of numerous proteins. Mutations affecting any of these gene products may result in loss or impairment of motility. Flagellins are immunogenic and constitute a group of protein antigens called the H antigens, which are characteristic of a given species, strain, or variant of an organism.

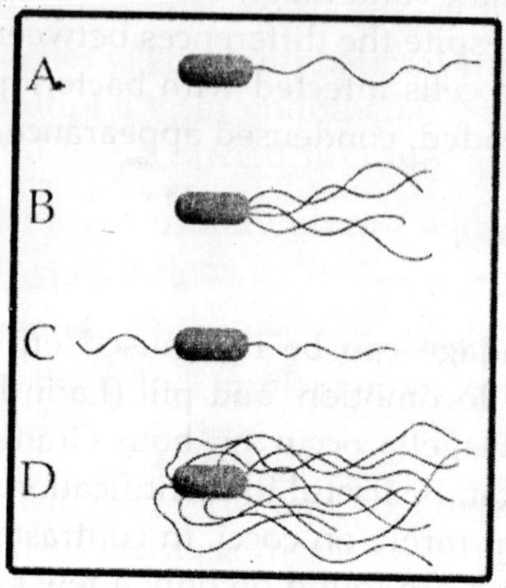

Figure 3. Bacterial falgella arrangements: (a) monotrichous; (b) lophotrichous; (c) amphitrichous; (d) peritrichous

The species specificity of the flagellins reflects differences in the primary structures of the proteins. Antigenic changes of the flagella known as the phase variation of H1 and H2 occurs in *Salmonella typhimurium*. The number and distribution of flagella on the bacterial surface are characteristic for a given species and hence are useful in identifying and classifying bacteria. V cholerae has a single flagellum at one pole of the cell, whereas Proteus vulgaris and E coli have many flagella distributed over the entire cell surface. The flagella of a peritrichous bacterium must aggregate as a posterior bundle to propel the cell in a forward direction. Flagella can be sheared from the cell surface without affecting the viability of the cell. The cell then becomes temporarily nonmotile. In time it synthesizes new flagella and regains motility. The protein synthesis inhibitor chloramphenicol, however, blocks regeneration of flagella.

Pili

The terms pili and fimbriae are usually used interchangeably to describe the thin, hairlike appendages on the surface of many Gram-negative bacteria and proteins of pili are referred to as pilins. Pili are more rigid in appearance than flagella. In some organisms, such as Shigella species and *E coli*, pili are distributed profusely over the cell surface, with as many as 200 per cell. As is easily recognised in strains of *E coli*, pili can come in two types: short, abundant common pili, and a small number of very long pili known as sex pili.

Sex pili can be distinguished by their ability to bind male-specific bacteriophages. The sex pili attach male to female bacteria during conjugation. Pili in many enteric bacteria confer adhesive properties on the bacterial cells, enabling them to adhere to various epithelial surfaces, to red blood cells, and to surfaces of yeast and fungal cells. These adhesive properties of piliated cells play an important role in bacterial colonisation of epithelial surfaces and are therefore referred to as colonisation factors. The common pili found on *E coli* exhibit a sugar specificity analogous to that of phytohemagglutinins and lectins, in that adhesion and hemagglutinating capacities of the organism are inhibited specifically by mannose. Organisms possessing this type of hemagglutination are called mannose-sensitive organisms. Other piliated organisms, such as gonococci, are adhesive and hemagglutinating, but are insensitive to the inhibitory effects of mannose. Extensive antigenic variations in pilins of gonococci are well known.

Surface Layers

The surface layers of the bacterial cell have been identified by various techniques: light microscopy and staining; electron microscopy of thin-sectioned, freeze-fractured, and negatively stained cells; and isolation and biochemical characterisation of individual morphologic components of the cell. The principal surface layers are capsules and loose slime, the cell wall of Gram-positive bacteria and the complex cell envelope of Gram-negative bacteria, plasma (cytoplasmic) membranes, and mesosomal membrane vesicles, which arise from invaginations of the plasma membrane. In bacteria, the cell wall forms a rigid structure of uniform thickness around the cell and is responsible for the characteristic shape of the cell (rod, coccus, or spiral). Inside the cell wall (or rigid peptidoglycan layer) is the plasma (cytoplasmic) membrane; this is usually closely apposed to the wall layer.

Capsules and Loose Slime

Some bacteria form capsules, which constitute the outermost layer of the bacterial cell and surround it with a relatively thick layer of viscous gel. Capsules may be up to 10 μm thick. Some organisms lack a well-defined capsule but have loose, amorphous slime layers external to the cell wall or cell envelope. The a hemolytic Streptococcus mutans, the primary organism found in dental plaque is able to synthesis a large extracellular mucoid glucans from sucrose. Not all bacterial species produce capsules; however, the capsules of encapsulated pathogens are often important determinants of virulence. Encapsulated species are found among both Gram-positive and Gram-negative bacteria. In both groups, most capsules are composed of highmolecular-weight viscous polysaccharides that are retained as a thick gel outside the cell wall or envelope. The capsule of Bacillus anthracis is unusual in that it is composed of a g-glutamyl polypeptide. A plasma membrane stage is involved in the biosynthesis and assembly of the capsular substances, which are extruded or secreted through the outer wall or envelope structures. Mutational loss of enzymes involved in the biosynthesis of the capsular polysaccharides can result in the smooth-to-rough variation seen in the pneumococci. The capsule is not essential for viability. Viability is not affected when capsular polysaccharides are removed enzymatically from the cell surface. The exact functions of capsules are not fully understood, but they do confer resistance to phagocytosis and hence provide the bacterial cell with protection against host defenses to invasion.

Cell Wall and Gram-Negative Cell Envelope

The Gram stain broadly differentiates bacteria into Gram-positive and Gram-negative groups; a few organisms are consistently Gram-variable. Gram-positive and Gram-negative organisms differ drastically in the organisation of the structures outside the plasma membrane but below the capsule: in Gram-negative organisms these structures constitute the cell envelope, whereas in Gram-positive organisms they are called a cell wall.

Most Gram-positive bacteria have a relatively thick (about 20 to 80 nm), continuous cell wall, which is composed largely of peptidoglycan. In thick cell walls, other cell wall polymers are covalently attached to the peptidoglycan. In contrast, the peptidoglycan layer in Gram-negative bacteria is thin; in E coli, the peptidoglycan is probably only a monolayer thick. Outside the peptidoglycan layer in the Gram-negative envelope is an outer membrane structure. In most Gram-negative bacteria, this membrane structure is anchored noncovalently to lipoprotein molecules, which, in turn, are covalently linked to the peptidoglycan. The lipopolysaccharides of the Gram-negative cell envelope form part of the outer leaflet of the outer membrane structure. The organisation and overall dimensions of the outer membrane of the Gram-negative cell envelope are similar to those of the plasma membrane. Moreover, in Gram-negative bacteria such as E coli, the outer and inner membranes adhere to each other at several hundred sites; these sites can break up the continuity of the peptidoglycan layer.

The basic differences in surface structures of Gram-positive and Gram-negative bacteria explain the results of Gram staining. Both Gram-positive and Gram-negative bacteria take up the same amounts of crystal violet (CV) and iodine (I). The CV-I complex, however, is trapped inside the Gram-positive cell by the dehydration and reduced porosity of the thick cell wall as a result of the differential washing step with 95 percent ethanol or other solvent mixture. In contrast, the thin peptidoglycan layer and probable discontinuities at the membrane adhesion sites do not impede solvent extraction of the CV-I complex from the Gram-negative cell. The above mechanism of the Gram stain based on the structural differences between the two groups has been confirmed by sophisticated methods of electron microscopy. Moreover, mechanical disruption of the cell wall of Gram-positive organisms or its enzymatic removal with lysozyme results in complete extraction of the CV-I complex and conversion to a Gram-negative reaction. Therefore, autolytic wall-degrading enzymes that cause cell wall breakage may account for Gram-negative or variable reactions in cultures of Gram-positive organisms.

Peptidoglycan

Unique features of almost all prokaryotic cells are cell wall peptidoglycan and the specific enzymes involved in its biosynthesis. These enzymes are target sites for inhibition of peptidoglycan synthesis by specific antibiotics. The primary chemical structures of peptidoglycans of both Gram-positive and Gram-negative bacteria have been established; they consist of a glycan backbone of repeating groups of b1, 4-linked disaccharides of b1,4-N-acetylmuramyl-N-acetylglucosamine.

Tetrapeptides of L-alanine-D-isoglutamic acid-L-lysine -n-alanine are linked through the carboxyl group by amide linkage of muramic acid residues of the glycan chains; the D-alanine residues are directly cross-linked to the e-amino group of lysine or diaminopimelic acid on a neighboring tetrapeptide, or they are linked by a peptide bridge. In S aureus peptidoglycan, a glycine pentapeptide bridge links the two adjacent peptide structures. The extent of direct or peptide-bridge cross-linking varies from one peptidoglycan to another.

The staphylococcal peptidoglycan is highly cross-linked, whereas that of E coli is much less so, and has a more open peptidoglycan mesh. The diamino acid providing the e-amino group for cross-linking is lysine or diaminopimelic acid, the latter being uniformly present in Gram-negative peptidoglycans. A peptidoglycan with a chemical structure substantially different from that of all eubacteria has been discovered in certain archaebacteria. Instead of muramic acid, this peptidoglycan contains talosaminuronic acid and lacks the D-amino acids found in the eubacterial peptidoglycans. Interestingly, organisms containing this wall polymer are insensitive to penicillin, an inhibitor of the transpeptidases involved in peptidoglycan biosynthesis in eubacteria.

The ß-1,4 glycosidic bond between N-acetylmuramic acid and N-acetylglucosamine is specifically cleaved by the bacteriolytic enzyme lysozyme. Widely distributed in nature, this enzyme is present in human tissues and secretions and can cause complete digestion of the peptidoglycan walls of sensitive organisms. When lysozyme is allowed to digest the cell wall of Gram-positive bacteria suspended in an osmotic stabilizer, protoplasts are formed. These protoplasts are able to survive and continue to grow on suitable media in the wall-less state. Gram-negative bacteria treated similarly produce spheroplasts, which retain much of the outer membrane structure. The dependence of bacterial shape on the peptidoglycan is shown by the transformation of rod-shaped bacteria to spherical protoplasts (spheroplasts) after enzymatic breakdown of the peptidoglycan. The mechanical protection afforded by the wall peptidoglycan layer is evident in the osmotic fragility of both protoplasts and spheroplasts.

There are two groups of bacteria that lack the protective cell wall peptidoglycan structure, the Mycoplasma species, one of which causes atypical pneumonia and some genitourinary tract infections and the L-forms, which originate from Gram-positive or Gram-negative bacteria and are so designated because of their discovery and description at the Lister Institute, London. The mycoplasmas and L-forms are all Gram-negative and insensitive to penicillin and are bounded by a surface membrane structure. L-forms arising "spontaneously" in cultures or isolated from infections are structurally related to protoplasts and spheroplasts; all three forms revert infrequently and only under special conditions.

Teichoic Acids

Wall teichoic acids are found only in certain Gram-positive bacteria; so far, they have not been found in gram- negative organisms. Substituent groups on the polyol chains can include D-alanine, N-acetylglucosamine, N-acetylgalactosamine, and glucose; the substituent is characteristic for the teichoic acid from a particular bacterial species and can act as a specific antigenic determinant. Teichoic acids are covalently linked to the peptidoglycan. These highly negatively charged polymers of the bacterial wall can serve as a cation-sequestering mechanism.

Accessory Wall Polymers

In addition to the principal cell wall polymers, the walls of certain Gram-positive bacteria possess polysaccharide molecules linked to the peptidoglycan. For example, the C polysaccharide of streptococci confers group specificity. Acidic polysaccharides attached to the peptidoglycan are called teichuronic acids. Mycobacteria have peptidoglycolipids, glycolipids, and waxes associated with the cell wall.

Lipopolysaccharides

A characteristic feature of Gram-negative bacteria is possession of various types of complex macromolecular lipopolysaccharide (LPS). So far, only one Gram-positive organism, Listeria monocytogenes, has been found to contain an authentic LPS. The LPS of this bacterium and those of all Gram-negative species are also called endotoxins, thereby distinguishing these cell-bound, heat-stable toxins from heat-labile, protein exotoxins secreted into culture media. Endotoxins possess an array of powerful biologic activities and play an important role in the pathogenesis of many Gram-negative bacterial infections. In addition to causing endotoxic shock, LPS is pyrogenic, can activate macrophages and complement, is mitogenic for B lymphocytes, induces interferon production, causes tissue necrosis and tumor regression, and has adjuvant properties. The endotoxic properties of LPS reside largely in the lipid A components. Usually, the LPS molecules have three regions: the lipid A structure required for insertion in the outer leaflet of the outer membrane bilayer; a covalently attached core composed of 2-keto-3deoxyoctonic acid (KDO), heptose, ethanolamine, N-acetylglucosamine, glucose, and galactose; and polysaccharide chains linked to the core.

The polysaccharide chains constitute the O-antigens of the Gram-negative bacteria, and the individual monosaccharide constituents confer serologic specificity on these components. The demonstration of the structure of lipid A of LPS of a heptoseless mutant of Salmonella typhimurium has established that amide-linked hydroxymyristoyl and lauroxymyristoyl groups are attached to the nitrogen of the 2- and 2'-carbons, respectively, and that hydroxymyristoyl and myristoxymyristoyl groups are attached to the oxygen of the 3- and 3'-carbons of the disaccharide, respectively. Therefore, only position 6' is left for attachment of KDO units.

LPS and phospholipids help confer asymmetry to the outer membrane of the Gram-negative bacteria, with the hydrophilic polysaccharide chains outermost. Each LPS is held in the outer membrane by relatively weak cohesive forces and can be dissociated from the cell surface with surface-active agents. As in peptidoglycan biosynthesis, LPS molecules are assembled at the plasma or inner membrane. These newly formed molecules are initially inserted into the outer-inner membrane adhesion sites.

The outer membranes of Gram-negative bacteria appear broadly similar to the plasma or inner membranes; however, they differ from the inner membranes and walls of Gram-positive bacteria in numerous respects. The lipid A of LPS is inserted with phospholipids to create the outer leaflet of the bilayer structure; the lipid portion of the lipoprotein and phospholipid form the inner leaflet of the outer membrane bilayer of most Gram-negative bacteria. In addition to these components, the outer membrane possesses several major outer membrane proteins; the most abundant is called porin. The assembled subunits of porin form a channel that limits the passage of hydrophilic molecules across the outer membrane barrier to those having molecular weights that are usually less than 600 to 700. Evidence also suggests that hydrophobic pathways exist across the outer membrane and are partly responsible for the differential penetration and effectiveness of certain b-lactam antibiotics that are active against various Gram-negative bacteria.

Although the outer membranes act as a permeability barrier or molecular sieve, they do not appear to possess energy-transducing systems to drive active transport. Several outer membrane proteins, however, are involved in the specific uptake of metabolites and iron from the medium. Thus, outer membranes of the Gram-negative bacteria provide a selective barrier to external molecules and thereby prevent the loss of metabolite-binding proteins and hydrolytic enzymes found in the periplasmic space. The periplasmic space is the region between the outer surface of the inner membrane and the inner surface of the outer membrane. Thus, Gram-negative bacteria have a cellular compartment that has no equivalent in Gram-positive organisms. In addition to the hydrolytic enzymes, the periplasmic space holds binding proteins involved in membrane transport and chemotactic receptor activities. Moreover, plasmid-encoded b-lactamases and aminoglycoside-modifying enzymes in the periplasmic space produce antibiotic resistance by degrading or modifying an antibiotic in transit to its target sites on the membrane or on the ribosomes.

INTRACELLULAR COMPONENTS

Plasma (Cytoplasmic) Membranes

Bacterial plasma membranes, the functional equivalents of eukaryotic plasma membranes, are referred to variously as cytoplasmic, protoplast, or (in Gram-negative organisms) inner membranes. Similar in overall dimensions and appearance in thin sections to biomembranes from eukaryotic cells, they are composed primarily of proteins and lipids. Protein-to-lipid ratios of bacterial plasma membranes are approximately 3: 1, close to those for mitochondrial membranes. Unlike eukaryotic cell membranes, the bacterial membrane has no sterols, and bacteria lack the enzymes required for sterol biosynthesis.

Although their composition is similar to that of inner membranes of Gram-negative species, cytoplasmic membranes from Gram-positive bacteria possess a class of macromolecules not present in the Gram-negative membranes. Many Gram-positive bacterial membranes contain membrane-bound lipoteichoic acid, and species lacking this component contain an analogous membrane-bound succinylated lipomannan. Lipoteichoic acids are structurally similar to the cell wall glycerol teichoic acids in that they have basal polyglycerol phosphodiester 1-3 linked chains. These chains terminate with the phosphomonoester end of the polymer, which is linked covalently to either a glycolipid or a phosphatidyl glycolipid moiety. Thus, a hydrophobic tail is provided for anchoring in the membrane lipid layers.

As in the cell wall glycerol teichoic acid, the lipoteichoic acids can have glycosidic and D-alanyl ester substituents on the C-2 position of the glycerol. Both membrane-bound lipoteichoic acid and membrane-bound succinylated lipomannan can be detected as antigens on the cell surface, and the glycerol-phosphate and succinylated mannan chains appear to extend through the cell wall structure. This class of polymer has not yet been found in the cytoplasmic membranes of Gram-negative organisms. In both instances, the lipoteichoic acids and the lipomannans are negatively charged components and can sequester positively charged substances. They have been implicated in adhesion to host cells, but their functions remain to be elucidated. Multiple functions are performed by the plasma membranes of both Gram-positive and Gram-negative bacteria. Plasma membranes are the site of active transport, respiratory chain components, energy-transducing systems, the H^+-ATPase of the proton pump, and membrane stages in the biosynthesis of phospholipids, peptidoglycan, LPS, and capsular polysaccharides. In essence, the bacterial cytoplasmic membrane is a multifunction structure that combines the mitochondrial transport and biosynthetic functions that are usually compartmentalised in discrete membranous organelles in eukaryotic cells.

Mesosomes

Gram-positive bacteria reveal the presence of vesicular or tubular-vesicular membrane structures called mesosomes, which are apparently formed by an invagination of the plasma membrane. These structures are much more prominent in Gram-positive than in Gram-negative organisms. At one time, the mesosomal vesicles were thought to be equivalent to bacterial mitochondria; however, many other membrane functions have also been attributed to the mesosomes. At present, there is no satisfactory evidence to suggest that they have a unique biochemical or physiologic function. Indeed, electron-microscopic studies have suggested that the mesosomes, as usually seen in thin sections, may arise from membrane perturbation and fixation artifacts. No general agreement exists about this theory, however, and some evidence indicates that mesosomes may be related to events in the cell division cycle.

OTHER INTRACELLULAR COMPONENTS

In addition to the nucleoid and cytoplasm (cytosol), the intracellular compartment of the bacterial cell is densely packed with ribosomes of the 70S type. These ribonucleoprotein particles, which have a diameter of 18 nm, are not arranged on a membranous rough endoplasmic reticulum as they are in eukaryotic cells. Other granular inclusions randomly distributed in the cytoplasm of various species include metabolic reserve particles such as poly-b-hydroxybutyrate (PHB), polysaccharide and glycogen-like granules, and polymetaphosphate or metachromatic granules. Endospores are highly heat-resistant, dehydrated resting cells formed intracellularly in members of the genera Bacillus and Clostridium. Sporulation, the process of forming endospores, is an unusual property of certain bacteria. The series of biochemical and morphologic changes that occur during sporulation represent true differentiation within the cycle of the bacterial cell. The process, which usually begins in the stationary phase of the vegetative cell cycle, is initiated by depletion of nutrients. The cell then undergoes a highly complex, well-defined sequence of morphologic and biochemical events that ultimately lead to the formation of mature endospores.

As many as seven distinct stages have been recognised by morphologic and biochemical studies of sporulating Bacillus species: stage 0, vegetative cells with two chromosomes at the end of exponential growth; stage I, formation of axial chromatin filament and excretion of exoenzymes, including proteases; stage II, forespore septum formation and segregation of nuclear material into two compartments; stage III, spore protoplast formation and elevation of tricarboxylic acid and glyoxylate cycle enzyme levels; stage IV, cortex formation and refractile appearance of spore; stage V, spore coat protein formation; stage VI, spore maturation, modification of cortical peptidoglycan, uptake of dipicolinic acid and calcium, and development of resistance to heat and organic solvents; and stage VII, final maturation and liberation of endospores from mother cells (in some species).

When newly formed, endospores appear as round, highly refractile cells within the vegetative cell wall, or sporangium. Some strains produce autolysins that digest the walls and liberate free endospores. The spore protoplast, or core, contains a complete nucleus, ribosomes, and energy generating components that are enclosed within a modified cytoplasmic membrane. The peptidoglycan spore wall surrounds the spore membrane; on germination, this wall becomes the vegetative cell wall. Surrounding the spore wall is a thick cortex that contains an unusual type of peptidoglycan, which is rapidly released on germination. A spore coat of keratinlike protein encases the spore contained within a membrane (the exosporium). During maturation, the spore protoplast dehydrates and the spore becomes refractile and resistant to heat, radiation, pressure, desiccation, and chemicals; these properties correlate with the cortical peptidoglycan and the presence of large amounts of calcium dipicolinate. Recent evidence indicated that the spores of Bacillus spharicus were revived which had been preserved in amber for more than 25 million years. Their claims need to be reevaluated.

REFERENCES

Alcamo IE. 2001. *Fundamentals of microbiology*. Boston: Jones and Bartlett.

Atlas RM. 1995. *Principles of microbiology*. St. Louis: Mosby.

Martinko JM, Madigan MT.2005. *Brock Biology c Aicroorganisms* (11th ed. ed.). Englewood Cliffs, N.J: Prentice Hall.

Holt JC, Bergey DH. 1994. *Bergey's manual of leterminative bacteriology* (9th ed. ed.). Baltimore: Williams & Wilkins.

Hugenholtz P, Goebel BM, Pace NR. 1998. "Impact of culture-independent studies on the emerging phylogenetic view of bacterial diversity". *J Bacteriol* 180 (18): 4765–74.

Funke BR, Tortora GJ, Case CL. 2004. *Microbiology: an introduction* (8th ed ed.). San Francisco: Benjamin Cummings.

Shively, Jessup M. 2006. *Complex Intracellular Structures in Prokaryotes* (Microbiology Monographs). Berlin: Springer.

4

Classification of Bacteria

Bacteria are classified and identified to distinguish one organism from another and to group similar organisms by criteria of interest to microbiologists or other scientists. The most important level of this type of classification is the species level. A species name should mean the same thing to everyone. Within one species, strains and subgroups can differ by the disease they produce, their environmental habitat, and many other characteristics. Formerly, species were created on the basis of such criteria, which may be extremely important for clinical microbiologists and physicians but which are not a sufficient basis for establishing a species. Verification of existing species and creation of new species should involve biochemical and other phenotypic criteria as well as DNA relatedness. In numerical or phenetic approaches to classification, strains are grouped on the basis of a large number of phenotypic characteristics. DNA relatedness is used to group strains on the basis of overall genetic similarity.

Species are identified in the clinical laboratory by morphological traits and biochemical tests, some of which are supplemented by serologic assessments (e.g., identification of *Salmonella* and *Shigella* species). Because of differences in pathogenicity *(Escherichia coli)* or the necessity to characterise a disease outbreak (*Vibrio cholerae*, methicillin-resistant *Staphylococcus aureus)*, strains of medical interest are often classified below the species level by serology or identification of toxins. Pathogenic or epidemic strains also can be classified by the presence of a specific plasmid, by their plasmid profile (the number and sizes of plasmids), or by bacteriophage susceptibility patterns (phage typing). Newer molecular biologic techniques have enabled scientists to identify some species and strains (without the use of biochemical tests) by identifying a specific gene or genetic sequence, sometimes directly from the clinical specimen.

Laboratories have no difficulty in identifying typical strains of common bacteria using commonly available test systems. Problems do arise, however, when atypical strains or rare or newly described species are not in the data base. Such difficulties are compounded when the strains are misidentified rather than unidentified, and so laboratory personnel and physicians (at least infectious diseases specialists) should be familiar with taxonomic reference texts and journals that publish papers on new species. Bacterial nomenclature at the genus and species level changes often, based primarily on the use of newer genetic techniques. A species may acquire more than one name. In some cases the recognition of a new species results in a unique correlation with specific clinical problems. For example, recognition of *Porphyromonas gingivalis* as a unique species, separate from its previous inclusion within *Bacteroides melaninogenicus* (now known to be composed of several taxonomic groups of black-pigmenting anaerobic gram-negative bacilli), elucidated its role as a key pathogen in adult periodontitis. It is important to understand why these changes and synonyms exist in taxonomy.

Taxonomy is the science of classification, identification, and nomenclature. For classification purposes, organisms are usually organised into subspecies, species, genera, families, and higher orders. For eukaryotes, the definition of the species usually stresses the ability of similar organisms to reproduce sexually with the formation of a zygote and to produce fertile offspring. However, bacteria do not undergo sexual reproduction in the eukaryotic sense. Other criteria are used for their classification.

Classification is the orderly arrangement of bacteria into groups. There is nothing inherently scientific about classification, and different groups of scientists may classify the same organisms differently. For example, clinical microbiologists are interested in the serotype, antimicrobial resistance pattern, and toxin and invasiveness factors in *Escherichia coli,* whereas geneticists are concerned with specific mutations and plasmids.

Nomenclature (naming) is the means by which the characteristics of a species are defined and communicated among microbiologists. A species name should mean the same thing to all microbiologists, yet some definitions vary in different countries or microbiologic specialty groups. For example, the organism known as *Clostridium perfringens* in the United States is called *Clostridium welchii* in England.

A bacterial species is a distinct organism with certain characteristic features, or a group of organisms that resemble one another closely in the most important features of their organisation. In the past, unfortunately, there was little agreement about these criteria or about the number of features necessary to distinguish a species. Species were often defined solely by such criteria as host range, pathogenicity, or ability to produce gas during the fermentation of a given sugar. Without a universal consensus, criteria reflected the interests of the investigators who described a particular species. For example, bacteria that caused plant diseases were often defined by the plant from which they were isolated; also, each new *Salmonella* serotype that was discovered was given species status. These practices have been replaced by generally accepted genetic criteria that can be used to define species in all groups of bacteria.

APPROACHES TO TAXONOMY

Numerical Approach

In their studies on members of the family Enterobacteriaceae, Edwards and Ewing established the following principles to characterise, classify, and identify organisms:

Classification and identification of an organism should be based on its overall morphologic and biochemical pattern. A single characteristic (pathogenicity, host range, or biochemical reaction), regardless of its importance, is not a sufficient basis for classifying or identifying an organism.

A large and diverse strain sample must be tested to determine accurately the biochemical characteristics used to distinguish a given species.

Atypical strains often are perfectly typical members of a given biogroup within an existing species, but sometimes they are typical members of an unrecognised new species.

In numerical taxonomy (also called computer or phenetic taxonomy) many (50 to 200) biochemical, morphological, and cultural characteristics, as well as susceptibilities to antibiotics and inorganic compounds, are used to determine the degree of similarity between organisms. In numerical studies, investigators often calculate the coefficient of similarity or percentage of similarity between strains (where strain indicates a single isolate from a specimen). A dendrogram or a similarity matrix is constructed that joins individual strains into groups and places one group with other groups on the basis of their percentage of similarity.

In some cases, certain characteristics may be weighted more heavily; for example, the presence of spores in *Clostridium* might be weighted more heavily than the organism's ability to use a specific carbon source. A given level of similarity can be equated with relatedness at the genus, species, and, sometimes, subspecies levels. For instance, strains of a given species may cluster at a 90% similarity level, species within a given genus may cluster at the 70 percent level, and different genera in the same family may cluster at the 50 percent or lower level.

When this approach is the only basis for defining a species, it is difficult to know how many and which tests should be chosen; whether and how the tests should be weighted; and what level of similarity should be chosen to reflect relatedness at the genus and species levels.

Most bacteria have enough DNA to specify some 1,500 to 6,000 average-sized genes. Therefore, even a battery of 300 tests would assay only 5 to 20 percent of the genetic potential of a bacterium. Tests that are comparatively simple to conduct (such as those for carbohydrate utilisation and for enzymes, presence of which can be assayed colorimetrically) are performed more often than tests for structural, reproductive, and regulatory genes, presence of which is difficult to assay. Thus, major differences may go undetected.

Other types of errors may occur when species are classified solely on the basis of phenotype. For example, different enzymes (specified by different genes) may catalyse the same reaction. Also, even if a metabolic gene is functional, negative reactions can occur because of the inability of the substrate to enter the cell, because of a mutation in a regulatory gene, or by production of an inactive protein. There is not necessarily a one-to-one correlation between a reaction and the number of genes needed to carry out that reaction. For instance, six enzymatic steps may be involved in a given pathway. If an assay for the end product is performed, a positive reaction indicates the presence of all six enzymes, whereas a negative reaction can mean the absence or nonfunction of one to six enzymes. Several other strain characteristics can affect phenotypic characterisation; these include growth rate, incubation temperature, salt requirement, and pH. Plasmids that carry metabolic genes can enable strains to carry out reactions atypical for strains of that species.

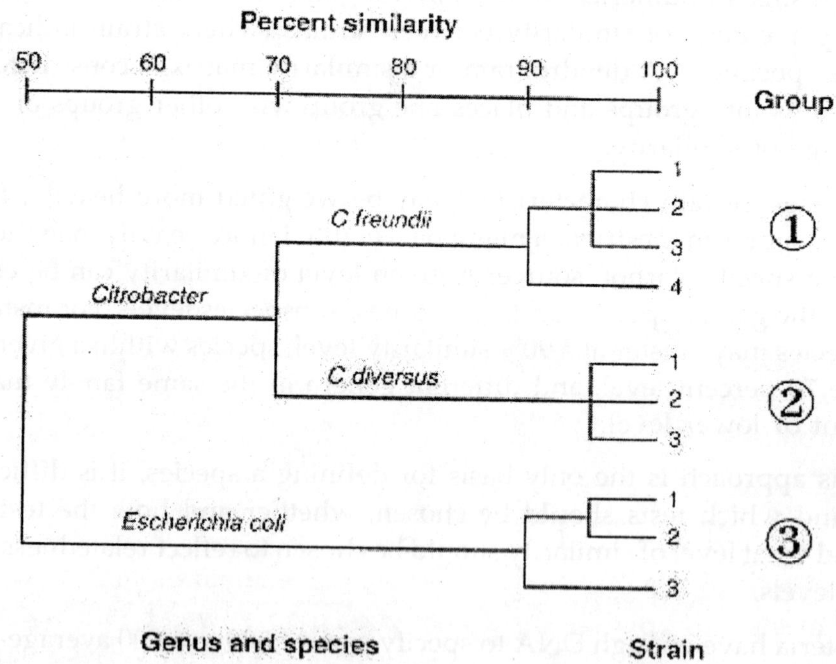

Figure 1. Example of Dendrogram

The same set of "definitive" reactions cannot be used to classify all groups of organisms, and there is no standard number of specific reactions that allows identification of a species. Organisms are identified on the basis of phenotype, but, from the taxonomic standpoint, definition of species solely on this basis is subject to error.

Phylogenetic Approach

The ideal means of identifying and classifying bacteria would be to compare each gene sequence in a given strain with the gene sequences for every known species. This cannot be done, but the total DNA of one organism can be compared with that of any other organism by a method called nucleic acid hybridisation or DNA hybridisation. This method can be used to measure the number of DNA sequences that any two organisms have in common and to estimate the percentage of divergence within DNA sequences that are related but not identical. DNA relatedness studies have been done for yeasts, viruses, bacteriophages, and many groups of bacteria.

Five factors can be used to determine DNA relatedness: genome size, guanine-plus-cytosine (G+C) content, DNA relatedness under conditions optimal for DNA reassociation, thermal stability of related DNA sequences, and DNA relatedness under conditions supraoptimal for DNA reassociation. Because it is not practical to conduct these genotypic or phylogenetic evaluations in clinical laboratories, the results of simpler tests usually must be correlated with known phylogenetic data. For example, yellow strains of *Enterobacter cloacae* were shown, by DNA relatedness, to form a separate species, *Enterobacter sakazakii,* but were not designated as such until results of practical tests were correlated with the DNA data to allow routine laboratories to identify the new species.

Genome Size

True bacterial DNAs have genome sizes (measured as molecular weight) between 1 X 109 and 8 X 109. Genome size determinations sometimes can distinguish between groups. They were used to distinguish *Legionella pneumophila* (the legionnaire's disease bacterium) from *Bartonella (Rickettsia) qui*[illegible], the agent of trench fever. *L pneumophila* has a genome size of about 3 X 109; that of *B quintana* is about 1 X 109.

Guanine-plus-Cytosine Content

The G+C content in bacterial DNA ranges from about 25 to 75 percent. This percentage is specific, but not exclusive, for a species; two strains with a similar G+C content may or may not belong to the same species. If the G+C contents are very different, however, the strains cannot be members of the same species.

DNA Relatedness

DNA relatedness is determined by allowing single-stranded DNA from one strain to reassociate with single-stranded DNA from a second strain, to form a double-stranded DNA molecule (Figure 2). This is a specific, temperature-dependent reaction. The optimal temperature for DNA reassociation is 25 to 30°C below the temperature at which native double-stranded DNA denatures into single strands. Many studies indicate that a bacterial species is composed of strains that are 70 to 100 percent related. In contrast, relatedness between different species is 0 to about 65 percent. It is important to emphasize that the term "related" does not mean "identical" or "homologous." Similar but nonidentical nucleic acid sequences can reassociate.

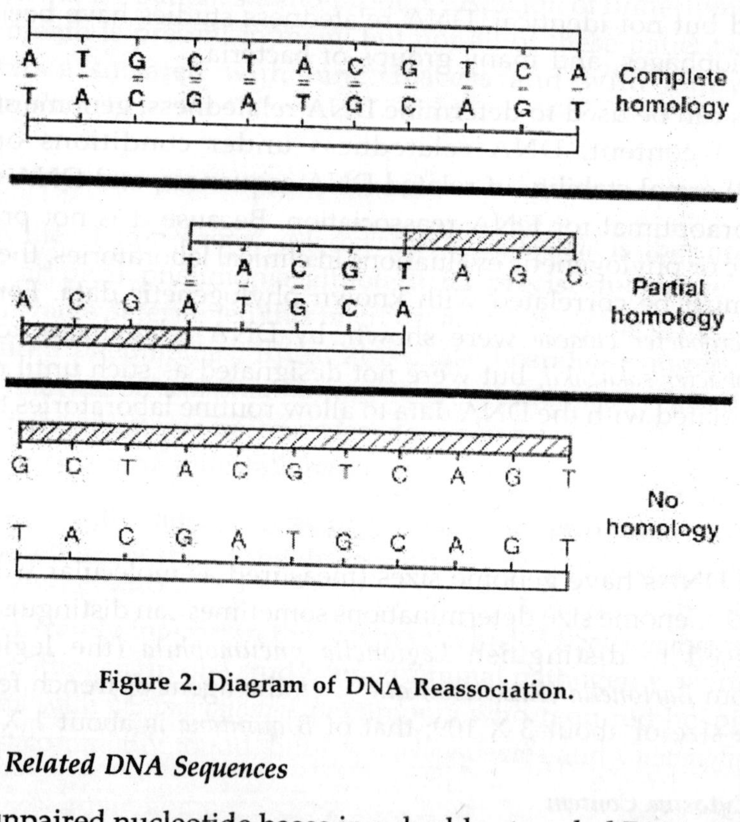

Figure 2. Diagram of DNA Reassociation.

Thermal Stability of Related DNA Sequences

Each 1 percent of unpaired nucleotide bases in a double-stranded DNA sequence causes a 1 percent decrease in the thermal stability of that DNA duplex. Therefore, a comparison between the thermal stability of a control double-stranded molecule (in which both strands of DNA are from the same organism) and that of a heteroduplex (DNA strands from two different organisms)' allows assessment of divergence between related nucleotide sequences.

When the incubation temperature used for DNA reassociation is raised from 25-30° C below the denaturation temperature to only 10-15° C below the denaturation temperature, only very closely related (and therefore highly thermally stable) DNA sequences can reassociate. Strains from the same species are 60 percent or more related at these supraoptimal incubation temperatures.

Defining Species on the Basis of DNA Relatedness

Use of these five factors allows a species definition based on DNA. Thus, *E coli* can be defined as a series of strains with a G+C content of 49 to 52 moles percent, a genome molecular weight of 2.3 X 109 to 3.0 X 109, relatedness of 70 percent or more at an optimal reassociation temperature with 0 to 4 percent divergence in related sequences, and relatedness of 60 percent or more at a supraoptimal reassociation temperature. Experience with more than 300 species has produced an arbitrary phylogenetic definition of a species to which most taxonomists subscribe: "strains with approximately 70% or greater DNA-DNA relatedness and with 5° C or less divergence in related sequences." When these two criteria are met, genome size and G+C content are always similar, and relatedness is almost always 60 percent or more at supraoptimal incubation temperatures. The 70 percent species relatedness rule has been ignored occasionally when the existing nomenclature is deeply ingrained, as is that for *E coli* and the four *Shigella* species. Because these organisms are all 70 percent or more related, DNA studies indicate that they should be grouped into a single species, instead of the present five species in two genera. This change has not been made because of the presumed confusion that would result.

DNA relatedness provides one species definition that can be applied equally to all organisms. Moreover, it cannot be affected by phenotypic variation, mutations, or the presence or absence of metabolic or other plasmids. It measures overall relatedness, and these factors affect only a very small percentage of the total DNA.

Polyphasic Approach

In practice, the approach to bacterial taxonomy should be polyphasic. The first step is phenotypic grouping of strains by morphological, biochemical and any other characteristics of interest. The phenotypic groups are then tested for DNA relatedness to determine whether the observed phenotypic homogeneity (or heterogeneity) is reflected by phylogenetic homogeneity or heterogeneity. The third and most important step is reexamination of the biochemical characteristics of the DNA relatedness groups. This allows determination of the biochemical borders of each group and determination of reactions of diagnostic value for the group. For identification of a given organism, the importance of specific tests is weighted on the basis of correlation with DNA results.

Occasionally, the reactions commonly used will not distinguish completely between two distinct DNA relatedness groups. In these cases, other biochemical tests of diagnostic value must be sought.

CHARACTERISTICS USEFUL IN CLASSIFICATION AND IDENTIFICATION

Morphologic Characteristics

Both wet-mounted and properly stained bacterial cell suspensions can yield a great deal of information. These simple tests can indicate the Gram reaction of the organism; whether it is acid-fast; its motility; the arrangement of its flagella; the presence of spores, capsules, and inclusion bodies; and, of course, its shape. This information often can allow identification of an organism to the genus level, or can minimise the possibility that it belongs to one or another group. Colony characteristics and pigmentation are also quite helpful. For example, colonies of several *Porphyromonas* species autofluoresce under long-wavelength ultraviolet light, and *Proteus* species swarm on appropriate media.

Growth Characteristics

A primary distinguishing characteristic is whether an organism grows aerobically, anaerobically, facultatively (i.e., in either the presence or absence of oxygen), or microaerobically (i.e., in the presence of a less than atmospheric partial pressure of oxygen). The proper atmospheric conditions are essential for isolating and identifying bacteria. Other important growth assessments include the incubation temperature, pH, nutrients required, and resistance to antibiotics. For example, one diarrheal disease agent, *Campylobacter jejuni,* grows well at 42° C in the presence of several antibiotics; another, *Y enterocolitica,* grows better than most other bacteria at 4° C. *Legionella, Haemophilus,* and some other pathogens require specific growth factors, whereas *E coli* and most other Enterobacteriaceae can grow on minimal media.

Antigens and Phage Susceptibility

Cell wall (O), flagellar (H), and capsular (K) antigens are used to aid in classifying certain organisms at the species level, to serotype strains of medically important species for epidemiologic purposes, or to identify serotypes of public health importance. Serotyping is also sometimes used to distinguish strains of exceptional virulence or public health importance, for example with *V cholerae* (O1 is the pandemic strain) and *E coli* (enterotoxigenic, enteroinvasive, enterohemorrhagic, and enteropathogenic serotypes).

Phage typing (determining the susceptibility pattern of an isolate to a set of specific bacteriophages) has been used primarily as an aid in epidemiologic surveillance of diseases caused by *Staphylococcus aureus,* mycobacteria, *P aeruginosa, V cholerae,* and *S typhi*. Susceptibility to bacteriocins has also been used as an epidemiologic strain marker. In most cases recently, phage and bacteriocin typing have been supplanted by molecular methods.

Biochemical Characteristics

Most bacteria are identified and classified largely on the basis of their reactions in a series of biochemical tests. Some tests are used routinely for many groups of bacteria (oxidase, nitrate reduction, amino acid degrading enzymes, fermentation or utilisation of carbohydrates); others are restricted to a single family, genus, or species (coagulase test for staphylococci, pyrrolidonyl arylamidase test for Gram-positive cocci).

Both the number of tests needed and the actual tests used for identification vary from one group of organisms to another. Therefore, the lengths to which a laboratory should go in detecting and identifying organisms must be decided in each laboratory on the basis of its function, the type of population it serves, and its resources. Clinical laboratories today base the extent of their work on the clinical relevance of an isolate to the particular patient from which it originated, the public health significance of complete identification, and the overall cost-benefit analysis of their procedures. For example, the Centers for Disease Control and Prevention (CDC) reference laboratory uses at least 46 tests to identify members of the Enterobacteriaceae, whereas most clinical laboratories, using commercial identification kits or simple rapid tests, identify isolates with far fewer criteria.

CLASSIFICATION BELOW AND ABOVE SPECIES LEVEL

Below the Species Level

Particularly for epidemiological purposes, clinical microbiologists must distinguish strains with particular traits from other strains in the same species. For example, serotype O157:H7 *E coli* are identified in stool specimens because of their association with bloody diarrhea and subsequent hemolytic uremic syndrome.

Below the species level, strains are designated as groups or types on the basis of common serologic or biochemical reactions, phage or bacteriocin sensitivity, pathogenicity, or other characteristics. Many of these characteristics are already used and accepted: serotype, phage type, colicin type, biotype, bioserotype (a group of strains from the same species with common biochemical and serologic characteristics that set them apart from other members of the species), and pathotype (e.g., toxigenic *Clostridium difficile,* invasive *E coli,* and toxigenic *Corynebacterium diphtheriae*).

Above Species Level

In addition to species and subspecies designations, clinical microbiologists must be familiar with genera and families. A genus is a group of related species, and a family is a group of related genera.

An ideal genus would be composed of species with similar phenotypic and phylogenetic characteristics. Some phenotypically homogeneous genera approach this criterion *(Citrobacter, Yersinia,* and *Serratia).* More often, however, the phenotypic similarity is present, but the genetic relatedness is not. *Bacillus, Clostridium,* and *Legionella* are examples of accepted phenotypic genera in which genetic relatedness between species is not 50 to 65 percent, but 0 to 65 percent. When phenotypic and genetic similarity are not both present, phenotypic similarity generally should be given priority in establishing genera. Identification practices are simplified by having the most phenotypically similar species in the same genus. The primary consideration for a genus is that it contain biochemically similar species that are convenient or important to consider as a group separate from other groups of organisms.

The sequencing of ribosomal RNA (rRNA) genes, which have been highly conserved through evolution, allows phylogenetic comparisons to be made between species whose total DNAs are essentially unrelated. It also allows phylogenetic classification at the genus, family, and higher taxonomic levels. The rRNA sequence data are usually not used to designate genera or families unless supported by similarities in phenotypic tests.

DESIGNATION OF NEW SPECIES AND NOMENCLATURAL CHANGES

Species are named according to principles and rules of nomenclature set forth in the Bacteriological Code. Scientific names are taken from Latin or Greek. The correct name of a species or higher taxon is determined by three criteria: valid publication, legitimacy of the name with regard to the rules of nomenclature, and priority of publication (that is, it must be the first validly published name for the taxon).

To be published validly, a new species proposal must contain the species name, a description of the species, and the designation of a type strain for the species, and the name must be published in the *International Journal for Systematic Bacteriology (IJSB).* Once proposed, a name does not go through a formal process to be accepted officially; in fact, the opposite is truea validly published name is assumed to be correct unless and until it is challenged officially. A challenge is initiated by publishing a request for an opinion (to the Judicial Commission of the International Association of Microbiological Societies) in the *IJSB.* This occurs only in cases in which the validity of a name is questioned with respect to compliance with the rules of the Bacteriological Code. A question of classification that is based on scientific data (for example, whether a species, on the basis of its biochemical or genetic characteristics, or both, should be placed in a new genus or an existing genus) is not settled by the Judicial Commission, but by the preference and usage of the scientific community. This is why there are pairs of names such as *Providencia rettgeri/Proteus rettgeri, Moraxella catarrhalis/Branhamella catarrhalis,* and *Legionella micdadei/Tatlockia micdadei.* More than one name may thus exist for a single organism. This is not, however, restricted to bacterial nomenclature. Multiple names exist for many antibiotics and other drugs and enzymes.

A number of genera have been divided into additional genera and species have been moved to new or existing genera, such as *Arcobacter* (new genus for former members of *Campylobacter*) and *Burkholderia* species (formerly species of *Pseudomonas*). Two former *Campylobacter* species (*cinaedi* and *fennelliae*) have been moved to the existing genus *Helicobacter* in another example.

Assessing Newly Described Bacteria

Since 1974, the number of genera in the family Enterobacteriaceae has increased from 12 to 28 and the number of species from 42 to more than 140, some of which have not yet been named. Similar explosions have occurred in other genera. In 1974, five species were listed in the genus *Vibrio* and four in *Campylobacter;* the genus *Legionella* was unknown. Today, there are at least 25 species in *Vibrio,* 12 *Campylobacter* species, and more than 40 species in *Legionella.* The total numbers of genera and species continue to increase dramatically.

The clinical significance of the agent of legionnaire's disease was well known long before it was isolated, characterised, and classified as *Legionella pneumophila.* In most cases, little is known about the clinical significance of a new species at the time it is first described. Assessments of clinical significance begin after clinical laboratories adopt the procedures needed to detect and identify the species and accumulate a body of data.

In fact, the detection and even the identification of uncultivatable microbes from different environments are now possible using standard molecular methods. The agents of cat scratch disease (*Bartonella henselae*) and Whipple's disease (*Tropheryma whippelii*) were elucidated in this manner. *Bartonella henselae* has since been cultured from several body sites from numerous patients; *T whippelii* remains uncultivated.

New species will continue to be described. Many will be able to infect humans and cause disease, especially in those individuals who are immunocompromised, burned, postsurgical, geriatric, and suffering from acquired immunodeficiency syndrome (AIDS). With today's severely immunocompromised patients, often the beneficiaries of advanced medical interventions, the concept of "pathogen" holds little meaning. Any organism is capable of causing disease in such patients under the appropriate conditions.

Role of the Clinical Laboratory

Clinical laboratory scientists should be able to isolate, identify, and determine the antimicrobial susceptibility pattern of the vast majority of human disease agents so that physicians can initiate appropriate treatment as soon as possible, and the source and means of transmission of outbreaks can be ascertained to control the disease and prevent its recurrence. The need to identify clinically relevant microorganisms both quickly and cost-effectively presents a considerable challenge.

To be effective, the professional clinical laboratory staff must interact with the infectious diseases staff. Laboratory scientists should attend infectious disease rounds. They must keep abreast of new technology, equipment, and classification and should communicate this information to their medical colleagues. They should interpret, qualify, or explain laboratory reports. If a bacterial name is changed or a new species reported, the laboratory should provide background information, including a reference.

The clinical laboratory must be efficient. A concerted effort must be made to eliminate or minimise inappropriate and contaminated specimens and the performance of procedures with little or no clinical relevance. Standards for the selection, collection, and transport of specimens should be developed for both laboratory and nursing procedure manuals and reviewed periodically by a committee composed of medical, nursing, and laboratory staff. Ongoing dialogues and continuous communication with other health care workers concerning topics such as specimen collection, test selection, results interpretation, and new technology are essential to maintaining high quality microbiological services.

Biochemical and Susceptibility Testing

Most laboratories today use either commercially available miniaturised biochemical test systems or automated instruments for biochemical tests and for susceptibility testing.

The kits usually contain 10 to 20 tests. The test results are converted to numerical biochemical profiles that are identified by using a codebook or a computer. Carbon source utilisation systems with up to 95 tests are also available. Most identification takes 4 to 24 hours. Biochemical and enzymatic test systems for which data bases have not been developed are used by some reference laboratories.

Automated instruments can be used to identify most Gram-negative fermenters, nonfermenters, and Gram-positive bacteria, but not for anaerobes. Antimicrobial susceptibility testing can be performed for some microorganisms with this equipment, with results expressed as approximate minimum inhibitory drug concentrations. Both tasks take 4 to 24 hours. If semiautomated instruments are used, some manipulation is done manually, and the cultures (in miniature cards or microdilution plates) are incubated outside of the instrument. The test containers are then read rapidly by the instrument, and the results are generated automatically. Instruments are also available for identification of bacteria by cell wall fatty acid profiles generated with gas-liquid chromatography (GLC), analysis of mycolic acids using high performance liquid chromatography (HPLC), and by protein-banding patterns generated by polyacrylamide gel electrophoresis (PAGE). Some other instruments designed to speed laboratory diagnosis of bacteria are those that detect (but do not identify) bacteria in blood cultures, usually faster than manual systems because of continuous monitoring. Also available are many rapid screening systems for detecting one or a series of specific bacteria, including certain streptococci, *N meningitidis*, salmonellae, *Chlamydia trachomatis*, and many others. These screening systems are based on fluorescent antibody, agglutination, or other rapid procedures.

It is important to inform physicians as soon as a presumptive identification of an etiologic agent is obtained so that appropriate therapy can be initiated as quickly as possible. Gram stain and colony morphology; acid-fast stains; and spot indole, oxidase, and other rapid enzymatic tests may allow presumptive identification of an isolate within minutes.

Role of Reference Laboratory

Despite recent advances, the armamentarium of the clinical laboratory is far from complete. Few laboratories can or should conduct the specialised tests that are often essential to distinguish virulent from avirulent strains. Serotyping is done only for a few species, and phage typing only rarely. Few pathogenicity tests are performed. Not many laboratories can conduct comprehensive biochemical tests on strains that cannot be identified readily by commercially available biochemical systems. Even fewer laboratories are equipped to perform plasmid profiles, gene probes, or DNA hybridisation. These and other specialised tests for the serologic or biochemical identification of some exotic bacteria, yeasts, molds, protozoans, and viruses are best done in regional reference laboratories. It is not cost-effective for smaller laboratories to store and control the quality of reagents and media for tests that are seldom run or quite complex. In addition, it is impossible to maintain proficiency when tests are performed rarely. Sensitive methods for the epidemiologic subtyping of isolates from disease outbreaks, such as electrophoretic enzyme typing, rRNA fingerprinting, whole-cell protein electrophoretic patterns, and restriction endonuclease analysis of whole-cell or plasmid DNA, are used only in reference laboratories and a few large medical centers.

Specific genetic probes are now available commercially for identifying virulence factors and many bacteria and viruses. Genetic probes are among the most common methods used for identification of *Mycobacterium tuberculosis* and *M avium* complex in the U.S. today. Probes for *Neisseria gonorrhoeae* and *Chlamydia trachomatis* are now being used directly on clinical specimens with excellent sensitivity and almost universal specificity with same-day results. Mycobacterial probes are also being evaluated for direct specimen testing.

Interfacing with Public Health

Hospital and local clinical laboratories interact with district, state, and federal public health laboratories in several important ways. The clinical laboratories participate in quality control and proficiency testing programs that are conducted by federally regulated agencies. The government reference laboratories supply cultures and often reagents for use in quality control, and they conduct training programs for clinical laboratory personnel.

All types of laboratories should interact closely to provide diagnostic services and epidemic surveillance. The primary concern of the clinical laboratory is identifying infectious disease agents and studying nosocomial and local outbreaks of disease. When the situation warrants, the local laboratory may ask the state laboratory for help in identifying an unusual organism, discovering the cause or mode of transmission in a disease outbreak, or performing specialised tests not done routinely in clinical laboratories. Cultures should be pure and should be sent on appropriate media following appropriate procedures for transport of biohazardous materials. Pertinent information, including the type of specimen; patient name (or number), date of birth, and sex; clinical diagnosis, associated illness, date of onset, and present condition; specific agent suspected, and any other organisms isolated; relevant epidemiologic and clinical data; treatment of patient; previous laboratory results (biochemical or serologic tests); and necessary information about the submitting party must accompany each request.

These data allow the state laboratory to test the specimen properly and quickly, and they provide information about occurrences within the state. For example, a food-borne outbreak might extend to many parts of the state (or beyond its boundaries). The state laboratory can alert local physicians to the possibility of such outbreaks.

REFERENCES

Fleming DO, Richardson JH, Tulis JI, Vesley D (eds). 1995. *Laboratory Safety: Principles and Practices.* 2nd Ed. ASM Press, Washington, D.C.

Lennette EH, et al (eds). 1985. *Manual of Clinical Microbiology* 4th Ed. American Society for Microbiology, Washington, D.C.

Murray PR, Baron EJ, Pfaller MA, Tenover FC, Yolken RH (eds). 1995. *Manual of Clinical Microbiology* 6th Ed. ASM Press, Washington, D.C.

Sneath, PHA (ed). 1990. *International Code of Nomenclature of Bacteria: Bacteriological Code,* 1990 Revision. American Society for Microbiology, Washington, D.C.,

5

Human Microbial Flora

In a healthy animal, the internal tissues, e.g. blood, brain, muscle, etc., are normally free of microorganisms. However, the surface tissues, i.e., skin and mucous membranes, are constantly in contact with environmental organisms and become readily colonized by various microbial species. The mixture of organisms regularly found at any anatomical site is referred to as the normal flora, except by researchers in the field who prefer the term "indigenous microbiota". The normal flora of humans consists of a few eucaryotic fungi and protists, but bacteria are the most numerous and obvious microbial components of the normal flora.

Incredibly, our bodies are actually composed of more bacterial cells than human cells; while the human body is made up of about 1013 human cells, we harbor near 1014 bacteria. This group of organisms, traditionally referred to as "normal flora" (although they are not plants) is composed of a fairly stable set of genera, mostly anaerobes. While each person has a relatively unique set of normal flora, members of the *Streptococcus* and *Bacteroides* make up a large percentage of the inhabitants. These organisms contribute to our existence in several ways. These normal flora may:

— Help us by competing with pathogens such as *Salmonella*

— Help us by providing vitamins or eliminating toxins (e.g. *Bacteroides*)

— Harm us by promoting disease (e.g. dental caries)

— Cause neither help nor harm (e.g. "commensals").

One of the most important functions of our normal flora is to protect us from highly pathogenic organisms. For example, in a normal (bacterially inhabited animal), about 106 *Salmonella* must be ingested in order to cause disease. However, when an animal has been maintained in a sterile environment all of its life (a "gnotobiotic" animal), the same level of disease can be produced by as few as 10 *Salmonella*. This dramatic difference is simply due to competition.

To a microorganism, the human body seems very much like the planet Earth seems to us. Just like our planet, our bodies contain numerous different environments, ranging from dry deserts (e.g. the forearm) to tropical forests (e.g. the perineum) to extremely hostile regions (e.g. the intestinal tract). Each environment possesses certain advantages and disadvantages and different microorganisms have adapted to certain regions of the body for their particular needs.

The surface of the skin itself comprises several distinct environments. Areas such as the axilla (armpit), the perineum (groin) and the toe webs provide typically moister regions for bacterial growth. These "tropical forest" environments often harbor the largest diversity amongst the skin flora. Typical organisms include *Staphylococcus aureus*, *Corynebacterium* and some Gram-negative bacteria. The bulk of the human skin surface, however, is much drier and is predominantly inhabited by *Staphylococcus epidermidis* and *Propionobacterium*.

Streptococci predominate in the oral cavity and nasopharyngeal regions but one can also find other anaerobes and species of *Neisseria*. Many potential pathogens may also be found in the nasopharynx of a healthy individual, providing a reservoir for infection of others. These pathogens include *Streptococcus pneumoniae, Neisseria meningitidis* and *Haemophilus influenzae*.

The intestinal tract is a rather hostile environment for microorganisms yet the bulk of our normal flora inhabit this region of the body. In fact, the colon may contain 109 to 1011 bacteria per gram of material. Most (95 - 99.9%) of these are anaerobes, represented by *Bacteroides, Bifidobacterium*, anaerobic streptococci and *Clostridium*. These organisms inhibit the growth of other pathogens but some can be opportunistic (e.g. *C. difficile* can produce pseudomembranous colitis).

The urogenital tract is normally sterile with the exception of the vagina and the distal 1 cm of the urethra. Various members of the genus *Lactobacillus* predominate in the vagina. These organisms generally lower the pH to around 4-5, which is optimal for the lactobacilli but inhibitory for the growth of many other bacteria. Loss of this protective effect by antibiotic therapy can lead to infection by *Candida* ("yeast infection"). The urethra may contain predominantly skin microorganisms including staphylococci, streptococci and diphtheroids.

SIGNIFICANCE OF MICROBIAL FLORA

A diverse microbial flora is associated with the skin and mucous membranes of every human being from shortly after birth until death. The human body, which contains about 10^{13} cells, routinely harbors about 10^{14} bacteria. This bacterial population constitutes the normal microbial flora. The normal microbial flora is relatively stable, with specific genera populating various body regions during particular periods in an individual's life.

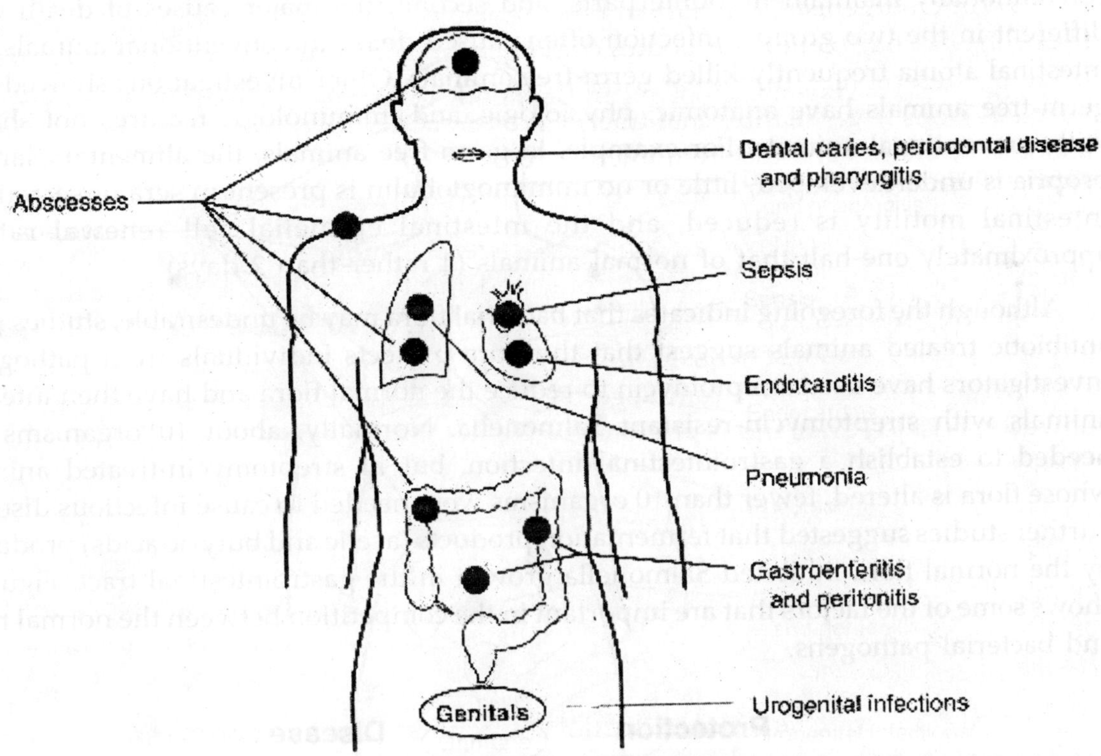

Figure 1. Normal flora in human body

Microorganisms of the normal flora may aid the host (by competing for microenvironments more effectively than such pathogens as Salmonella spp or by producing nutrients the host can use), may harm the host (by causing dental caries, abscesses, or other infectious diseases), or may exist as commensals (inhabiting the host for long periods without causing detectable harm or benefit). Even though most elements of the normal microbial flora inhabiting the human skin, nails, eyes, oropharynx, genitalia, and gastrointestinal tract are harmless in healthy individuals, these organisms frequently cause disease in compromised hosts. Viruses and parasites are not considered members of the normal microbial flora by most investigators because they are not commensals and do not aid the host.

The fact that the normal flora substantially influences the well-being of the host was not well understood until germ-free animals became available. Germ-free animals were obtained by cesarean section and maintained in special isolators; this allowed the investigator to raise them in an environment free from detectable viruses, bacteria, and other organisms. Two interesting observations were made about animals raised under germ-free conditions. First, the germ-free animals lived almost twice as long as their conventionally maintained counterparts, and second, the major causes of death were different in the two groups. Infection often caused death in conventional animals, but intestinal atonia frequently killed germ-free animals. Other investigations showed that germ-free animals have anatomic, physiologic, and immunologic features not shared with conventional animals. For example, in germ-free animals, the alimentary lamina propria is underdeveloped, little or no immunoglobulin is present in sera or secretions, intestinal motility is reduced, and the intestinal epithelial cell renewal rate is approximately one-half that of normal animals (4 rather than 2 days).

Although the foregoing indicates that bacterial flora may be undesirable, studies with antibiotic treated animals suggest that the flora protects individuals from pathogens. Investigators have used streptomycin to reduce the normal flora and have then infected animals with streptomycin-resistant Salmonella. Normally, about 10^6 organisms are needed to establish a gastrointestinal infection, but in streptomycin-treated animals whose flora is altered, fewer than 10 organisms were needed to cause infectious disease. Further studies suggested that fermentation products (acetic and butyric acids) produced by the normal flora inhibited Salmonella growth in the gastrointestinal tract. Figure 2 shows some of the factors that are important in the competition between the normal flora and bacterial pathogens.

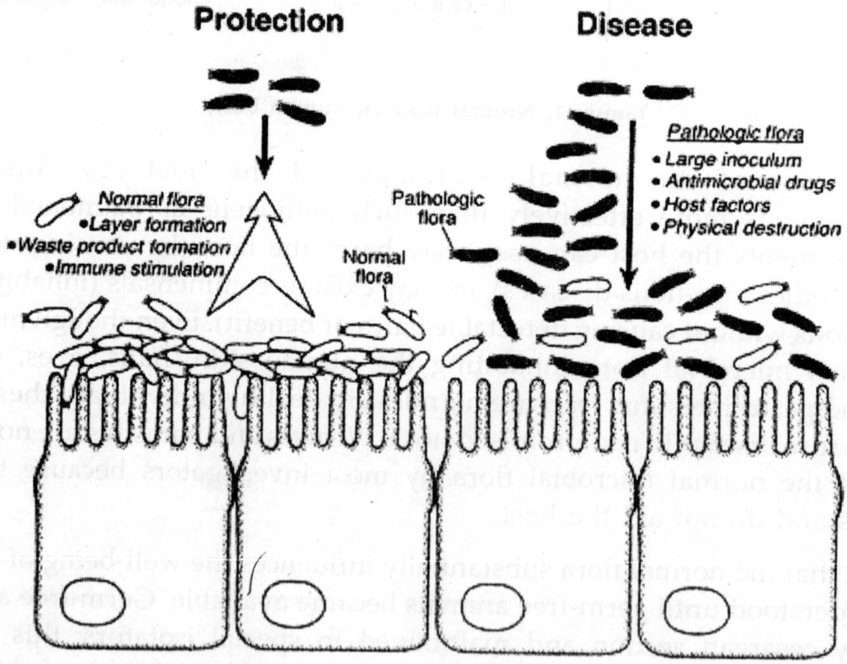

Figure 2. Mechanisms by which the normal flora competes with invading pathogens.

The normal flora in humans usually develops in an orderly sequence, or succession, after birth, leading to the stable populations of bacteria that make up the normal adult flora. The main factor determining the composition of the normal flora in a body region is the nature of the local environment, which is determined by pH, temperature, redox potential, and oxygen, water, and nutrient levels. Other factors such as peristalsis, saliva, lysozyme secretion, and secretion of immunoglobulins also play roles in flora control. The local environment is like a concerto in which one principal instrument usually dominates.

For example, an infant begins to contact organisms as it moves through the birth canal. A Gram-positive population (bifidobacteria arid lactobacilli) predominates in the gastrointestinal tract early in life if the infant is breast-fed. This bacterial population is reduced and displaced somewhat by a Gram-negative flora (Enterobacteriaceae) when the baby is bottle-fed. The type of liquid diet provided to the infant is the principal instrument of this flora control; immunoglobulins and, perhaps, other elements in breast milk may also be important.

What, then, is the significance of the normal flora? Animal and some human studies suggest that the flora influences human anatomy, physiology, lifespan, and, ultimately, cause of death. Although the causal relationship of flora to death and disease in humans is accepted, of her roles of the human microflora need further study.

NORMAL FLORA OF SKIN

Skin provides good examples of various microenvironments. Skin regions have been compared to geographic regions of Earth: the desert of the forearm, the cool woods of the scalp, and the tropical forest of the armpit. The composition of the dermal microflora varies from site to site according to the character of the microenvironment. A different bacterial flora characterises each of three regions of skin:

(1) axilla, perineum, and toe webs;

(2) hand, face and trunk; and

(3) upper arms and legs.

Skin sites with partial occlusion (axilla, perineum, and toe webs) harbor more microorganisms than do less occluded areas (legs, arms, and trunk). These quantitative differences may relate to increased amount of moisture, higher body temperature, and greater concentrations of skin surface lipids. The axilla, perineum, and toe webs are more frequently colonised by Gram-negative bacilli than are drier areas of the skin.

The number of bacteria on an individual's skin remains relatively constant; bacterial survival and the extent of colonisation probably depend partly on the exposure of skin to a particular environment and partly on the innate and species-specific bactericidal activity in skin. Also, a high degree of specificity is involved in the adherence of bacteria to epithelial surfaces. Not all bacteria attach to skin; staphylococci, which are the major element of the nasal flora, possess a distinct advantage over viridans streptococci in colonising the nasal mucosa. Conversely, viridans streptococci are not seen in large numbers on the skin or in the nose but dominate the oral flora.

The microbiology literature is inconsistent about the density of bacteria on the skin; one reason for this is the variety of methods used to collect skin bacteria. The scrub method yields the highest and most accurate counts for a given skin area. Most microorganisms live in the superficial layers of the stratum corneum and in the upper parts of the hair follicles. Some bacteria, however, reside in the deeper areas of the hair follicles and are beyond the reach of ordinary disinfection procedures. These bacteria are a reservoir for recolonisation after the surface bacteria are removed.

Staphylococcus Epidermidis

S epidermidis is a major inhabitant of the skin, and in some areas it makes up more than 90 percent of the resident aerobic flora.

Staphylococcus Aureus

The nose and perineum are the most common sites for S aureus colonisation, which is present in 10 percent to more than 40 percent of normal adults. S aureus is prevalent (67 percent) on vulvar skin. Its occurrence in the nasal passages varies with age, being greater in the newborn, less in adults. S aureus is extremely common (80 to 100 percent) on the skin of patients with certain dermatologic diseases such as atopic dermatitis, but the reason for this finding is unclear.

Micrococci

Micrococci are not as common as staphylococci and diphtheroids; however, they are frequently present on normal skin. Micrococcus luteus, the predominant species, usually accounts for 20 to 80 percent of the micrococci isolated from the skin.

Diphtheroids (Coryneforms)

The term diphtheroid denotes a wide range of bacteria belonging to the genus Corynebacterium. Classification of diphtheroids remains unsatisfactory; for convenience, cutaneous diphtheroids have been categorised into the following four groups: lipophilic or nonlipophilic diphtheroids; anaerobic diphtheroids; diphtheroids producing porphyrins (coral red fluorescence when viewed under ultraviolet light); and those that possess some keratinolytic enzymes and are associated with trichomycosis axillaris (infection of axillary hair). Lipophilic diphtheroids are extremely common in the axilla, whereas nonlipophilic strains are found more commonly on glabrous skin.

Anaerobic diphtheroids are most common in areas rich in sebaceous glands. Although the name Corynebacterium acnes was originally used to describe skin anaerobic diphtheroids, these are now classified as Propionibacterium acnes and as P granulosum. P acnes is seen eight times more frequently than P granulosum in acne lesions and is probably involved in acne pathogenesis. Children younger than 10 years are rarely colonised with P acnes. The appearance of this organism on the skin is probably related to the onset of secretion of sebum (a semi-fluid substance composed of fatty acids and epithelial debris secreted from sebaceous glands) at puberty. P avidum, the third species of cutaneous anaerobic diphtheroids, is rare in acne lesions and is more often isolated from the axilla.

Streptococci

Streptococci, especially ß-hemolytic streptococci, are rarely seen on normal skin. The paucity of ß-hemolytic streptococci on the skin is attributed at least in part to the presence of lipids on the skin, as these lipids are lethal to streptococci. Other groups of streptococci, such as a-hemolytic streptococci, exist primarily in the mouth, from where they may, in rare instances, spread to the skin.

Gram-negative bacilli

Gram-negative bacteria make up a small proportion of the skin flora. In view of their extraordinary numbers in the gut and in the natural environment, their scarcity on skin is striking. They are seen in moist intertriginous areas, such as the toe webs and axilla, and not on dry skin. Desiccation is the major factor preventing the multiplication of Gram-negative bacteria on intact skin. Enterobacter, Klebsiella, Escherichia coli, and Proteus spp are the predominant Gram-negative organisms found on the skin. Acinetobacter spp also occurs on th[illegible]n of normal individuals and, like other Gram-negative bacteria, is more common in the moist intertriginous areas.

Nail flora

The microbiology of a normal nail is generally similar to that of the skin. Dust particles and other extraneous materials may get trapped under the nail, depending on what the nail contacts. In addition to resident skin flora, these dust particles may carry fungi and bacilli. Aspergillus, Penicillium, Cladosporium, and Mucor are the major types of fungi found under the nails.

ORAL AND UPPER RESPIRATORY TRACT FLORA

The oral flora is involved in dental caries and periodontal disease, which affect about 80 percent. of the population in the Western world. Anaerobes in the oral flora are responsible for many of the brain, face, and lung infections that are frequently manifested by abscess formation.

The pharynx and trachea contain primarily those bacterial genera found in the normal oral cavity (for example, alpha-and ß-hemolytic streptococci); however, anaerobes, staphylococci, neisseriae, diphtheroids, and others are also present. Potentially pathogenic organisms such as Haemophilus, mycoplasmas, and pneumococci may also be found in the pharynx. Anaerobic organisms also are reported frequently. The upper respiratory tract is so often the site of initial colonisation by pathogens (Neisseria meningitides, C diphtheriae, Bordetella pertussis, and many others) and could be considered the first region of attack for such organisms. In contrast, the lower respiratory tract (small bronchi and alveoli) is usually sterile, because particles the size of bacteria do not readily reach it. If bacteria do reach these regions, they encounter host defense mechanisms, such as alveolar macrophages, that are not present in the pharynx.

GASTROINTESTINAL TRACT FLORA

The stomach is a relatively hostile environment for bacteria. It contains bacteria swallowed with the food and those dislodged from the mouth. Acidity lowers the bacterial count, which is highest (approximately 10^3 to 10^6 organisms/g of contents) after meals and lowest (frequently undetectable) after digestion. Some Helicobacter species can colonise the stomach and are associated with type B gastritis and peptic ulcer disease. Aspirates of duodenal or jejunal fluid contain approximately 10^3 organisms/ml in most individuals. Most of the bacteria cultured (streptococci, lactobacilli, Bacteroides) are thought to be transients. Levels of 10^5 to about 10^7 bacteria/ml in such aspirates usually indicate an abnormality in the digestive system (for example, achlorhydria or malabsorption syndrome). Rapid peristalsis and the presence of bile may explain in part the paucity of organisms in the upper gastrointestinal tract. Further along the jejunum and into the ileum, bacterial populations begin to increase, and at the ileocecal junction they reach levels of 10^6 to 10^8 organisms/ml, with streptococci, lactobacilli, Bacteroides, and bifidobacteria predominating.

Concentrations of 10^9 to 10^{11} bacteria/g of contents are frequently found in human colon and feces. This flora includes a bewildering array of bacteria (more than 400 species have been identified); nonetheless, 95 to 99 percent belong to anaerobic genera such as Bacteroides, Bifidobacterium, Eubacterium, Peptostreptococcus, and Clostridium. In this highly anaerobic region of the intestine, these genera proliferate, occupy most available niches, and produce metabolic waste products such as acetic, butyric, and lactic acids.

The strict anaerobic conditions, physical exclusion (as is shown in many animal studies), and bacterial waste products are factors that inhibit the growth of other bacteria in the large bowel.

Although the normal flora can inhibit pathogens, many of its members can produce disease in humans. Anaerobes in the intestinal tract are the primary agents of intra-abdominal abscesses and peritonitis. Bowel perforations produced by appendicitis, cancer, infarction, surgery, or gunshot wounds almost always seed the peritoneal cavity and adjacent organs with the normal flora. Anaerobes can also cause problems within the gastrointestinal lumen. Treatment with antibiotics may allow certain anaerobic species to become predominant and cause disease. For example, *Clostridium difficile*, which can remain viable in a patient undergoing antimicrobial therapy, may produce pseudomembranous colitis. Other intestinal pathologic conditions or surgery can cause bacterial overgrowth in the upper small intestine. Anaerobic bacteria can then deconjugate bile acids in this region and bind available vitamin B12 so that the vitamin and fats are malabsorbed. In these situations, the patient usually has been compromised in some way; therefore, the infection caused by the normal intestinal flora is secondary to another problem.

More information is available on the animal than the human microflora. Research on animals has revealed that unusual filamentous microorganisms attach to ileal epithelial cells and modify host membranes with few or no harmful effects. Microorganisms have been observed in thick layers on gastrointestinal surfaces and in the crypts of Lieberkuhn. Other studies indicate that the immune response can be modulated by the intestinal flora. Studies of the role of the intestinal flora in biosynthesis of vitamin K and other host-utilisable products, conversion of bile acids (perhaps to cocarcinogens), and ammonia production (which can play a role in hepatic coma) show the dual role of the microbial flora in influencing the health of the host. More basic studies of the human bowel flora are necessary to define their effect on humans.

UROGENITAL FLORA

The type of bacterial flora found in the vagina depends on the age, pH, and hormonal levels of the host. Lactobacillus spp predominate in female infants (vaginal pH, approximately 5) during the first month of life. Glycogen secretion seems to cease from about I month of age to puberty. During this time, diphtheroids, *S epidermidis*, streptococci, and *E coli* predominate at a higher pH (approximately pH 7). At puberty, glycogen secretion resumes, the pH drops, and women acquire an adult flora in which L acidophilus, corynebacteria, peptostreptococci, staphylococci, streptococci, and Bacteroides predominate. After menopause, pH again rises, less glycogen is secreted, and the flora returns to that found in prepubescent females. Yeasts (Torulopsis and Candida) are occasionally found in the vagina (10 to 30 percent of women); these sometimes increase and cause vaginitis. In the anterior urethra of humans, *S epidermidis*, enterococci, and diphtheroids are found frequently; *E coli*, Proteus, and Neisseria (nonpathogenic species) are reported occasionally (10 to 30 percent). Because of the normal flora residing in the urethra, care must be taken in clinically interpreting urine cultures; urine samples may contain these organisms at a level of 10^4/ml if a midstream (clean-catch) specimen is not obtained.

CONJUNCTIVAL FLORA

The conjunctival flora is sparse. Approximately 17 to 49 percent of culture samples are negative. Lysozyme, secreted in tears, may play a role in controlling the bacteria by interfering with their cell wall formation. When positive samples show bacteria, corynebacteria, neisseriae, and moraxellae are cultured. Staphylococci and streptococci are also present, and recent reports indicate that Haemophilus parainfluenzae is present in 25 percent of conjunctival samples.

REFERENCES

Bitton G, Marshall KC. 1980. *Adsorption of Microorganisms to Surfaces*. John Wiley & Sons, New York,

Draser BS, Hill MJ. 1974. *Human Intestinal Flora*. Academic Press, London.

Freter R, Brickner J, Botney M, et al. 1983. "Survival and implantation of Escherichia coli in the intestinal tract". *Infect Immun* 39:686,

Herthelius M, Gorbach SL, Mollby R, et al. 1989. "Elimination of vaginal colonization with Escherichia coli by administration of indigenous flora". *Infect Immun* 57:2447.

Maibach H, Aly R. 1969. *Skin Microbiology: Relevance to Clinical Infection*. Springer-Verlag, New York, 1981

Marples MJ: Life in the skin. Sci Am 220:108.

Savage DC. 1977. "Microbial ecology of the gastrointestinal tract". *Annu Rev Microbiol* 31:107.

Tannock GW. 1995. *Normal Microflora*. Chapman and Hall,London, UK.

6

Viruses: Structure and Functions

Epidemiologic studies show that viral infections in developed countries are the most common cause of acute disease that does not require hospitalisation. In developing countries, viral diseases also exact a heavy toll in mortality and permanent disability, especially among infants and children. Emerging viral diseases such as those due to HIV, ebolavirus and hantavirus, appear regularly. Now that antibiotics effectively control most bacterial infections, viral infections pose a relatively greater and less controlled threat to human health.

Viruses can infect all forms of life (bacteria, plants, protozoa, fungi, insects, fish, reptiles, birds, and mammals); however, this section covers only viruses capable of causing human infections. Like other microorganisms, viruses may have played a role in the natural selection of animal species. A documented example is the natural selection of rabbits resistant to virulent myxoma virus during several epidemics deliberately induced to control the rabbit population in Australia.

Indirect evidence suggests that the same selective role was played by smallpox virus in humans. Another possible, though unproved, mechanism by which viruses may affect evolution is by introducing viral genetic material into animal cells by mechanisms similar to those that govern gene transfer by bacteriophages. Genes from avirulent retrovirus integrated into genomes of chickens or mice produce resistance to reinfection by related, virulent retroviruses. The same relationship may exist for human retroviruses, since human leukemia-causing retroviruses have been reported. Viruses are small, subcellular agents that are unable to multiply outside a host cell (intracellular, obligate parasitism).

The assembled virus (virion) is formed to include only one type of nucleic acid (RNA or DNA) and, in the simplest viruses, a protective protein coat. The nucleic acid contains the genetic information necessary to program the synthetic machinery of the host cell for viral replication. The protein coat serves two main functions: first, it protects the nucleic acid from extracellular environmental insults such as nucleases; second, it permits attachment of the virion to the membrane of the host cell, the negative charge of which would repel a naked nucleic acid. Once the viral genome has penetrated and thereby infected the host cell, virus replication mainly depends on host cell machinery for energy and synthetic requirements.

The various virion components are synthesized separately within the cell and then assembled to form progeny particles. This assembly type of replication is unique to viruses and distinguishes them from all other small, obligate, intracellular parasites. The basic structure of viruses may permit them to be simultaneously adaptable and selective.

Many viral genomes are so adaptable that once they have penetrated the cell membrane under experimental conditions, viral replication can occur in almost any cell. On the other hand, intact viruses are so selective that most virions can infect only a limited range of cell types. This selectivity exists largely because penetration of the nucleic acid usually requires a specific reaction for the coat to attach to the host cell membrane and some specific intracellular components.

Although some viruses may establish some forms of silent infection of cells, their multiplication usually causes cell damage or death; however, since viruses must depend on host survival for their own survival, they tend to establish mild infections in which death of the host is more an aberration than a regular outcome. Notable exceptions are HIV, ebolavirus, hantavirus and rabiesvirus. Viruses are distinct among microorganisms in their extreme dependence on the host cell. Since a virus must grow within a host cell, the virus must be viewed together with its host in any consideration of pathogenesis, epidemiology, host defenses, or therapy.

The bilateral association between the virus and its host imposes specific conditions for pathogenesis. For example, rhinoviruses require a temperature not exceeding 34°C; this requirement restricts their growth to only those cells in the cool outer layer of the nasal mucosa, thereby preventing spread to deeper cells where temperatures are higher. The intracellular location of the virus often protects the virus against some of the host's immune mechanisms; at the same time, this location makes the virus vulnerable because of its dependence on the host cell's synthetic machinery, which may be altered by even subtle physical and chemical changes produced by the viral infection (inflammation, fever, circulatory alterations, and interferon).

Epidemiologic properties depend greatly on the characteristics of the virus-host association. For example, some arthropod-borne viruses require a narrow range of temperature to multiply in insects; as a result, these viruses are found only under certain seasonal and geographic conditions. Other environmental conditions determine the transmissibility of viruses in aerosols and in food.

Viruses are difficult targets for chemotherapy because they replicate only within host cells, mainly utilising many of the host cell's biosynthetic processes. The similarity of host-directed and virus-directed processes makes it difficult to find antiviral agents specific enough to exert a greater effect on viral replication in infected cells than on functions in uninfected host cells. It is becoming increasingly apparent, however, that each virus may have a few specific steps of replication that may be used as targets for highly selective, carefully aimed chemotherapeutic agents.

Therefore, proper use of such drugs requires a thorough knowledge of the suitable targets, based on a correct diagnosis and a precise understanding of the replicative mechanisms for the offending virus. Knowledge of the pathogenetic mechanisms by which virus enters, spreads within, and exits from the body also is critical for correct diagnosis and treatment of disease and for prevention of spread in the environment.

Effective treatment with antibody-containing immunoglobulin requires knowing when virus is susceptible to antibody and when virus reaches target organs where antibody is less effective. Many successful vaccines have been based on knowledge of pathogenesis and immune defenses. Comparable considerations govern treatment with interferon. Clearly, viral infections are among the most difficult and demanding problems a physician must face. Unfortunately, some of these problems still lack satisfactory solutions, although tremendous progress has been made during the last several decades.

Many aspects of medical virology are now understood, others are being clarified gradually, and many more are still obscure. Knowledge of the properties of viruses and the relationships they establish with their hosts is crucial to successful investigation and clinical management of their pathologic processes. Our plan for conveying this knowledge is to present, first, concepts of viral structure, and then relate them to principles of viral multiplication. Together these concepts form the basis for understanding how viruses are classified, how they affect cells, and how their genetic system functions. These molecular and cellular mechanisms are combined with the concepts of immunology to explain viral pathogenesis, nonspecific defenses, persistent infections, epidemiology, evolution, and control.

The important virus families are then discussed individually. Having studied the virology section, the reader should be able to use many principles of virology to explain individual manifestations of virus infection and the processes that bring them about.

ORIGIN OF VIRUSES

Viruses are found wherever there is life and have probably existed since living cells first evolved. The origin of viruses is unclear because they do not form fossils, so molecular techniques have been the most useful means of investigating how they arose. These techniques rely on the availability of ancient viral DNA or RNA, but, unfortunately, most of the viruses that have been preserved and stored in laboratories are less than 90 years old. There are three main theories of the origins of viruses:

— *Regressive theory*: Viruses may have once been small cells that parasitised larger cells. Over time, genes not required by their parasitism were lost. The bacteria rickettsia and chlamydia are living cells that, like viruses, can reproduce only inside host cells. They lend credence to this theory, as their dependence on parasitism is likely to have caused the loss of genes that enabled them to survive outside a cell. This is also called the degeneracy theory.

— *Cellular origin theory (sometimes called the vagrancy theory):* Some viruses may have evolved from bits of DNA or RNA that "escaped" from the genes of a larger organism. The escaped DNA could have come from plasmids—pieces of naked DNA that can move between cells or transposons. These are molecules of DNA that replicate and move around to different positions within the genes of the cell. Once called "jumping genes", these are examples of mobile genetic elements and could be the origin of some viruses. Transposons were discovered in maize by Barbara McClintock in 1950.

— *Coevolution theory*: Viruses may have evolved from complex molecules of protein and nucleic acid at the same time as cells first appeared on earth and would have been dependent on cellular life for many millions of years. Viroids are molecules of RNA that are not classified as viruses because they lack a protein coat. However, they have characteristics that are common to several viruses and are often called subviral agents. Viroids are important pathogens of plants. They do not code for proteins but interact with the host cell and use the host machinery for their replication. The hepatitis delta virus of humans has an RNA genome similar to viroids but has protein coat derived from hepatitis B virus and cannot produce one of its own. It is therefore a defective virus and cannot replicate without the help of hepatitis B virus.

The Virophage 'sputnik' infects the Mimivirus and the related Mamavirus which in turn infect the protozooan Acanthamoeba castellanii. These viruses that are dependent on other virus species are called satellites and may represent evolutionary intermediates of viroids and viruses. Prions are infectious protein molecules that do not contain DNA or RNA. They cause an infection in sheep called scrapie and cattle bovine spongiform encephalopathy ("mad cow" disease). In humans they cause kuru and Creutzfeld-Jacob disease. They are able to replicate because some proteins can exist in two different shapes and the prion changes the normal shape of a host protein into the prion shape. This starts a chain reaction where each prion protein converts many host proteins into more prions, and these new prions then go on to convert even more protein into prions. Although they are fundamentally different from viruses and viroids, their discovery gives credence to theory that viruses could have evolved from self-replicating molecules.

Computer analysis of viral and host DNA sequences is giving a better understanding of the evolutionary relationships between different viruses and may help identify the ancestors of modern viruses. To date, such analyses have not helped to decide on which of the theories are correct. However, it seems unlikely that all currently known viruses have a common ancestor and viruses have probably arisen numerous times in the past by one or more mechanisms.

Opinions differ on whether viruses are a form of life, or organic structures that interact with living organisms. They have been described as "organisms at the edge of life", since they resemble organisms in that they possess genes and evolve by natural selection, and reproduce by creating multiple copies of themselves through self-assembly. However, although they have genes, they do not have a cellular structure, which is often seen as the basic unit of life. Additionally, viruses do not have their own metabolism, and require a host cell to make new products. They therefore cannot reproduce outside a host cell (though bacterial species such as rickettsia and chlamydia are considered living organisms despite the same limitation). Accepted forms of life use cell division to reproduce, whereas viruses spontaneously assemble within cells, which is analogous to the autonomous growth of crystals. Virus self-assembly within host cells has implications for the study of the origin of life, as it lends further credence to the hypothesis that life could have started as self-assembling organic molecules.

STRUCTURE AND FUNCTION

Viruses are small obligate intracellular parasites, which by definition contain either a RNA or DNA genome surrounded by a protective, virus-coded protein coat. Viruses may be viewed as mobile genetic elements, most probably of cellular origin and characterised by a long co-evolution of virus and host. For propagation viruses depend on specialised host cells supplying the complex metabolic and biosynthetic machinery of eukaryotic or prokaryotic cells.

A complete virus particle is called a virion. The main function of the virion is to deliver its DNA or RNA genome into the host cell so that the genome can be expressed (transcribed and translated) by the host cell. The viral genome, often with associated basic proteins, is packaged inside a symmetric protein capsid.

The nucleic acid-associated protein, called nucleoprotein, together with the genome, forms the nucleocapsid. In enveloped viruses, the nucleocapsid is surrounded by a lipid bilayer derived from the modified host cell membrane and studded with an outer layer of virus envelope glycoproteins.

CLASSIFICATION OF VIRUSES

Morphology: Viruses are grouped on the basis of size and shape, chemical composition and structure of the genome, and mode of replication. Helical morphology is seen in nucleocapsids of many filamentous and pleomorphic viruses. Helical nucleocapsids consist of a helical array of capsid proteins wrapped around a helical filament of nucleic acid. Icosahedral morphology is characteristic of the nucleocapsids of many "spherical" viruses. The number and arrangement of the capsomeres are useful in identification and classification. Many viruses also have an outer envelope.

Chemical composition and mode of replication: The genome of a virus may consist of DNA or RNA, which may be single stranded (ss) or double stranded (ds), linear or circular. The entire genome may occupy either one nucleic acid molecule or several nucleic acid segments. The different types of genome necessitate different replication strategies.

NOMENCLATURE

Aside from physical data, genome structure and mode of replication are criteria applied in the classification and nomenclature of viruses, including the chemical composition and configuration of the nucleic acid, whether the genome is monopartite or multipartite. The genomic RNA strand of single-stranded RNA viruses is called sense (positive sense, plus sense) in orientation if it can serve as mRNA, and antisense (negative sense, minus sense) if a complementary strand synthesized by a viral RNA transcriptase serves as mRNA. Also considered in viral classification is the site of capsid assembly and, in enveloped viruses, the site of envelopment.

Viruses are inert outside the host cell. Small viruses, e.g., polio and tobacco mosaic virus, can even be crystallised. Viruses are unable to generate energy. As obligate intracellular parasites, during replication, they fully depend on the complicated biochemical machinery of eukaryotic or prokaryotic cells.

The main purpose of a virus is to deliver its genome into the host cell to allow its expression (transcription and translation) by the host cell. A fully assembled infectious virus is called a virion. The simplest virions consist of two basic components: nucleic acid (single- or double-stranded RNA or DNA) and a protein coat, the capsid, which functions as a shell to protect the viral genome from nucleases and which during infection attaches the virion to specific receptors exposed on the prospective host cell.

Capsid proteins are coded for by the virus genome. Because of its limited size the genome codes for only a few structural proteins. Capsids are formed as single or double protein shells and consist of only one or a few structural protein species. Therefore, multiple protein copies must self assemble to form the continuous three-dimensional capsid structure. Self assembly of virus capsids follows two basic patterns: helical symmetry, in which the protein subunits and the nucleic acid are arranged in a helix, and icosahedral symmetry, in which the protein subunits assemble into a symmetric shell that covers the nucleic acid-containing core.

Some virus families have an additional covering, called the envelope, which is usually derived in part from modified host cell membranes. Viral envelopes consist of a lipid bilayer that closely surrounds a shell of virus-encoded membrane-associated proteins. The exterior of the bilayer is studded with virus-coded, glycosylated (trans-) membrane proteins. Therefore, enveloped viruses often exhibit a fringe of glycoprotein spikes or knobs, also called peplomers. In viruses that acquire their envelope by budding through the plasma or another intracellular cell membrane, the lipid composition of the viral envelope closely reflects that of the particular host membrane.

The outer capsid and the envelope proteins of viruses are glycosylated and important in determining the host range and antigenic composition of the virion. In addition to virus-specified envelope proteins, budding viruses carry also certain host cell proteins as integral constituents of the viral envelope.

Virus envelopes can be considered an additional protective coat. Larger viruses often have a complex architecture consisting of both helical and isometric symmetries confined to different structural components. Small viruses, e.g., hepatitis B virus or the members of the picornavirus or parvovirus family, are orders of magnitude more resistant than are the larger complex viruses, e.g. members of the herpes or retrovirus families.

Viruses are classified on the basis of morphology, chemical composition, and mode of replication. The viruses that infect humans are currently grouped into 21 families, reflecting only a small part of the spectrum of the multitude of different viruses whose host ranges extend from vertebrates to protozoa and from plants and fungi to bacteria.

In the replication of viruses with helical symmetry, identical protein subunits (protomers) self-assemble into a helical array surrounding the nucleic acid, which follows a similar spiral path. Such nucleocapsids form rigid, highly elongated rods or flexible filaments; in either case, details of the capsid structure are often discernible by electron microscopy. In addition to classification as flexible or rigid and as naked or enveloped, helical nucleocapsids are characterised by length, width, pitch of the helix, and number of protomers per helical turn. The most extensively studied helical virus is tobacco mosaic virus (Fig. 1). Many important structural features of this plant virus have been detected by x-ray diffraction studies.

ICOSAHEDRAL SYMMETRY

An icosahedron is a polyhedron having 20 equilateral triangular faces and 12 vertices (Fig. 2). Lines through opposite vertices define axes of fivefold rotational symmetry: all structural features of the polyhedron repeat five times within each 360° of rotation about any of the fivefold axes. Lines through the centers of opposite triangular faces form axes of threefold rotational symmetry; twofold rotational symmetry axes are formed by lines through midpoints of opposite edges. An icosaheron (polyhedral or spherical) with fivefold, threefold, and twofold axes of rotational symmetry (Fig. 2) is defined as having 532 symmetry.

Viruses were first found to have 532 symmetry by x-ray diffraction studies and subsequently by electron microscopy with negative-staining techniques. In most icosahedral viruses, the protomers, i.e. the structural polypeptide chains, are arranged in oligomeric clusters called capsomeres, which are readily delineated by negative staining ele(

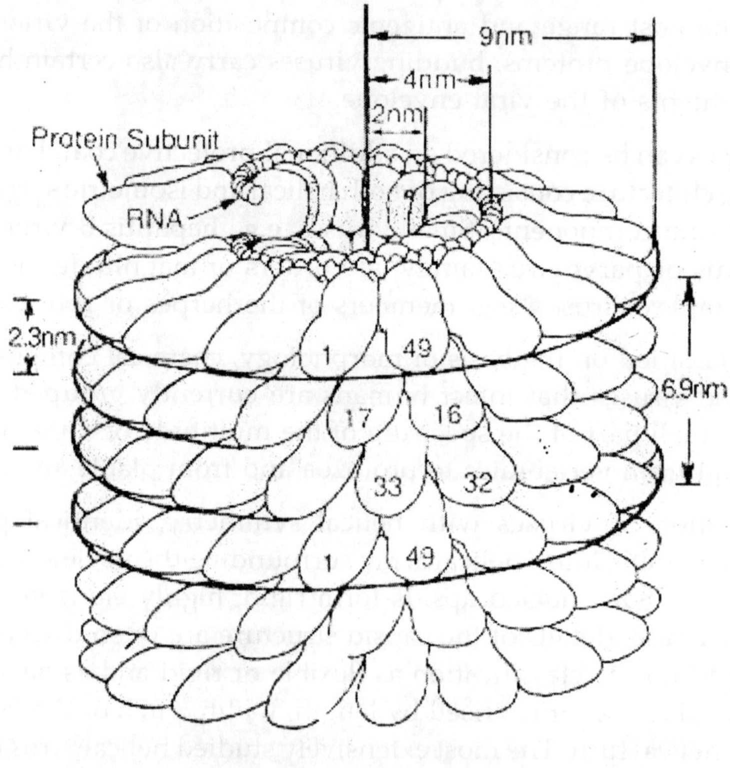

Figure 1. The helical structure of the rigid tobacco mosaic virus rod.

The arrangement of capsomeres into an icosahedral shell (compare Fig. 3 with the upper right model in Fig. 2) permits the classification of such viruses by capsomere number and pattern. This requires the identification of the nearest pair of vertex capsomeres (called penton: those through which the fivefold symmetry axes pass) and the distribution of capsomeres between them.

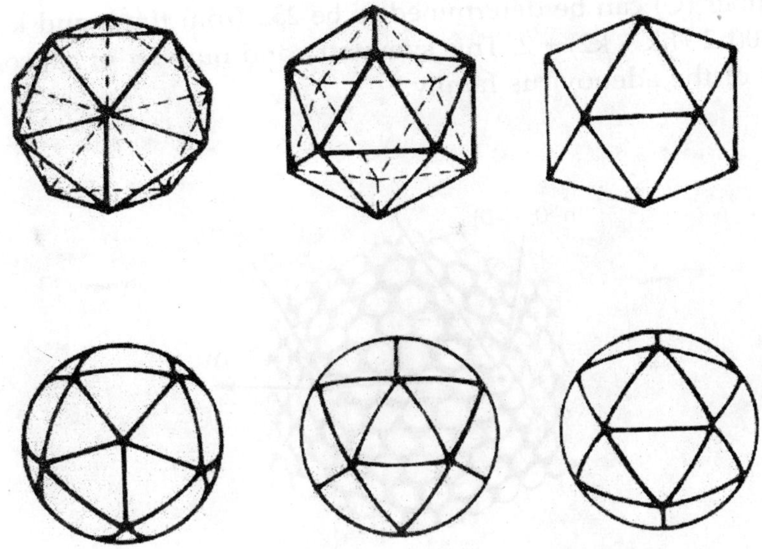

Figure 2. Icosahedral models seen, left to right, on fivefold, threefold, and twofold axes of rotational symmetry.

In the adenovirus model in Figure 3, one of the penton capsomeres is arbitrarily assigned the indices h = 0, k = 0 (origin), where h and k are the indicated axes of the inclined (60°) net of capsomeres. The net axes are formed by lines of the closest-packed neighboring capsomeres. In adenoviruses, the h and k axes also coincide with the edges of the triangular faces.

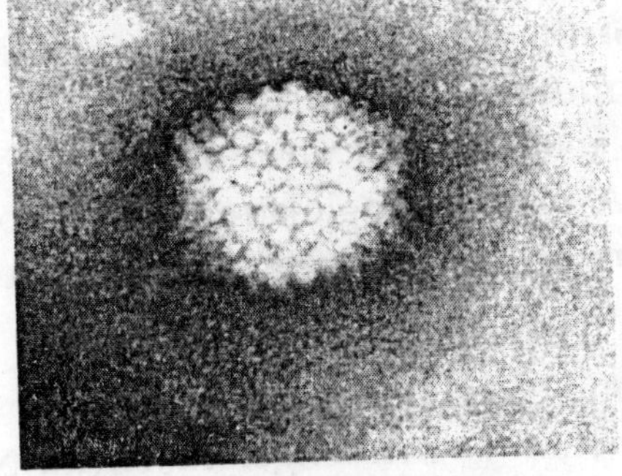

Figure 3a. Adenovirus after Negative Stain Electron Microscopy.

Any second neighboring vertex capsomere has indices h = 5, k = 0 (or h = 0, k = 5). The capsomere number (C) can be determined to be 252 from the h and k indices and the equation: C = 10(h2 +hk + k2) + 2. This symmetry and number of capsomeres is typical of all members of the adenovirus family.

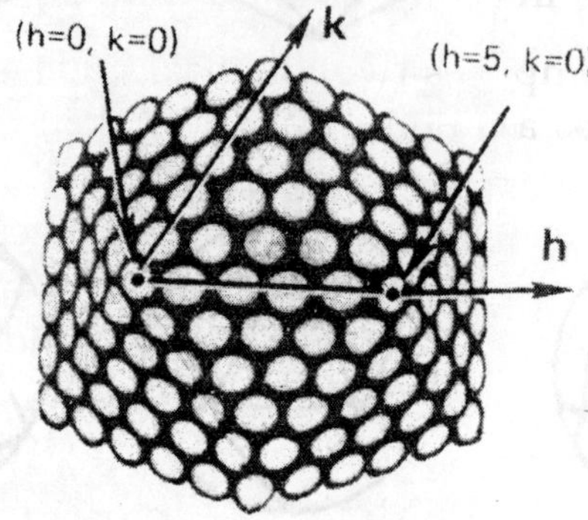

Figure 3b. Adenovirus Model

VIRUS CORE STRUCTURE

Except in helical nucleocapsids, little is known about the packaging or organisation of the viral genome within the core. Small virions are simple nucleocapsids containing 1 to 2 protein species. The larger viruses contain in a core the nucleic acid genome complexed with basic protein(s) and protected by a single- or double layered capsid (consisting of more than one species of protein) or by an envelope.

RNA VIRUS

RNA viruses, comprising 70% of all viruses, vary remarkably in genome structure. Because of the error rate of the enzymes involved in RNA replication, these viruses usually show much higher mutation rates than do the DNA viruses. Mutation rates of 10-4 lead to the continuous generation of virus variants which show great adaptability to new hosts. The viral RNA may be single-stranded (ss) or double-stranded (ds), and the genome may occupy a single RNA segment or be distributed on two or more separate segments.

In addition, the RNA strand of a single-stranded genome may be either a sense strand (plus strand), which can function as messenger RNA (mRNA), or an antisense strand, which is complementary to the sense strand and cannot function as mRNA protein translation. Sense viral RNA alone can replicate if injected into cells, since it can function as mRNA and initiate translation of virus-encoded proteins.

Antisense RNA, on the other hand, has no translational function and cannot per se produce viral components. DsRNA viruses, e.g., members of the reovirus family, contain 10, 11 or 12 separate genome segments coding for 3 enzymes involved in RNA replication, 3 major capsid proteins and a number of smaller structural proteins. Each segment consists of a complementary sense and antisense strand that is hydrogen bonded into a linear ds molecule. The replication of these viruses is complex; only the sense RNA strands are released from the infecting virion to initiate replication.

The retrovirus genome comprises two identical, plus-sense ssRNA molecules, each monomer 7-11 kb in size, that are noncovalently linked over a short terminal region. Retroviruses contain 2 envelope proteins encoded by the env-gene, 4-6 nonglycosylated core proteins and 3 non-structural functional proteins specified by the gag-gene. The RT transcribes the viral ssRNA into double-stranded, circular proviral DNA. This DNA, mediated by the viral integrase, becomes covalently bonded into the DNA of the host cell to make possible the subsequent transcription of the sense strands that eventually give rise to retrovirus progeny.

After assembly and budding, retroviruses show structural and functional maturation. In immature virions the structural proteins of the core are present as a large precursor protein shell. After proteolytic processing by the viral protease the proteins of the mature virion are rearranged and form the dense isometric or cone-shaped core typical of the mature virion, and the particle becomes infectious.

DNA VIRUS

Most DNA viruses contain a single genome of linear dsDNA. The papovaviruses, comprising the polyoma- and papillomaviruses, however, have circular DNA genomes, about 5.1 and 7.8 kb pairs in size. DsDNA serves as a template both for mRNA and for self-transcription. Three or 2 structural proteins make up the papovavirus capsid: in addition, 5-6 nonstructural proteins are encoded that are functional in virus transcription, DNA replication and cell transformation.

Single-stranded linear DNA, 4-6 kb in size, is found with the members of the Parvovirus family that comprises the parvo-, the erythro- and the dependoviruses. The virion contains 2-4 structural protein species which are differently derived from the same gene product. The adeno-associated virus is incapable of producing progeny virions except in the presence of helper viruses. It is therefore said to be replication defective.

Circular single-stranded DNA of only 1.7 to 2.3 kb is found in members of the Circovirus family which comprise the smallest autonomously propagated viruses. The isometric capsid measures 17 nm and is composed of 2 protein species only.

On the basis of shared properties viruses are grouped at different hierarchical levels of order, family, subfamily, genus and species. More than 30,000 different virus isolates are known today and grouped in more than 3,600 species, in 164 genera and 71 families. Viral morphology provides the basis for grouping viruses into families. A virus family may consist of members that replicate only in vertebrates, only in invertebrates, only in plants, or only in bacteria. Certain families contain viruses that replicate in more than one of these hosts. This section concerns only the 21 families and genera of medical importance.

Besides physical properties, several factors pertaining to the mode of replication play a role in classification: the configuration of the nucleic acid (ss or ds, linear or circular), whether the genome consists of one molecule of nucleic acid or is segmented, and whether the strand of ss RNA is sense or antisense. Also considered in classification is the site of viral capsid assembly and, in enveloped viruses, the site of nucleocapsid envelopment.

The use of Latinized names ending in -viridae for virus families and ending in -virus for viral genera has gained wide acceptance. The names of subfamilies end in -virinae. Vernacular names continue to be used to describe the viruses within a genus. In this text, Latinized endings for families and subfamilies usually are not used. In the early days of virology, viruses were named according to common pathogenic properties, e.g. organ tropism and/or modes of transmission, and often also after their discoverers. From the early 1950s until the mid-1960s, when many new viruses were being discovered, it was popular to compose virus names by using sigla.

Thus the name Picornaviridae is derived from pico (small) and RNA; the name Reoviridae is derived from respiratory, enteric, and orphan viruses because the agents were found in both respiratory and enteric specimens and were not related to other classified viruses; Papovaviridae is from papilloma, polyoma, and vacuolating agent; retrovirus is from reverse transcriptase; Hepadnaviridae is from the replication of the virus in hepatocytes and their DNA genomes, as seen in hepatitis B virus.

Hepatitis A virus is classified now in the family Picornaviridae, genus Hepatovirus. Although the current rules for nomenclature do not prohibit the introduction of new sigla, they require that the siglum be meaningful to workers in the field and be recognised by international study groups. The names of the other families that contain viruses pathogenic for humans are derived as follows: Adenoviridae; Astroviridae; Arenaviridae describes the sandy appearance of the virion. Bunyaviridae; Calicivirus; Coronaviridae describes the appearance of the peplomers protruding from the viral surface; Filoviridae describes the morphology of these viruses. Herpesviridae describes the nature of the lesions; Orthomyxoviridae ortho, "true," plus myxo "mucus," a substance for which the viruses have an affinity; Paramyxoviridae derived from para, "closely resembling" and myxo; Parvoviridae; Poxviridae; Rhabdoviridae "rod" describes the shape of the viruses and Togaviridae refers to the tight viral envelope.

Several viruses of medical importance still remain unclassified. Some are difficult or impossible to propagate in standard laboratory host systems and thus cannot be obtained in sufficient quantity to permit more precise characterisation. Hepatitis E virus, the Norwalk virus and similar agents that cause nonbacterial gastroenteritis in humans are now assigned to the calicivirus family.

The fatal transmissible dementias in humans and other animals are caused by the accumulation of non-soluble amyloid fibrils in the central nervous systems. The agents causing transmissible subacute spongiform encephalopathies have been linked to viroids or virinos because of their resistance to chemical and physical agents.

According to an alternative theory, the term "prion" has been coined to point to an essential nonviral infectious cause for these fatal encephalopathiesprion standing for self-replicating proteinaceous agent devoid of demonstrable nucleic acid.

Some of the transmissible amyloidoses show a familial pattern and can be explained by defined mutations which render a primary soluble glycoprotein insoluble, which in turn leads to the pathognomonic accumulation of amyloid fibers and plaques. The pathogenesis of the sporadic amyloidoses, however, is still a matter of highly ambitious research.

GENETIC CHANGE

Viruses undergo genetic change by several mechanisms. These include a process called genetic drift where individual bases in the DNA or RNA mutate to other bases. Most of these point mutations are silent in that they do not change the protein that the gene encodes, but others can confer evolutionary advantages such as resistance to antiviral drugs. Antigenic shift is where there is a major change in the genome of the virus. This occurs as a result of recombination or reassortment. When this happens with influenza viruses, pandemics may result. RNA viruses often exist as quasispecies or swarms of viruses of the same species but with slightly different genome nucleoside sequences. Such quasispecies are a prime target for natural selection.

Segmented genomes confer evolutionary advantages; different strains of a virus with a segmented genome can shuffle and combine genes and produce progeny viruses or (offspring) that have unique characteristics. This is called reassortment or viral sex.

Genetic recombination is the process by which a strand of DNA is broken and then joined to the end of a different DNA molecule. This can occur when viruses infect cells simultaneously and studies of viral evolution have shown that recombination has been rampant in the species studied. Recombination is common to both RNA and DNA viruses.

REFERENCES

Collier, Leslie; Balows, Albert; Sussman, Max. 1998. *Topley and Wilson's Microbiology and Microbial Infections* (ninth edition), Volume 1, *Virology*, volume editors: Mahy, Brian and Collier, Leslie. Arnold.

Dimmock, N.J; Easton, Andrew J; Leppard, Keith. 2007. *Introduction to Modern Virology* (sixth edition), Blackwell Publishing.

Knipe, David M; Howley, Peter M; Griffin, Diane E; Lamb, Robert A; Martin, Malcolm A; Roizman, Bernard; Straus Stephen E. 2007 *Fields Virology*. Lippincott Williams & Wilkins.

Shors, Teri . 2008. *Understanding Viruses*. Jones and Bartlett Publishers.

7

Viral Effects on Cells

Mostly, disturbances of bodily function that are manifested as the signs and symptoms of viral disease result from the direct effects of viruses on cells. Knowledge of the morphologic, physiologic, biochemical, and immunologic effects of viruses on cells is essential in understanding the pathophysiology of viral disease and in developing accurate diagnostic procedures and effective treatment.

Virus-host cell interactions may produce either:

1) cytocidal (cytolytic) infections, in which production of new infectious virus kills the cell;
2) persistent infections, in which the virus or its genome resides in some or all of the cells without killing most of them; or
3) transformation, in which the virus does not kill the cell, but produces genetic, biochemical, physiologic, and morphologic changes that may result in the acquisition of malignant properties.

The type of virus infection and the virus-induced effects on cells are dependent on the virus, the cell type and species, and often the physiologic state of the cell.

CYTOCIDAL INFECTIONS

Morphological Effects

Infection of permissive cells with virus leads to productive infection and often results in cell death (cytocidal, cytolytic infection). The first effects of the replication of cytocidal viruses to be described were the morphologic changes known as cytopathic effects. Cultured cells that are infected by most viruses undergo morphologic changes, which can be observed easily in unfixed, unstained cells by a light microscope. Some viruses cause characteristic cytopathic effects; thus, observation of the cytopathic effect is an important tool for virologists concerned with isolating and identifying viruses from infected animals or humans.

Many types of cytopathic effects occur. Often the first sign of viral infections is rounding of the cells. In some diseased tissues, intracellular structures called inclusion bodies appear in the nucleus and/or cytoplasm of infected cells. Inclusion bodies were first identified by light microscopy in smears and stained sections of infected tissues. Their composition can often be clarified by electron microscopy. In an adenovirus infection, for example, crystalline arrays of adenovirus capsids accumulate in the nucleus to form an inclusion body.

Inclusions may alternatively be host cell structures altered by the virus. For example, in reovirus-infected cells, virions associate with the microtubules, giving rise to a crescent-shaped perinuclear inclusion. Infection of cells by other viruses causes specific alterations in the cytoskeleton of cells. For example, extensive changes in cellular intermediate filaments in relation to formation of viral inclusions may be observed after cytomegalovirus infection.

A particularly striking cytopathic effect of some viral infections is the formation of syncytia, or polykaryocytes, which are large cytoplasmic masses that contain many nuclei (*poly*, many; *karyon*, nucleus) and are usually produced by fusion of infected cells. The mechanism of cell fusion during viral infection probably results from the interaction between viral gene products and host cell membranes. Cell fusion may be a mechanism by which virus spreads from infected to uninfected cells.

Effects on Cell Physiology

Research into the pathogenesis of virus infections suggests a close correlation between cellular physiologic responses and the replication of some viruses. In other words, the physiological state of living cells has a significant effect on the outcome of the virus infection, since the host cell provides the synthetic machinery, key regulatory molecules, and precursors for the newly synthesized viral proteins and nucleic acids. The optimal intracellular environment for virus replication develops through events that begin to take place with attachment of virus to the cell membrane. Binding of virus to the cell membrane receptor(s) may be followed by cascades of events that are associated with biochemical, physiological and morphological changes in the cells. The virus receptor is a cell membrane component that participates in virus binding, facilitates viral infection, and is a determinant of virus host range, as well as tissue tropism. Some viruses recognise more than one cellular receptor (e.g., HIV, adenoviruses) and the binding is a multistage process. Multiple receptors may act together either to modulate each other's activity or to contribute complementary functions.

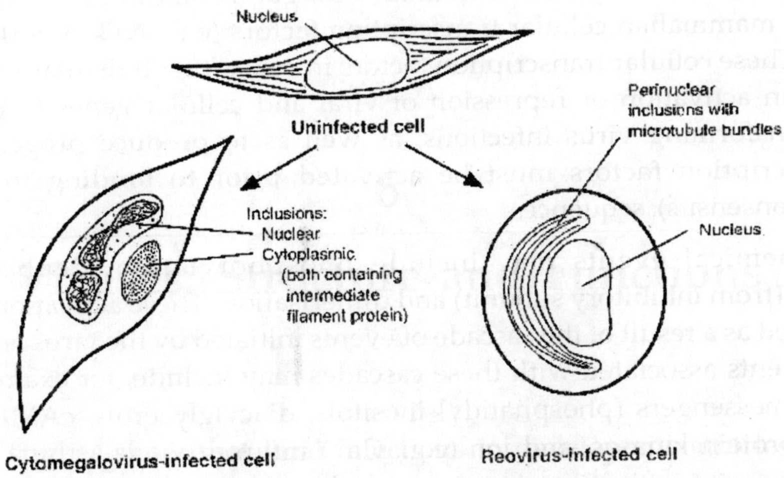

Figure 1. Alteration of cytoskeleton organization by virus infection

Other virus-associated alterations in cell physiology are related to insertion of viral proteins or other changes in the cell membrane. One example is the leaky cell membrane that appears after infection with picornaviruses or Sindbis virus; the change in intracellular ion concentrations that results from the leaky membrane may favor translation of the more salt-stable (e.g., Na^+ or K^+) viral mRNA over cellular, mRNA. These and other effects may be maintained or modified by immediate early and/or early viral gene products (e.g., changes in transcription and protein levels of cell cycle regulatory molecules).

Effects on Cell Biochemistry

Virus binding to the cell membrane in concert with immediate early (e.g., IE proteins of herpesviruses), early non-structural proteins (e.g., E-6, E-7 of HPVs) or virion components (e.g., ICP0, ICP4, VP16 of herpes simplex virus, penton protein of adenoviruses) may mediate a series of biochemical changes that optimise the intracellular milieu for use of cellular synthetic machinery, low molecular weight precursors for productive virus replication or to achieve latency, chronic, slow or transforming infection. For example, studies of transcriptional regulation of viral genes and post-transcriptional modification of gene products (splicing, polyadenylation of RNA) demonstrate that the nature of the basic biochemical processes for virus replication are similar to the mechanisms used to regulate expression of cellular genes.

Viruses have sequence motifs in their nucleic acid for binding of known transcriptional regulators of cellular origin. Thus, promoter regions of regulatory and structural proteins for many viruses contain contiguous binding sites for a large array of identifiable mammalian cellular transcription factors (e.g., NFk B, Sp1, CRE/B, AP-1, Oct-1, NF-1). These cellular transcription factors in concert with regulatory viral proteins are involved in activation or repression of viral and cellular genes to develop latent, persistent, transforming virus infections, as well as to produce progeny virus. Most cellular transcription factors must be activated prior to binding to their specific recognition (consensus) sequences.

The biochemical events may include phosphorylation, dephosphorylation, disassociation (from inhibitory subunit) and dimerisation. These activation processes can be accomplished as a result of the cascade of events initiated by the virus and cell receptor interaction. Events associated with these cascades may include, for example, formation of secondary messengers (phosphatidyl inositols, diacylglycerols, cAMP, cGMP, etc.), activation of protein kinases, and ion (e.g., Ca^{2+}) influxes.

To maintain cell activation processes, viruses have evolved unique mechanisms to regulate these cellular processes, adapting their proteins to interact with cellular proteins. Examples include the association of early virus gene products (e.g., E-6, E-7 of papillomaviruses; IE proteins of herpesviruses; SV40 T antigen) with the Rb tumor suppresser protein which results in liberation of the E2F transcription factor that is required for modification (activation/inhibition) of cellular biochemical pathways, for synthesis of viral DNA, or initiation of cellular apoptotic processes (programmed cell death).

In some cases the virus directly incorporates cellular biochemical regulatory strategies by triggering the cells to overproduce and excrete regulatory molecules (e.g., transforming growth factors, tumor necrosis factors, interleukins), which may activate in an autocrine fashion cellular biochemical cascades involved in virus (e.g., HIV, herpesviruses, papillomaviruses) replication, maintenance of or reactivation from a latent state, or maintaining a transformed phenotype. On the other hand, these soluble cellular regulatory molecules may inhibit biochemical reactions of immune cells in a paracrine manner to compromise elimination of infected cells.

Inhibition of cellular macromolecule synthesis may result from virus infection and provide an advantage for synthesis of virus proteins and nucleic acids in the absence of competing synthesis of cellular products. This inhibition occurs in characteristic ways. In poliovirus or herpes simplex infections, for example, selective inhibition of host protein synthesis occurs prior to the maximal synthesis of viral proteins. In some cases, viral products inhibit both protein and nucleic acid synthesis. Purified adenovirus penton fibers significantly decrease the synthesis of host protein, RNA, and DNA.

Total inhibition of host macromolecular synthesis also may occur when excess viral products accumulate in the cell late in the viral replicative cycle. Some picornaviruses specify a protein that causes cell damage independent of the viral proteins that inhibit cell macromolecular synthesis. Cellular mRNA may be degraded. For example, in influenza virus and herpes simplex virus infections, cellular mRNA stops binding with ribosomes to form polyribosomes; only virus-specific mRNA is bound, giving viral mRNAs a selective advantage. Cell DNA synthesis is inhibited in most cytolytic virus infections. This may be achieved by virus-induced apoptosis or by a decrease in cellular protein synthesis. Reoviruses and some herpesviruses may be exceptions in that they cause a decrease in cell DNA synthesis before a substantial decline in cellular protein synthesis occurs. Direct degradation of host DNA is seen in vaccinia virus infections due to a virion-associated DNase.

Genotoxic Effects

Chromosome damage may be caused directly by the virus particle or indirectly by events occurring during synthesis of new viral macromolecules (RNA, DNA, protein). The chromosome damage may or may not be faithfully repaired, and in either case, it may or may not be compatible with survival of the infected cell. When the cell survives, the virus genome may persist within the cell, possibly leading to continued instability of cellular genomic material or to altered expression of cellular genes (e.g., cellular oncogenes). Virus-induced genomic instability appears to be associated with accumulation of mutations and related to the process of cell immortalisation and oncogenic transformation.

Biologic Effects

The biologic consequences of virus infection results from the aforementioned biochemical, physiological, structural, morphological and genetic changes. In productive infections virus-induced biological modifications of the cell may be closely related to the efficiency of virus replication or to the recognition of these cells by the immune system. For cells that are persistently infected, the cellular changes caused by the virus could lead to disease (e.g., subacute sclerosing panencephalitis after measles infection), cellular genetic damage (e.g., hepatitis B virus), immortalisation (e.g., Epstein-Barr virus), or malignant transformation (e.g., HTLV-1, HTLV-2, hepatitis B virus). The wide variety of these effects of virus infection points to the complex interaction between the viruses and their host cell.

Relation of Cellular Effects to Viral Pathogenesis

Although most of the events that damage or modify the host cell during lytic infection are difficult to separate from viral replication, the effects are not always linked directly to the production of progeny virions. For example, changes in cell size, shape, and physiologic parameters may occur before progeny virions or even many virus proteins, are produced. These alterations in cell structure and function may be important aspects of the pathogenesis of a number of viral infections. For example, through their cellular effects many viruses (e.g., rotaviruses, caliciviruses, Norwalk viruses) induce gastrointestinal symptoms (ranging from mild alteration in absorption of ions to severe watery diarrhea). Cytocidal viral infections (e.g., herpesviruses, togaviruses, flaviviruses, bunyaviruses) of the central nervous system are related to necrosis, inflammation or phagocytosis by supporting cells. Rubella virus infections are associated with demyelination without neural degeneration. The long-term effects of persistent virus infections may also be related to such progressive diseases as atherosclerosis and demyelination in multiple sclerosis.

PERSISTENT INFECTIONS

In a persistent infection the virus is not eliminated from all of the host tissues after initial infection or the acute phase of disease. The several types of persistent infection [chronic, slow, latent, and transforming] differ in the mechanisms controlling their pathogenesis. In chronic infections, a limited number of cells (in the target organs) are infected. These infected cells may demonstrate a cytopathic effect, synthesize virus macromolecules, and release infectious virus.

The spread of infection is limited by host factors such as humoral and cell-mediated immune responses, interferons and other nonspecific inhibitors. Slow infections induced by conventional (e.g., measles) or unconventional viruses (e.g., prions) are characterised by long incubation periods (years), preceding the onset of clinical symptoms. In latent infections, infectious virus is seldom detected between clinical episodes of disease. Few cells are infected, and virus expression and replication are extensively restricted.

Common features of latent infection are their ability to reactivate in response to various environmental stimuli (e.g., heat, ultraviolet irradiation), and immune suppression brought on by heterologous virus infection (e.g., HIV) or chemotherapy, often associated with organ transplantation. In transforming persistent infections, infectious virus (RNA tumor viruses) may or may not be released, but as a result of heritable genetic changes to cellular genes, acquisition of a viral oncogene, or the effect of integrated viral sequences, the cells undergo alterations that result in malignant phenotypes.

Responses to Persistent Infections

Autoimmune injury and other forms of cell damage may occur during persistent infections. Budding virions and viral peptides associated with the cell membrane change the antigenic characteristics of the cell so that the immune system may recognise it as foreign. The cell then may be attacked by the humoral and cellular immune systems of the host and may die, even if it was infected by a noncytocidal virus.

The immune response also may cause formation of circulating antigen-antibody complexes involving viral antigens. These complexes may deposit (e.g., in the glomeruli) and elicit inflammation by activating the classical pathway of complement. The long-term association of the virus with specific target cells may lead to altered function or responses; this type of mechanism is thought to be responsible for the progressive neurologic disease associated with slow virus infections such as kuru, Creutzfeldt-Jakob disease, or subacute sclerosing panencephalitis.

TRANSFORMING INFECTIONS

The term oncogenic transformation refers to the process through which control of cell proliferation is genetically modified, so that the cell becomes cancerous. In the context of virus-cell interactions, the cells can also undergo various types of heritable changes, that result in biochemical, antigenic, morphologic, and physiologic alterations, called non-oncogenic transformation.

Transforming Virus Host Cell Interactions

DNA viruses induce transformation only under conditions that restrict virus replication and permit survival of infected cells (e.g., in noncytolytic infections of selected cell types or animal hosts, or in infection with incomplete virions). Under such conditions, immediate early (e.g., EBNAs of Epstein-Barr virus) or early proteins (e.g., SV40 "T" or polyoma middle "t" antigen) are usually present, but infectious progeny virions are seldom produced. In contrast, because RNA tumor virus replication is usually noncytocidal, they can cause oncogenic transformation in permissive cells or in their natural hosts, and viral products may be produced whether or not virus is released.

Stages and Mechanisms of Cellular Transformation

Current data indicate that transformation of a cell involves at least two components: first, the cell gains the capacity for unlimited cell division (immortalisation), and second, the immortalised cells acquire the ability to produce a tumor in an appropriate host. Some viral genes that can immortalise cells include, for example, T antigens of papovaviruses (e.g., polyoma, JC, SV40), early proteins (e.g., E6, E7) of papillomaviruses, and Epstein-Barr virus (e.g. EBNA-5). In these cases, the viral proteins may interact and inactivate one or more cellular tumor suppresser proteins, resulting in a significantly impaired cell cycle regulation. During the perturbed cell cycling, accumulation of mutations may occur either spontaneously or as an effect of other agents (virus, chemical, radiation) in cellular oncogenes (e.g., H-*ras*, K-*ras*; c-*myc*), in anti-oncogenes (e.g., p53, Rb), or in other cellular genes. *In vivo*, the history of malignancies also suggests a multiple process of cellular evolution, involving cumulative genetic changes, selection of rare cells that have the ability to invade, metastasize, and avoid immune surveillance.

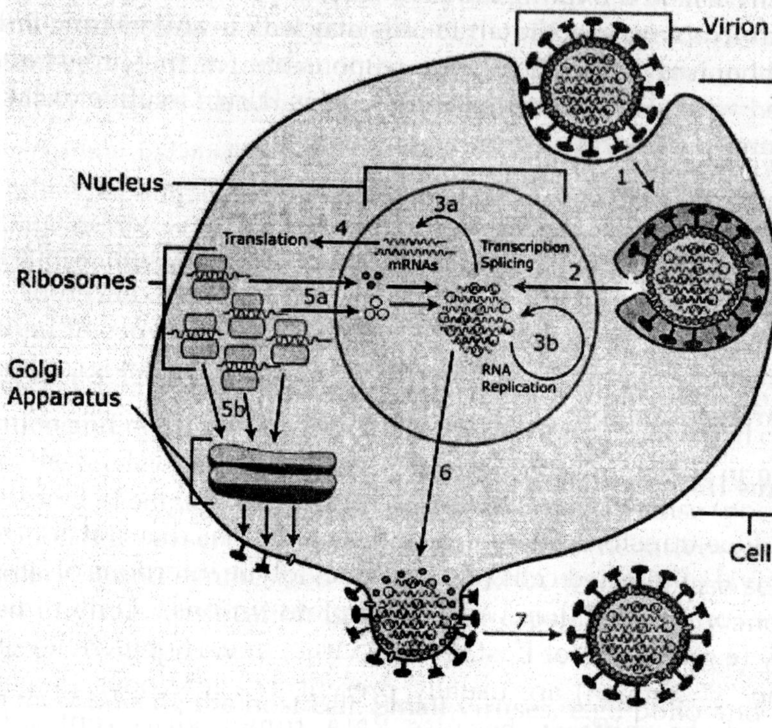

Figure 2. Virus Host Cell Interactions

The number of mutations supporting an oncogenic transformation is estimated to range from six to more. Virus infections may contribute both to immortalization and to the accumulation of mutations in growth related cellular genes during oncogenic processes.

VIRUSES AND TUMOURS

Viruses whose association with human oncogenic disease is considered to be causative should ideally fulfill certain criteria:

1. The virus or part of its genome should be closely associated with the oncogenic disease (e.g., should be present in tumor tissues).
2. The virus should persist throughout the disease.

3. A prospective study should show that infection with the virus precedes disease.
4. Prevention of virus infection (e.g., by vaccine) should prevent disease.
5. The location of the virus should be appropriate to account for disease.

In the absence of the ability to test whether the virus will induce a tumor, several of these criteria are good evidence for some causative role for the virus in the development of the tumor. Almost always, further cofactors are essential. Viruses from three DNA and one RNA genera fulfill at least some of these criteria, and their association with human tumors deserves further study.

Human Oncogenic RNA Viruses

Although several groups of DNA viruses are oncogenic in their natural hosts, only one group of RNA viruses, the retroviruses, has this property. The retroviruses are classified into three major subfamilies.

1. The oncoviruses contain the oncogenic retroviruses and are divided into type B, type C, and type D viruses on the basis of their morphology and genome structure.
2. The lentiviruses contain viruses (e.g., human immunodeficiency virus [HIV]) associated with slowly progressive, usually fatal conditions.
3. The spumaviruses are believed to be apathogenic.

Oncovirus Groups

Type B Oncoviruses

Only one member of the type B oncovirusesmouse mammary tumor virus (MMTV)has been clearly identified. This virus has a distinctive morphology and is produced when a preformed core buds through a cytoplasmic membrane.

Type C Oncoviruses

The type C oncoviruses include the human and animal leukemogenic retroviruses. Neoplastic transformation is induced by these viruses through direct interaction with cellular genes called proto-oncogenes.

One group of type C oncoviruses includes human T-cell leukemia viruses types I and 2 (HTLV-l and HTLV-2), as well as bovine (BLV) and simian (STLV-l) leukemogenic viruses. These viruses establish predominantly latent infections, often with the concurrent production of antiviral antibody. They produce a narrower disease spectrum than the other type C oncoviruses (such as avian leukemia virus); for HTLV-l, the disease spectrum includes adult T-cell leukemia, lymphoma, and tropical spastic paraparesis.

The HTLV group of viruses carry a viral gene, *tax*, which is probably involved in neoplastic transformation

Type D Oncoviruses

Type D oncoviruses have been isolated only from nonhuman primates, in which they induce both immunosuppression and proliferative syndromes.

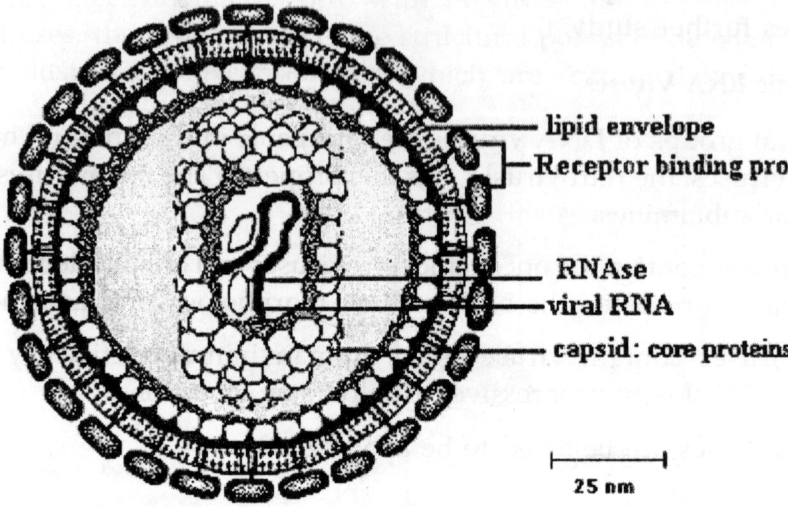

Figure 3. Human Oncogenic RNA Viruses

Replication

Although there are important differences in the replication strategies among the oncogenic retroviruses, there is a strong common theme. Retroviruses are enveloped RNA viruses, and two copies of the viral genome are enclosed within a core, together with the virion-associated enzyme reverse transcriptase. Surrounding the core is an envelope containing the viral glycoprc ins, which serves as the anti-receptor for the virus to bind to its target cell. Once bo ınd, the core is released into the cytoplasm and the RNA genome of the virus is transcribed into double-stranded DNA by reverse transcriptase. The double-stranded DNA copy of the viral genomecalled the provirusmigrates in a nucleoprotein complex to the nucleus, where it becomes covalently integrated into the chromosomal DNA. Integration of the provirus is dependent on an integrase function of reverse transcriptase.

At the gross level, integration appears to be a random process occurring anywhere within open chromatin, although fine-structural features of the chromosomal target may influence the exact integration point. In HTLV infection, the proviral copy number usually ranges from one to three proviruses per cell. The HTLV replication cycle may be halted at this point, and then a latent infection is established within the cell.

Once transcription is initiated, new genomic RNA and mRNA encoding viral proteins are transcribed from the provirus. Viral glycoproteins become substituted into the plasma membrane, and in the type C viruses internal core proteins assemble beneath this region, forming a nascent virion, which is released by a budding process. In contrast to infection with HIV, release of the type C oncoviruses is not usually a cytopathic process; hence, a cell can continue to divide and function normally while releasing virions. These features, coupled with the integration of the provirus into chromosomal DNA, lead to persistent, usually life-long infections with these viruses.

Oncogenes

Oncogenes are cellular and viral genes that influence cell growth and differentiation and may lead to oncogenic disease. The discovery of the human oncogenic and immunosuppressive retroviruses has been of major importance in human medicine; however, study of animal retroviruses has led to the discovery of the cellular oncogenes, whose mutation or dysregulation leads to oncogenic disease.

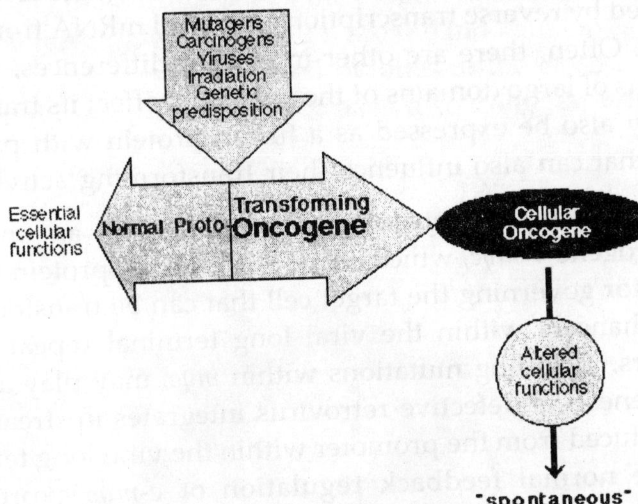

Figure 4. Diagram showing the dual nature of oncogene

When avian leukemia virus infects a chicken, many months may pass before tumors occur, if they occur at all. However, when viruses are isolated from tumor tissue, they may contain acutely transforming viruses (e.g., Rous sarcoma virus [RSV]) that are capable of rapidly reproducing the same type of tumor. Rous sarcoma virus was found to contain an additional gene, src, which is responsible for the transforming activity of the virus. This gene is of cellular origin and becomes incorporated into the virus by recombination.

Subsequently, other acutely transforming retroviruses were characterised, and the generic term viral oncogene (v-onc) became used to describe the transduced cellular gene within the viral genome. The cellular homologs of the viral oncogenes are referred to as c-oncs or proto-oncogenes, to indicate that the cellular genes fulfill normal functions governing signal transduction, cell proliferation and differentiation. Rous sarcoma virus is unique in that it is replication competent, which most acutely transforming retroviruses are not. The other viral oncogene-containing retroviruses are defective, having lost viral sequences and require the presence of a conventional leukemia virus to complement their defectiveness and permit their replication. Most of the evidence suggests that viral oncogene-containing viruses are generated de novo in each individual and are not transmitted from animal to animal.

Viral Oncogenes are Derived from Processed Cellular RNA

Examination of viral oncogenes indicates that they differ from their normal cellular counterparts. The most obvious difference is that viral oncogenes lack introns, indicating that they were derived by reverse transcription of spliced mRNA from which the introns have been removed. Often, there are other important differences, ranging from point mutations to deletions of large domains of the gene, that affect its transforming capacity. Viral oncogenes may also be expressed as a fusion protein with part of the viral gag sequences, a factor that can also influence their transforming activity.

Features that influence the transforming properties of an oncogene have been described for the oncogene v-*myc*, which encodes a phosphoprotein that localises in the nucleus. A major factor governing the target cell that can be transformed resides in the properties of the enhancers within the viral long terminal repeat sequences (LTRs), although other factors, including mutations within *myc*, may play a part. In one form of insertional mutagenesis, a defective retrovirus integrates upstream of exon 2 and a hybrid mRNA is produced from the promoter within the viral long terminal repeat. This process disrupts the normal feedback regulation of c-*myc* expression, leading to uncoordinated production of the myc gene product. Hybrid virus-*myc* RNAs of this form are probably precursors of v-*myc*-containing viruses.

Nature of Oncogenes

A wide range of genes are activated or transduced by retroviruses, and where their gene products have been identified, it has been possible to assign them to putative functional groups.

Growth factor receptors have been identified as viral oncogenes, and genes for other cell surface structures, like the beta chain of the T-cell antigen receptor, have also been found in retroviruses. For instance, the oncogene v-*erb*-B is the homolog of the epidermal growth factor (EGF) receptor. In this case, there has been extensive modification of the protein, including deletion of the extracellular domain, forming the putative epidermal growth factor-binding site. Other changes in the cytoplasmic domain include the loss of a site for autophosphorylation. Phosphorylation of proteins is an important method of regulating protein activity and of communicating signals intracellularly. Consequently, the net effect of these changes is to produce a receptor that is constitutively active in the absence of its natural ligand.

Another large group of oncogenes, which includes the *src* gene of Rous sarcoma virus, also have a tyrosine-protein kinase activity (i.e., they phosphorylate proteins on tyrosine residues). The products of this group of oncogenes appear to be associated with the plasma membrane and may increase levels of second messengers, transducing signals from cell surface receptors.

A distinct group of oncogenes, the *ras* oncogenes, appears to be related to the signal-transducing G proteins, which are regulated by binding to GTP and GDP, the GTP-bound forms being active and the GDP-bound forms inactive. The active form of *ras* bound to GTP is regulated by another protein, GAP, which has a GTPase activity, converting GTP to GDP. However, both v-*ras* genes within retroviruses and *ras* genes activated in nonviral cancers possess point mutations that prevent GAP-mediated GTPase activity. Consequently, these *ras* gene products remain in their GTP-bound activated state.

Other oncogene products, such as those of *myc, myb, fos, ski,* and *erb*-A genes, are located within the nucleus. The c-*erb*-A gene is the thyroid hormone receptor gene. Like other receptors of this class, the normal thyroid hormone receptor activates transcription through the binding of a hormone receptor complex to enhancer elements of their target genes. The v-*erb*-A gene is unable to bind thyroid hormone and is likely to be constitutively active.

Role of Multiple Genetic Events in Transformation

A common theme in viral oncogenesis is the concept that viruses act as initiators of transformation, with secondary genetic events being required for progression to the full neoplastic phenotype. Many of the oncogenes identified as retrovirus-activated genes are also activated in nonviral cancers. Moreover, different groups of oncogenes appear to complement one another in transformation. For example, genes such as *myc* exert an immortalising function on primary embryo fibroblasts, which are normally capable of only a restricted number of divisions in vitro. Such immortalised cells are not tumorigenic, but become so when other genes such as *ras* are activated.

Recently, attention has been directed to a new, important class of repressor genes that exert an anti-oncogenic effect. The best-characterised example is the p53 protein. The anti-oncogene or tumor suppressor gene p53 is bound by the transforming proteins of papova, papilloma and adenoviruses, a state which inhibits normal cell function. Mutations in p53 similarly destroy the anti-oncogene activity in non-virus induced tumors.

HTLV Replication

HTLV-1 and HTLV-2, as well as the bovine and simian leukemia viruses, have a genomic organisation and replication pattern that distinguish them from other animal leukemia viruses. In addition to the *gag, pol,* and *env* genes, these viruses contain a 3' px region encoding two genes, *tax* and *rex,* which are generated from a double spliced message.

HTLV-1 appears to remain as a latent infection in most infected cells in vivo. However, when the virus is activated, *tax* and *rex* transcripts are produced. The *tax* gene product acts as a transactivator, up-regulating the transcription of the provirus and thus creating a positive feedback loop. The effects of the *tax* protein are not direct but are mediated through cellular activating transcription factors namely the cAMP response-element binding proteins which bind to three 21bp repeat elements in U3 of the virus LTR; and ATF/CREB, AP2 and ETS transcription factors. These factors act through protein interactions, the modification of pre-existing proteins or by the induction of new cellular proteins.

The Rex nuclear phosphoprotein of MW 27Kd has a different regulatory function from *tax.* It favors the accumulation of unspliced and single-spliced mRNA encoding structural proteins over the double-spliced mRNAs encoding the regulatory proteins. The mode of action of Rex is not fully resolved but it has been implicated in stabilising certain cellular mRNAs or the protein may be involved in the nuclear export of the structural mRNAs or play a role in the polyadenylation of transcripts. The balance between Rex and Tax production leads to an early/late switch in replication. When the provirus is first activated, all of the mRNA will be double spliced because of the low level of Rex. As Tax, which encodes two nuclear phosphoproteins of MWs 37 and 40, increases the level of transcription, Rex also increases, switching the balance toward the production of structural proteins. Thus Tax may regulate the switch from latency to virus production which might be achieved by increasing the stabilisation of the transcription factor CREB binding to the virus promoter.

Mode of Action of Tax

When HTLV-1 infects cord blood lymphocytes, a polyclonal proliferation of mainly CD4+ but also CD8+ T cells occurs, and these events are associated with the expression of IL-2 and the IL-2Rd receptor by these cells. The genes for both interleukin-2 and the interleukin-2 receptor are among several cellular genes transactivated by Tax, indicating that HTLV- 1 can initiate an autocrine stimulatory pathway. Tax also transactivates c-fos, c-myc, c-jun and c-sis (encoding PDGF) MHC class 1, AP1, ERG-1, GM-CSF and stimulates the overgrowth of transfected 3T3 and rat 1 cells as well as inducing anchorage independence in transfected cells.

However, the pathogenesis of HTLV- 1 -associated T-cell leukemia and lymphoma remains an enigma. Although clonal HTLV-1 proviruses can be detected in the neoplastic cells, viral RNA transcripts are not usually detected. This situation is somewhat similar to the role of Epstein-Barr virus (EBV), a herpesvirus which remains latent in Burkitt's lymphoma, and suggests that HTLV-1 is necessary for initiation of disease but secondary genetic events are important in the development of HTLV-1-associated T-cell leukemia and lymphoma. These events remain to be elucidated. Tax induces N-F Kappa B transcription factors via degradation of I Kappa B-alpha and involving signalling pathways which converge with those used by the tumor promoter phorbol 12 merysterate, ionomycin or TNF. The possibility that Tax may also modulate signals from the cell surface receptors such as CD4 and the T-cell receptor remains an interesting speculation.

Animal Model

Recent studies show that severe combined immunodeficient (SCID) mice can support HTLV replication. Furthermore human lymphomas can arise when PBL from ATL patients are injected intraperitoneally. This suggests the SCID mouse may prove useful in the future for studying the development of ATL.

Control

As it is important to prevent HTLV-1 infection of very young children to avoid the onset in adult life of ATL, the effort in Japan has gone into publicising the need for hygiene, the control of breast feeding to only three months and the use of condoms. Chemotherapy for ATL has only recently shown any success using a combination of AZT with interferon. Vaccine developments include efforts to include peptides to interfere with the binding of HLV-1 to its cell receptor and a recombinant adenovirus vector in which the envelope glycoprotein (env) can be expressed in Hela cells. This latter may be tested in countries with a high prevalence rate (New Guinea) and used for oral administration.

Human Hepatitis B Virus

Beasley and coworkers showed in 1981 that 40% of Chinese males infected with hepatitis B virus (HBV) will die owing to HBV related hepatocellular carcinoma (HCC). In common with many other tumors, primary hepatocellular carcinoma (PHC) is prevalent in certain geographic areas, principally in Africa and Asia. Its study has produced a convincing association of HBV with PHC. PHC is usually associated with chronic liver disease and with chronic liver injury and regeneration. Infection with HBV often goes from mother to child and within family groups living together.

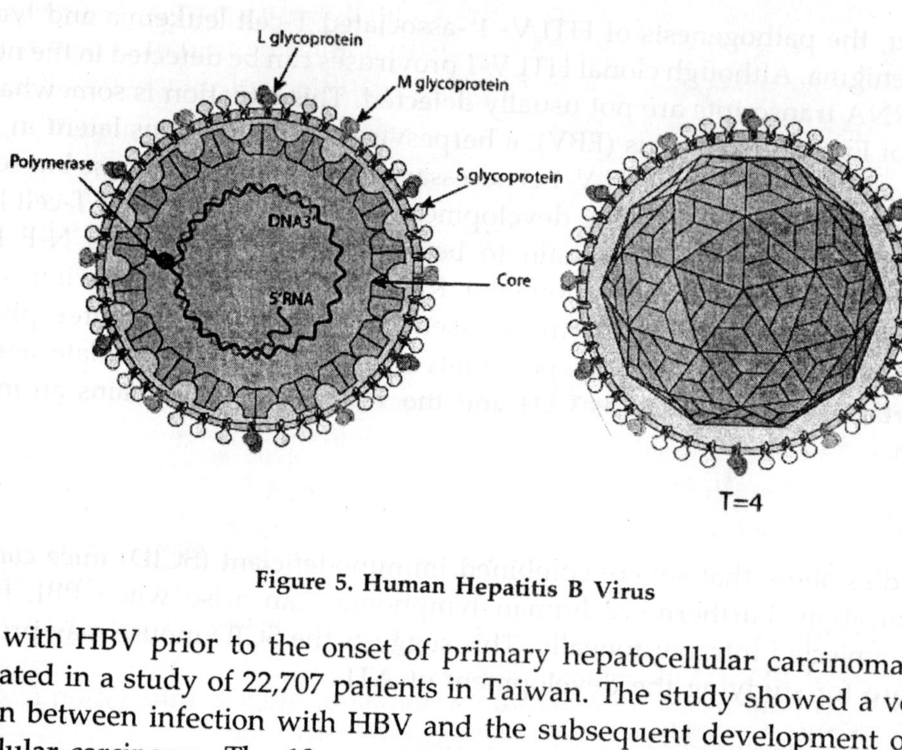

Figure 5. Human Hepatitis B Virus

Infection with HBV prior to the onset of primary hepatocellular carcinoma has been demonstrated in a study of 22,707 patients in Taiwan. The study showed a very strong association between infection with HBV and the subsequent development of primary hepatocellular carcinoma. The 10 percent of individuals in this series who were HBV carriers had a 223-fold higher risk of developing primary hepatocellular carcinoma than did the individuals not carrying the virus. That this is a very high risk is shown by the fact that moderate cigarette smoking increases the risk of developing lung cancer by 10-fold.

Cofactors

Infection with HBV at an early age, the development of chronic hepatitis and impaired immunity are risk factors in the development of HBV-associated primary hepatocellular carcinoma. Another risk factor is alcohol-associated hepatic cirrhosis. A potent fungal carcinogen, aflatoxin, is frequently found in the areas in which primary hepatocellular carcinoma is prevalent which may act synergistically to increase the risk of carcinogenesis.

Host Response and Viral Persistence

Because HBV cannot yet be cultured in vitro, another method is necessary to diagnose the continuing presence of the virus in patients. Therefore, the presence of HBV antigen in serum is monitored; in particular, the presence of HBV surface antigen (HBsAg) is used to indicate persistent infection. Patients with a poor antibody response (HBsAb) to HBsAg are those who frequently progress to develop primary hepatocellular carcinoma. Low levels of HBsAb indicate an impaired immune response. Another factor implicated in primary hepatocellular carcinoma is early age of infection of the patient before maturation of the immune system. The virus is often spread from mother to infant, and close family members also spread the virus. Low standards of living are also a contributing factor. Early infection may occur before maturation of the immune system.. In more developed Western countries, where primary hepatocellular carcinoma is not common, the initial infection takes place later in life (i.e., in the teens).

Persistence of Hepatitis B Virus DNA

Biopsy material from patients with primary hepatocellular carcinoma has been tested for the presence of HBV DNA. HBV DNA is integrated into the host genome, and isolation of restriction fragments larger than the HBV genome confirms HBV integration within the cell genome, as do DNA sequencing studies, which show that integration is usually random. The total genome is frequently not retained, and deletions, rearrangements, insertions and mutations are observed, indicating that the integrated HBV genome is probably not the source of the persistent infection. This finding also suggests that the primary role of HBV in carcinogenesis does not require the continuing expression of the integrated HBV genome (i.e., a hit-and-run mechanism). Indeed, integration frequently occurs during chronic infection with no evidence of HCC.

Molecular Mechanisms of Transformation

Despite numerous studies, no viral oncogene has been detected in HBV. One possible mechanism for the oncogenicity of HBV is integration of the HBV genome in or near a proto-oncogene. Such an integration event has been observed: in this case integration took place near a gene with homology to v-*erb*-A, the putative DNA-binding domain of the human glucocorticoid and estrogen receptors. This gene might subsequently be inappropriately transcribed. In addition, integration of HBV DNA within the DNA sequences coding for human cyclin A has been recorded. Cyclins are important proteins which control progression through the cell cycle. Integration of HBV at such a site could alter the growth control of the normal cell. However, there is no general pattern in tumors to suggest that this is a regular mechanism. The fact that HBV encodes a reverse transcriptase and replicates via an RNA intermediate similar to that of the retroviruses suggests that the mechanism of oncogenicity may be similar in the two virus groups. The X open reading frame of HBV DNA corresponds in position to the HTLV *tax* gene discussed above. Experimental evidence has shown that the X gene can transactivate promoters and enhancers and stimulate itself and other genes in the presence of cellular factors. Integrated HBV retains the X gene, in contrast to other HBV genes, which are frequently lost. Another possible mechanism of HBV oncogenesis is repression of the cellular interferon beta promoter by a *trans* mechanism. This event could promote persistent infection by HBV, leading to cirrhosis and ultimately to cancer. Infection with HBV correlates with deletion of p53, suggesting that one mechanism by which the virus achieves oncogenicity is deletion of an anti-oncogene.

HBV is an almost complete double stranded DNA genome which codes for its proteins in overlapping frames. Two transactivating gene products, pre-S and X, may be responsible for oncogenicity by transactivation of cellular genes important in growth control.

Receptors for HBV are present only in human and chimpanzee cells. Other viruses of this genus have been clearly demonstrated to be tumorigenic (e.g., woodchuck hepatitis virus, which induces tumors in 90 percent of infected woodchucks within 1 to 2 years, and duck hepatitis virus, which is oncogenic in ducks).

Control

Effective vaccination has shown a decrease of hepatocellular carcinoma in Alaskan Eskimos. However, it is essential that targets are met for the vaccination of neonates and young people in third world countries and vaccination of teenagers in the west where the incidence of HBV in heterosexual teenagers is increasing dramatically. Use of condoms and other hygiene precautions should decrease the risk of spread.

REFERENCES

Albrecht T, Fons M, Boldogh I, et al. 1991. "Metabolic and cellular effects of human cytomegalovirus infection". *Transplant Proc* 23:48.

Antoni BA, Stein SB, Rabson AB. 1994. "Regulation of human immunodeficiency virus infection: Implication for pathogenesis". *Adv Virus Res* 43:53,

Boldogh I, Abubakar S, Albrecht T. 1990. "Activation of protooncogenes: an immediate early event in human cytomegalovirus infection". *Science* 247:561,

Haywood, AM. 1994. "Virus receptors: binding, adhesion strengthening, and changes in virus structure". *J Virol* 68:1

Moran E. 1993. "DNA tumor virus transforming proteins and the cell cycle". *Curr Opin Gen Dev* 3:63.

8

Microbial Ecology

Microbial ecology is the relationship of microorganisms with one another and with their environment. It concerns the three major domains of life—Eukaryota, Archaea, and Bacteria—as well as viruses. Microorganisms, by their omnipresence, impact the entire biosphere. They are present in virtually all of our planet's environments, including some of the most extreme, from acidic lakes to the deepest ocean, and from frozen environments to hydrothermal vents.

Microbes, especially bacteria, often engage in symbiotic relationships with other organisms, and these relationships affect the ecosystem. One example of these fundamental symbioses are chloroplasts, which allow eukaryotes to conduct photosynthesis. Chloroplasts are considered to be endosymbiotic cyanobacteria, a group of bacteria that are thought to be the origins of aerobic photosynthesis.

Some theories state that this invention coincides with a major shift in the early earth's atmosphere, from a reducing atmosphere to an oxygen-rich atmosphere. Some theories go as far as saying that this shift in the balance of gasses might have triggered a global ice-age known as the Snowball Earth. They are the backbone of all ecosystems, but even more so in the zones where light cannot approach and thus photosynthesis cannot be the basic means to collect energy. In such zones, chemosynthetic microbes provide energy and carbon to the other organisms.

Other microbes are decomposers, with the ability to recycle nutrients from other organisms' waste poducts. These microbes play a vital role in biogeochemical cycles. The nitrogen cycle, the phosphorus cycle and the carbon cycle all depend on microorganisms in one way or another. For example, nitrogen which makes up 78% of the planet's atmosphere is "indigestible" for most organisms, and the flow of nitrogen into the biosphere depends on a microbial process called fixation. Due to the high level of horizontal gene transfer among microbial communities, microbial ecology is also of importance to studies of evolution.

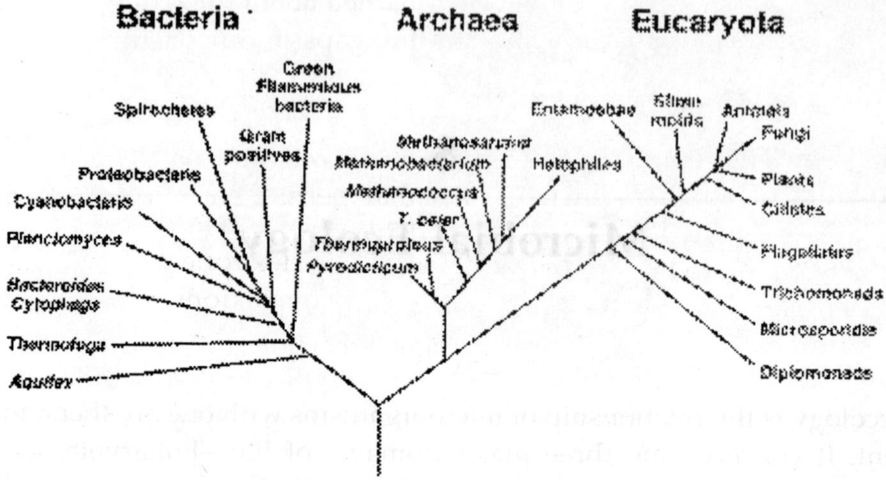

Figure 1. Phylogenic tree of microbes

DEVELOPMENT OF MICROBIAL ECOLOGY

Although plants and animals dominate our visual landscape of nature, microbes constitute a similar amount of biomass and play pivotal roles in maintaining the viability of Earth's biosphere. In their celebrated book *The Microbe's Contribution to Biology*, A.J. Kluyver and C.B. van Niel introduced the scientific community to microbes and their critical roles in sustaining life with the following argument:

> "Reliable estimates have shown that the amount of carbon dioxide consumed annually in the photosynthetic activity of the green plants is such that the quantity of this gas present in the atmosphere would be exhausted within some 30 years, if it were not replenished. Even taking into account the important carbon dioxide reservoir present in the oceans as a buffer system, such an exhaustion should have occurred within historic times. It is, therefore, clear that the green plants can continue to grow only because the assimilated carbon is in some way reconverted into carbon dioxide. At first sight this will not present any difficulties to our "macrobiologist"; he will refer to the slow combustion of the vegetable remains by man and animals. However, several independent estimates tend to show that the annual carbon dioxide production by this means amounts to only about 5 percent of the annual carbon dioxide consumption by the green plants. The conclusion seems inevitable that the remaining 95 percent is produced by the mineralising action of the microbe."

Although estimates of the magnitude of carbon flux and details of the processes that contribute to the cycling of carbon have been refined considerably in the past 50 years, microbes remain central catalytic agents in the global carbon cycle. Microbes also drive local and global cycles of nitrogen, sulfur, oxygen, phosphorus, and many of the transition metals: without the metabolic activity provided by microbes, plant or animal life could not be sustained. So what have we learned about the ecology of microbes since the publication of this book and where are the gaps in our scientific understanding?

METABOLIC DIVERSITY OF MICROBES

Kluyver and van Niel emphasized the core metabolic unity of life and described the impressive suite of auxiliary metabolisms that permit the harvest of energy while coupling the oxidation and reduction of a wide variety of chemical reactions. Reactions are arranged from top to bottom in the order of most electronegative to electropositive, providing a simple way to determine whether coupled reactions will yield energy under standard conditions: draw a line between the half reactions on the towers and if the slope of the line is negative, the reaction is exothermic and there are almost certainly microbes capable of capturing the energy released in the reaction.

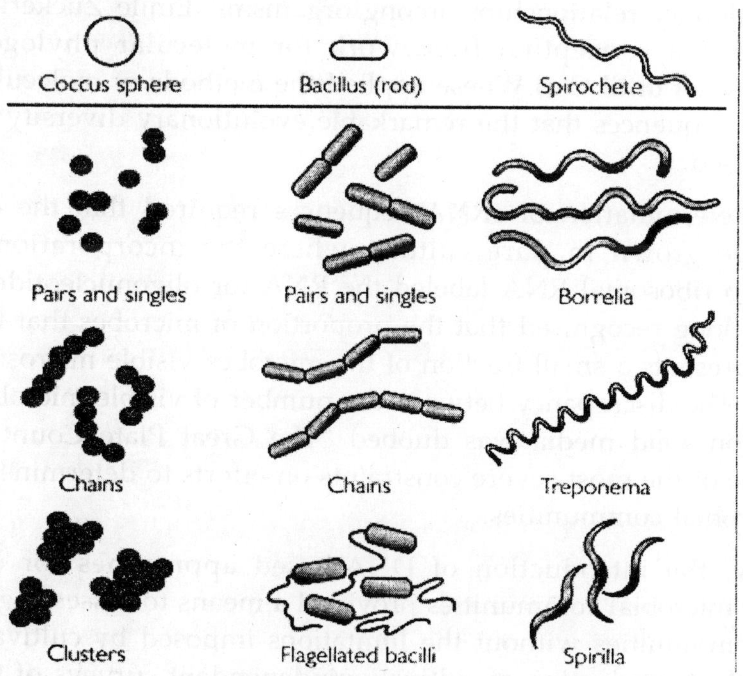

Figure 2. Various types of microbes

Although environmental conditions such as pH, eH, and the concentration of reactants and products, alter the free energy available in these redox reactions, as a first approximation energy-yielding reactions can be easily identified by a downward sloping line. In the past 50 years, there have been some spectacular discoveries of the types of energy-yielding reactions catalysed by microbes, including the capacity to use insoluble metal oxides as terminal electron acceptors and the use of anthropogenic hydrocarbons as electron acceptors.

The microbially catalysed, anaerobic oxidations of ammonia and methane have also expanded the types of microbial metabolisms in anoxic environments that contribute to global elemental cycles. It is in terms of their metabolism that microbes have revealed incredible diversity as compared to the uniform types of metabolism found throughout the plant and animal kingdoms. Nevertheless, understanding the evolutionary diversity underlying this metabolic diversity required a systematic means to determine the evolutionary relatedness of microbes.

EVOLUTIONARY DIVERSITY OF MICROBES

In a seminal article demonstrating that sequences of macromolecules could be used to infer the evolutionary relationships among organisms, Emile Zuckerkandl and Linus Pauling provided a conceptual framework for molecular phylogenetic analysis. However, it was not until Carl Woese applied the methods of molecular phylogeny to ribosomal RNA sequences that the remarkable evolutionary diversity of the microbial world was realised.

The early determination of rRNA sequences required that the organism under investigation be grown in pure culture, where the incorporation of radioactive phosphorus into ribosomal RNA labeled the RNA for oligonucleotide cataloging. But microbiologists long recognised that the proportion of microbes that form colonies on solid media represents a small fraction of the microbes visible microscopically in most environments. The discrepancy between the number of visible microbe and those that form colonies on solid media was dubbed "The Great Plate Count Anomaly", and represented one of the most severe constraints on efforts to determine the composition of natural microbial communities.

Fortunately, the introduction of DNA-based approaches for determining the composition of microbial communities provided a means for assessing the composition of microbial communities without the limitations imposed by cultivation. In the two decades since the introduction of cultivationindependent surveys of the environment, there have been dozens of new phyla discovered that lack cultured representatives.

Molecular surveys of microbial communities have become a routine dimension of studies of microbial communities, resulting in an explosion of 16S rRNA gene sequences derived directly from environmental samples. Analyses of molecular surveys now provide the opportunity to compare the composition of microbial communities and lead to hypotheses about how changes in composition result in changes in their function.

MICROBIAL COMMUNITIES

Over the past decade, molecular-based approaches have revealed a enormous phylogenetic diversity within the microbial world that is not yet represented in culture. This impression has come almost entirely by retrieval of small subunit (SSU) rRNA sequence information, which provides a phylogenetic context in which to quantify such a diversity, and typically involves cloning and sequencing SSU rRNAencoding genes. While phylogenetically informative, such clones do not always allow predictions to be made about the physiological properties of a source organism, unless the source was a member of a phylogenetic group exhibiting a distinctive type of metabolism, e.g., a cyanobacterium, a methanogenic archaeon. This is especially true for clones from phylogenetic groups that are seriously underrepresented in culture collections. How do we move beyond phylogenetic trees that reveal the evolutionary diversity of microbes present in an environment and advance our understanding of their metabolic potential?

In an effort to increase the information retrieved from environmentally cloned SSU rDNA, a number of investigators have exploited cloning vectors that can accommodate large fragments of genomic DNA that include numerous genes adjacent to the phylogenetic signature, i.e., the SSU rDNA. Stein and colleagues used a fosmid vector to obtain 35- to 45-kb fragments of genomic DNA from a marine picoplankton community containing an abundant, but yet-uncultured, marine crenarchaeote.

Sequencing of regions adjacent to the SSU rDNA locus revealed a number of genes, including an RNA helicase and one involved in heme biosynthesis, thereby providing a glimpse into the physiological potential of the source organism. Some used a bacterial artificial chromosome (BAC) vector to clone genomic DNA fragments up to 150 kb in size from marine planktonic assemblages. Among the genes flanking the SSU rDNA of one such clone, derived from a γ-proteobacterium, was a gene encoding a bacterial version of rhodopsin, implying a novel type of phototrophy occurring in the sea and suggesting obvious strategies that might be used to obtain such organisms in culture.

BAC vectors were also used to clone large fragments of DNA from soil microbes and revealed an assortment of genes encoding degradative and biosynthetic enzymes, as well as antibiotic production. Many of the applications of genomic sciences to microbial ecology have been pioneered in marine systems and are the focus of an accompanying article in this issue; only recently have they been applied to terrestrial ecosystems. This has been due in part to the difficulty of purifying nucleic acids from the soil matrix, but the effort will likely be worth the investment of time and energy because the biodiversity harbored in soils of the Earth is staggering. One gram of soil contains up to 1×10^{10} organisms, representing as many as one million bacterial species.

The combined acreage of soil contains 26×10^{28} Bacteria and Archaea that harbor 26 Pg of C, 6.2 Pg of N, and 0.65 Pg of P—huge amounts, even when compared to the total amount of these elements in terrestrial plants. The importance of microbes as a component of the Earth's terrestrial biomass, and their power to influence global ecosystems, is derived from their sheer numbers and the diverse array of biochemical reactions they catalyze. In addition to their impact on terrestrial habitats, soil microbes also affect atmospheric chemistry and global climate by influencing budgets of atmospheric gases, including CO_2, CH_4, H_2, N_2O, and NO.

Recent results from long-term studies suggest that varying management practices have major impacts on fluxes of greenhouse gases, i.e., CO_2, N_2O, and CH_4, and carbon sequestration. While it is generally accepted that microbial communities in soil are critical to the productivity and health of the biosphere, and the documented diversity in these communities is spectacular, the challenge of understanding how the function of a community is related to its structure remains. Given the magnitude of microbial diversity in both terrestrial and aquatic ecosystems and the integral role of microbes in virtually all environments, successful investigation of the structure and function of microbial communities requires enhanced opportunities in graduate education.

EDUCATION AND RESEARCH IN MICROBIAL ECOLOGY

It is a disturbing reality that we have only fragmentary understanding of the enormous microbial diversity that exists on our planet: This applies not merely to microbes living in extreme environments and which would be expected to possess unusual and perhaps not yet fully characterised properties, but also to those in mundane habitats—a gram of soil, a milliliter of seawater. Yet the metabolic activity of these microbes is essential to life on Earth: without the unique roles filled by microbes in the global recycling of carbon, sulfur, and nitrogen, plants and animals could not survive.

Despite their importance, many such microbes have continued to elude conventional isolation and cultivation techniques. Fortunately, the same developments in molecular biology presented above that have enabled us to recognise such untapped microbial diversity are now supplying the very tools necessary to quantify, monitor, dissect, and understand it.

However, the effective application of such technology has created a need for a 'new breed' of microbiologist: a hybrid individual who is as much at home annotating a genome sequence as evaluating stable isotope data; as conversant in microbial bioenergetics as in gene cloning strategies. Such individuals need a global perspective and experience in functioning within multidisciplinary, often international, teams, because the most important questions that confront us today are global in scope and often too complex to be addressed by individuals or small research teams. Several factors will likely drive much of the research in microbial ecology over the next decade, including:

(i) Recognition of the enormous biomass and diversity of microbial life on our planet.

(ii) The unsettling acknowledgement that most of this diversity is poorly understood, and yet almost certainly affects global-level changes and environmental stability on our planet.

(iii) The steadily growing facility with which microbial genomes and communities of microbes can be sequenced.

If this vision is accurate, then traditional graduate programs in microbiology must be re-tooled to educate the next generation of microbiologists, who must be broadly trained, versatile, and capable of working in multidisciplinary teams, and who can think holistically—from the genome level to global-level processes. Graduate students must also have the opportunity to gain research experience in the laboratory *and* in the field—ideally, in habitats representative of those that dominate the Earth, i.e., terrestrial, freshwater, and marine.

In this way, they can gain an appreciation for the various scales at which microbial processes are occurring, as well as an understanding of the most important questions and the methodologies used to address them. It would be ideal for graduate students to be exposed to nonlinear biological systems and mathematical methods for teasing out potentially deterministic relationships from "noise" as they begin to link the genomic and phylogenetic diversity of microbial life on Earth with its function and planetary impact.

Contemporary graduate training programs designed to accomplish this do not currently exist, and it is not difficult to understand why. It is rare that a single department, or university, has the blend of expertise and resources to provide such a robust training experience on its own. Nevertheless, atop a list of recommendations made in a recent report by the American Academy of Microbiology was a need to develop training opportunities to advance the emerging new field of "ecogenomics" as a means to explore the many levels of biological organisation that sustain the biosphere. Such training programs will require an integration of knowledge from a number of disciplines, including microbiology, ecology, evolutionary biology, genomic science, biogeochemistry, mathematics and bioinformatics.

As human civilisation leads to ever increasing perturbations of the environment, we need more than ever to understand the essential role of microbes in maintaining a habitable Earth. The recognition of such a need will ideally lead to new educational opportunities for scientists such that they will be in a position to advance our understanding of the microbial world and its effect on the Earth's biosphere, hydrosphere and atmosphere.

REFERENCES

Blum P (ed). 2008. Archaea: New Models for Prokaryotic Biology. Caister Academic Press.

Diaz E (ed). 2008. *Microbial Biodegradation: Genomics and Molecular Biology*. Caister Academic Press. ISBN 978-1-904455-17-2.

Garcia-Pichel F 2008. "Molecular Ecology and Environmental Genomics of Cyanobacteria". *The Cyanobacteria: Molecular Biology, Genomics and Evolution*. Caister Academic Press.

Heuner K, Swanson M (ed). 2008. *Legionella: Molecular Microbiology*. Caister Academic Press

Ljungh A, Wadstrom T (ed). 2009. *Lactobacillus Molecular Biology: From Genomics to Probiotics*. Caister Academic Press.

Verstraete W. 2007. "Microbial ecology and environmental biotechnology". *ISME J*. **1** (1): 4–8.

9

Microbial Genetics

Microorganisms include prokaryotes like bacteria, unicellular or mycelial eukaryotes e.g., yeasts and other fungi, and viruses, notably bacterial viruses (bacteriophages). Microbial genetics is concerned with the transmission of hereditary characters in microorganisms. Microbial genetics has played a unique role in developing the fields of molecular and cell biology and also has found applications in medicine, agriculture, and the food and pharmaceutical industries.

Hereditary processes in microorganisms are analogous to those in multicellular organisms. In both prokaryotic and eukaryotic microbes, the genetic material is DNA; the only known exceptions to this rule are the RNA viruses. Mutations, heritable changes in the DNA, occur spontaneously and the rate of mutation can be increased by mutagenic agents.

In practice, the susceptibility of bacteria to mutagenic agents has been used to identify potentially hazardous chemicals in the environment. For example, the Ames test was developed to evaluate the mutagenicity of a chemical in the following way. Plates containing a medium lacking in, for example, the nutrient histidine are inolculated with a histidine requiring strain of the bacterium *Salmonella typhimurium*. Thus, only cells that revert back to the wild type can grow on the medium. If plates are exposed to a mutagenic agent, the increase in the number of mutants compared with unexposed plates can be observed and a large number of revertants would indicate a strong mutagenic agent. For such studies, microorganisms offer the advantage that they have short mean generation times, are easily cultured in a small space under controlled conditions and have a relatively uncomplicated structure.

Microbes are ideally suited for combined biochemical and genetic studies, and have been successful in providing information on the genetic code and the regulation of gene activity. The many applications of microbial genetics in medicine and the pharmaceutical industry emerge from the fact that microbes are both the causes of disease and the producers of antibiotics. Genetic studies have been used to understand variation in pathogenic microbes and also to increase the yield of antibiotics from other microbes.

Microorganisms, and particularly bacteria, were generally ignored by the early geneticists because of their small in size and apparent lack of easily identifiable variable traits. Therefore, a method of identifying variation and mutation in microbes was fundamental for progress in microbial genetics. As many of the mutations manifest themselves as metabolic abnormalities, methods were developed by which microbial mutants could be detected by selecting or testing for altered phenotypes. Positive selection is defined as the detection of mutant cells and the rejection of unmutated cells. An example of this is the selection of penicillin resistant mutants, achieved by growing organisms in media containing penicillin such that only resistant colonies grow.

In contrast, negative selection detects cells that cannot perform a certain function and is used to select mutants that require one or more extra growth factors. Replica plating is used for negative selection and involves two identical prints of colony distributions being made on plates with and without the required nutrients. Those microbes that do not grow on the plate lacking the nutrient can then be selected from the identical plate, which does contain the nutrient.

GENETIC INFORMATION IN MICROBES

The genetic material of bacteria and plasmids is DNA. Bacterial viruses (bacteriophages or phages) have DNA or RNA as genetic material. The two essential functions of genetic material are replication and expression. Genetic material must replicate accurately so that progeny inherit all of the specific genetic determinants (the genotype) of the parental organism. Expression of specific genetic material under a particular set of growth conditions determines the observable characteristics (phenotype) of the organism. Bacteria have few structural or developmental features that can be observed easily, but they have a vast array of biochemical capabilities and patterns of susceptibility to antimicrobial agents or bacteriophages. These latter characteristics are often selected as the inherited traits to be analysed in studies of bacterial genetics.

Nucleic Acid Structure

Nucleic acids are large polymers consisting of repeating nucleotide units (Fig. 1). Each nucleotide contains one phosphate group, one pentose or deoxypentose sugar, and one purine or pyrimidine base. In DNA the sugar is D-2-deoxyribose; in RNA the sugar is D-ribose. In DNA the purine bases are adenine (A) and guanine (G), and the pyrimidine bases are thymine (T) and cytosine (C). In RNA, uracil (U) replaces thymine. Chemically modified purine and pyrimidine bases are found in some bacteria and bacteriophages.

The repeating structure of polynucleotides involves alternating sugar and phosphate residues, with phosphodiester bonds linking the 3'-hydroxyl group of one nucleotide sugar to the 5'-hydroxyl group of the adjacent nucleotide sugar.

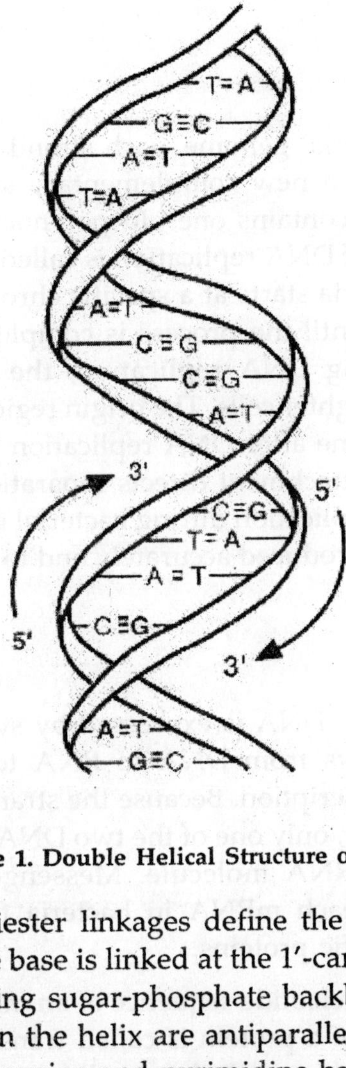

Figure 1. Double Helical Structure of DNA.

These asymmetric phosphodiester linkages define the polarity of the polynucleotide chain. A purine or pyrimidine base is linked at the 1'-carbon atom of each sugar residue and projects from the repeating sugar-phosphate backbone. Double-stranded DNA is helical, and the two strands in the helix are antiparallel. The double helix is stabilised by hydrogen bonds between purine and pyrimidine bases on the opposite strands. At each position, A on one strand pairs by two hydrogen bonds with T on the opposite strand, or G pairs by three hydrogen bonds with C.

The two strands of double-helical DNA are, therefore, complementary. Because of complementarity, double-stranded DNA contains equimolar amounts of purines (A + G) and pyrimidines (T + C), with A equal to T and G equal to C, but the mole fraction of

G + C in DNA varies widely among different bacteria. Information in nucleic acids is encoded by the ordered sequence of nucleotides along the polynucleotide chain, and in double-stranded DNA the sequence of each strand determines what the sequence of the complementary strand must be. The extent of sequence homology between DNAs from different microorganisms is the most stringent criterion for determining how closely they are related.

DNA Replication

During replication of the bacterial genome, each strand in double-helical DNA serves as a template for synthesis of a new complementary strand. Each daughter double-stranded DNA molecule thus contains one old polynucleotide strand and one newly synthesized strand. This type of DNA replication is called semiconservative. Replication of chromosomal DNA in bacteria starts at a specific chromosomal site called the origin and proceeds bidirectionally until the process is completed. When bacteria divide by binary fission after completing DNA replication, the replicated chromosomes are partitioned into each of the daughter cells. The origin regions specifically and transiently associate with the cell membrane after DNA replication has been intitiated, leading to a model whereby membrane attachment directs separation of daughter chromosomes. These characteristics of DNA replication during bacterial growth fulfill the requirements of the genetic material to be reproduced accurately and to be inherited by each daughter cell at the time of cell division.

Gene Expression

Genetic information encoded in DNA is expressed by synthesis of specific RNAs and proteins, and information flows from DNA to RNA to protein. The DNA-directed synthesis of RNA is called transcription. Because the strands of double-helical DNA are antiparallel and complementary, only one of the two DNA strands can serve as template for synthesis of a specific mRNA molecule. Messenger RNAs (mRNAs) transmit information from DNA, and each mRNA in bacteria functions as the template for synthesis of one or more specific proteins.

The process by which the nucleotide sequence of an mRNA molecule determines the primary amino acid sequence of a protein is called translation. Ribosomes, complexes of ribosomal RNAs (rRNAs) and several ribosomal proteins, translate each mRNA into the corresponding polypeptide sequence with the aid of transfer RNAs (tRNAs), amino-acyl tRNA synthesases, initiation factors and elongation factors. All of these components of the apparatus for protein synthesis function in the production of many different proteins. A gene is a DNA sequence that encodes a protein, rRNA, or tRNA molecule.

The genetic code determines how the nucleotides in mRNA specify the amino acids in a polypeptide. Because there are only 4 different nucleotides in mRNA (containing U, A, C and G), single nucleotides do not contain enough information to specify uniquely all 20 of the amino acids. In dinucleotides 16 (4 x 4) arrangements of the four nucleotides are possible, and in trinucleotides 64 (4 x 4 x 4) arrangements are possible. Thus, a minimum of three nucleotides is required to provide at least one unique sequence corresponding to each of the 20 amino acids. The "universal" genetic code employed by most organisms is a triplet code in which 61 of the 64 possible trinucleotides (codons) encode specific amino acids, and any of the three remaining codons (UAG, UAA or UGA) results in termination of translation.

The chain-terminating codons are also called nonsense codons because they do not specify any amino acids. The genetic code is described as degenerate, because several codons may be used for a single amino acid, and as nonoverlapping, because adjacent codons do not share any common nucleotides. Exceptions to the "universal" code include the use of UGA as a tryptophan codon in some species of Mycoplasma and in mitochondrial DNA, and a few additional codon differences in mitochondrial DNAs from yeasts, Drosophila, and mammals.

Translation of mRNA is usually initiated at an AUG codon for methionine, and adjacent codons are translated sequentially as the mRNA is read in the 5' to 3' direction. The corresponding polypeptide chain is assembled beginning at its amino terminus and proceeding toward its carboxy terminus. The sequence of amino acids in the polypeptide is, therefore, colinear with the sequence of nucleotides in the mRNA and the corresponding gene.

Expression of genetic determinants in bacteria involves the unidirectional flow of information from DNA to RNA to protein. In bacteriophages, either DNA or RNA can serve as genetic material. During infection of bacteria by RNA bacteriophages, RNA molecules serve as templates for RNA replication and as mRNAs. Studies with the retrovirus group of animal viruses reveal that DNA molecules can be synthesized from RNA templates by enzymes designated as RNA-dependent DNA polymerases (reverse transcriptases). This reversal of the usual direction for flow of genetic information, from RNA to DNA instead of from DNA to RNA, is an important mechanism for enabling information from retroviruses to be encoded in DNA and to become incorporated into the genomes of animal cells.

Genome Organisation

DNA molecules that replicate as discrete genetic units in bacteria are called replicons. In some Escherichia coli strains, the chromosome is the only replicon present in the cell. Other bacterial strains have additional replicons, such as plasmids and bacteriophages.

Chromosomal DNA

Bacterial genomes vary in size from about 0.4 x 109 to 8.6 x 10^9 daltons (Da), some of the smallest being obligate parasites (Mycoplasma) and the largest belonging to bacteria capable of complex differentiation such as Myxococcus. The amount of DNA in the genome determines the maximum amount of information that it can encode. Most bacteria have a haploid genome, a single chromosome consisting of a circular, double stranded DNA molecule. However linear chromosomes have been found in Gram-positive Borrelia and Streptomyces spp., and one linear and one circular chromosome is present in the Gram-negative bacterium Agrobacterium tumefaciens. The single chromosome of the common intestinal bacterium *E coli* is 3 x 109 Da (4,500 kilobase pairs [kbp]) in size, accounting for about 2 to 3 percent of the dry weight of the cell. The *E coli* genome is only about 0.1% as large as the human genome, but it is sufficient to code for several thousand polypeptides of average size (40 kDa or 360 amino acids).

The chromosome of *E coli* has a contour length of approximately 1.35 mm, several hundred times longer than the bacterial cell, but the DNA is supercoiled and tightly packaged in the bacterial nucleoid. The time required for replication of the entire chromosome is about 40 minutes, which is approximately twice the shortest division time for this bacterium.

DNA replication must be initiated as often as the cells divide, so in rapidly growing bacteria a new round of chromosomal replication begins before an earlier round is completed. At rapid growth rates there may be four chromosomes replicating to form eight at the time of cell division, which is coupled with completion of a round of chromosomal replication. Thus, the chromosome in rapidly growing bacteria is replicating at more than one point. The replication of chromosomal DNA in bacteria is complex and involves many different proteins.

Plasmids

Plasmids are replicons that are maintained as discrete, extrachromosomal genetic elements in bacteria. They are usually much smaller than the bacterial chromosome, varying from less than 5 to more than several hundred kbp, though plasmids as large as 2 Mbp occur in some bacteria. Plasmids usually encode traits that are not essential for bacterial viability, and replicate independently of the chromosome.

Most plasmids are supercoiled, circular, double-stranded DNA molecules, but linear plasmids have also been demonstrated in Borrelia and Streptomyces. Closely related or identical plasmids demonstrate incompatibility; they cannot be stably maintained in the same bacterial host. Classification of plasmids is based on incompatibility or on use of specific DNA probes in hybridisation tests to identify nucleotide sequences that are characteristic of specific plasmid replicons.

Some hybrid plasmids contain more than one replicon.Conjugative plasmids code for functions that promote transfer of the plasmid from the donor bacterium to other recipient bacteria, but nonconjugative plasmids do not. Conjugative plasmids that also promote transfer of the bacterial chromosome from the donor bacterium to other recipient bacteria are called fertility plasmids.

The average number of molecules of a given plasmid per bacterial chromosome is called its copy number. Large plasmids are often conjugative, have small copy numbers (1 to several per chromosome), code for all functions required for their replication, and partition themselves among daughter cells during cell division in a manner similar to the bacterial chromosome. Plasmids smaller than 7.5 kilobase pairs usually are nonconjugative, have high copy numbers (typically 10-20 per chromosome), rely on their bacterial host to provide some functions required for replication, and are distributed randomly between daughter cells at division.

Many plasmids control medically important properties of pathogenic bacteria, including resistance to one or several antibiotics, production of toxins, and synthesis of cell surface structures required for adherence or colonisation. Plasmids that determine resistance to antibiotics are often called R plasmids (or R factors). Representative toxins encoded by plasmids include heat-labile and heat-stable enterotoxins of *E coli*, exfoliative toxin of Staphylococcus aureus, and tetanus toxin of Clostridium tetani. Some plasmids are cryptic and have no recognisable effects on the bacterial cells that harbor them. Comparing plasmid profiles is a useful method for assessing possible relatedness of individual clinical isolates of a particular bacterial species for epidemiological studies

Bacteriophages

Bacteriophages (bacterial viruses, phages) are infectious agents that replicate as obligate intracellular parasites in bacteria. Extracellular phage particles are metabolically inert and consist principally of proteins plus nucleic acid. The proteins of the phage particle form a protective shell surrounding the tightly packaged nucleic acid genome. Phage genomes vary in size from approximately 2 to 200 kilobases per strand of nucleic acid and consist of double-stranded DNA, single-stranded DNA, or RNA. Phage genomes, like plasmids, encode functions required for replication in bacteria, but unlike plasmids they also encode capsid proteins and nonstructural proteins required for phage assembly. Several morphologically distinct types of phage have been described, including polyhedral, filamentous, and complex. Complex phages have polyhedral heads to which tails and sometimes other appendages are attached.

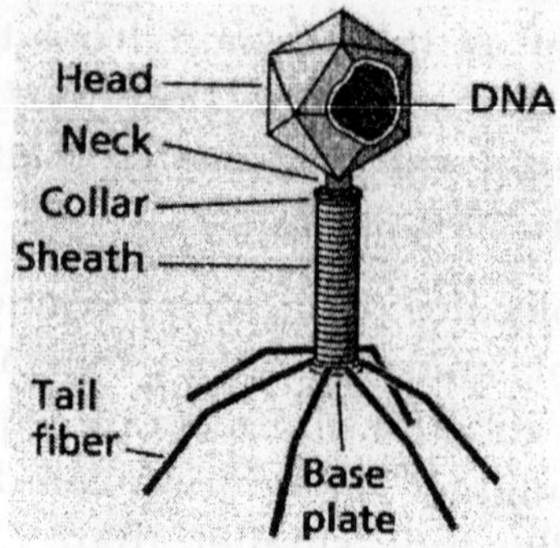

Figure 2. Bacteriophages

The capsids remain at the cell surface, and the DNA or RNA genomes enter the target cells. Because infectivity of genomic DNA or RNA is much less than that of mature virus, there is a time immediately after infection called the eclipse period during which intracellular infectious phage cannot be detected.

The infecting phage RNA or DNA is replicated to produce many new copies of the phage genome, and phage-specific proteins are produced. For most phages assembly of progeny occurs in the cytoplasm, and release of the progeny occurs by cell lysis. In contrast, filamentous phages are formed at the cell envelope and released without killing the host cells. The eclipse period ends when intracellular infectious progeny appear. The latent period is the interval from infection until extracellular progeny appear, and the rise period is the interval from the end of the latent period until all phage are extracellular. The average number of phage particles produced by each infected cell, called the burst size, is characteristic for each virus and often ranges between 50 and several hundred.

Phages are classified into two major groups: virulent and temperate. Growth of virulent phages in susceptible bacteria destroys the host cells. Infection of susceptible bacteria by temperate phages can have either of two outcomes: lytic growth or lysogeny. Lytic growth of temperate and virulent bacteriophages is similar, leading to production of phage progeny and death of the host bacteria. Lysogeny is a specific type of latent viral infection in which the phage genome replicates as a prophage in the bacterial cell.

In most lysogenic bacteria the genes required for lytic phage development are not expressed, and production of infectious phage does not occur. Furthermore, the lysogenic cells are immune to superinfection by the virus which they harbor as a prophage. The physical state of the prophage is not identical for all temperate viruses. For example, the prophage of bacteriophage l in *E coli* is integrated into the bacterial chromosome at a specific site and replicates as part of the bacterial chromosome, whereas the prophage of bacteriophage P1 in E coli replicates as an extrachromosomal plasmid.

Lytic phage growth occurs spontaneously in a small fraction of lysogenic cells, and a few extracellular phages are present in cultures of lysogenic bacteria. For some lysogenic bacteria, synchronous induction of lytic phage development occurs in the entire population of lysogenic bacteria when they are treated with agents that damage DNA, such as ultraviolet light or mitomycin C. The loss of prophage from a lysogenic bacterium, converting it to the nonlysogenic state and restoring susceptibility to infection by the phage that was originally present as prophage, is called curing. Some temperate phages contain genes for bacterial characteristics that are unrelated to lytic phage development or the lysogenic state, and expression of such genes is called phage conversion (or lysogenic conversion). Examples of phage conversion that are important for microbial virulence include production of diphtheria toxin by Corynebacterium diphtheriae,. erythrogenic toxin by Streptococcus pyogenes (group A b-hemolytic streptococci), botulinum toxin by Clostridium botulinum, and Shiga-like toxins by *E coli.*

In each of these examples the gene which encodes the bacterial toxin is present in a temperate phage genome. The specificity of O antigens in Salmonella can also be controlled by phage conversion. Phage typing is the testing of strains of a particular bacterial species for susceptibility to specific bacteriophages. The patterns of susceptibility to the set of typing phages provide information about the possible relatedness of individual clinical isolates. Such information is particularly useful for epidemiological investigations.

MUTATION AND SELECTION

Mutations are heritable changes in the genome. Spontaneous mutations in individual bacteria are rare. Some mutations cause changes in phenotypic characteristics; the occurrence of such mutations can be inferred from the effects they produce. In microbial genetics specific reference organisms are designated as wild-type strains, and descendants that have mutations in their genomes are called mutants. Thus, mutants are characterised by the inherited differences between them and their ancestral wild-type strains. Variant forms of a specific genetic determinant are called alleles. Genotypic symbols are lower case, italicised abbreviations that specify individual genes, with a (+) superscript indicating the wild type allele. Phenotypic symbols are capitalised and not italicised, to distinguish them from genotypic symbols. For example, the genotypic symbol for the ability to produce b-galactosidase, required to ferment lactose, is lacZ+, and mutants that cannot produce b-galactosidase are lacZ. The lactose-fermenting phenotype is designated Lac+, and inability to ferment lactose is Lac-.

Detection of Mutant Phenotypes

Selective and differential media are helpful for isolating bacterial mutants. Some selective media permit particular mutants to grow, but do not allow the wild-type strains to grow. Rare mutants can be isolated by using such selective media. Differential media permit wild-type and mutant bacteria to grow and form colonies that differ in appearance. Detection of rare mutants on differential media is limited by the total number of colonies that can be observed. Consider a wild-type strain of *E coli* that is susceptible to the antibiotic streptomycin (phenotype Strs) and can utilise lactose as the sole source of carbon (phenotype Lac+).

Spontaneously occurring Strr mutants are rare and are usually found at frequencies of less than one per 109 bacteria in cultures of wild-type *E coli*. Nevertheless, Strr mutants can be isolated easily by using selective media containing streptomycin, because the wild-type Strs bacteria are killed. Isolation of lactose-negative (phenotype Lac-) mutants of *E coli* poses a different problem.

On minimal media with lactose as the sole source of carbon, Lac+ wild-type strains will grow, but Lac- mutants cannot grow. On differential media such as MacConkey-lactose agar or eosin-methylene blue-lactose agar, Lac+ wild-type and Lac- mutant strains of *E coli* can be distinguished by their color, but spontaneous Lac- mutants are too rare to be isolated easily. Selective media for Lac- mutants of *E coli* can be made by incorporating chemical analogs of lactose that are converted into toxic metabolites by Lac+ bacteria but not by Lac- mutants.

The Lac- mutants can then grow on such media, but the Lac+ wild-type bacteria are killed. Mutations that inactivate essential genes in haploid organisms are usually lethal, but such potentially lethal mutations can often be studied if their expression is controlled by manipulation of experimental conditions. A mutation that increases the thermolability of an essential gene product may prevent bacterial growth at 42°C, although the mutant bacterium can still grow at 25°C. Conversely, cold-sensitive mutants express the mutant phenotype at low temperature, but not at high temperature. A conditional lethal phenotype indicates that the mutant gene is essential for viability.

Spontaneous and Induced Mutations

The mutation rate in bacteria is determined by the accuracy of DNA replication, the occurrence of damage to DNA, and the effectiveness of mechanisms for repair of damaged DNA. For a particular bacterial strain under defined growth conditions, the mutation rate for any specific gene is constant and is expressed as the probability of mutation per cell division.

In a population of bacteria grown from a small inoculum, the proportion of mutants usually increases progressively as the size of the bacterial population increases. Mutations in bacteria can occur spontaneously and independently of the experimental methods used to detect them. This principle was first demonstrated by the fluctuation test. The numbers of phage-resistant mutants of *E coli* in replicate cultures grown from small inocula were measured and compared with those in multiple samples taken from a single culture. If mutations to phage resistance occurred only after exposure to phage, the variability in numbers of mutants between cultures should be similar under both sets of conditions.

In contrast, if phage-resistant mutants occurred spontaneously before exposure of the bacteria to phage, the numbers of mutants should be more variable in the independently grown cultures, because differences in the size of the bacterial population when the first mutant appeared would contribute to the observed variability. The data indicated that the mutations to phage resistance in *E coli* occurred spontaneously with constant probability per cell division.

Replica plating confirmed that mutations in bacteria can occur spontaneously, without exposure of bacteria to selective agents. For replica plating, a flat, sterile, velveteen surface is used to pick up an inoculum from the surface of an agar master plate and transfer samples to other agar plates. In this manner, samples of the bacterial population from the master plate are transferred to the replica plates without distorting their spatial arrangement. If the replica plates contain selective medium and the master plates do not, the positions of selected mutant colonies on the replica plates can be noted, and bacteria that were not exposed to the selective conditions can be isolated from the same positions on the master plate. Mutants of *E coli* resistant to bacteriophage T1 or to streptomycin have been isolated in this way, without exposing the wild-type bacteria to the bacteriophage or the antibiotic.

Both environmental and genetic factors affect mutation rates. Exposure of bacteria to mutagenic agents causes mutation rates to increase, sometimes by several orders of magnitude. Many chemical and physical agents, including X-rays and ultraviolet light, have mutagenic activity. Chemicals that are carcinogenic for animals are often mutagenic for bacteria, or can be converted by animal tissues to metabolites that are mutagenic for bacteria. Standardised tests for mutagenicity in bacteria are used as screening procedures to identify environmental agents that may be carcinogenic in humans.

Mutator genes in bacteria cause an increase in spontaneous mutation rates for a wide variety of other genes. Expression of these genes, induced by DNA damage, enables the repair of DNA lesions that would otherwise be lethal, but by an error-prone mechanism that increases the rate of mutation.

The overall mutation ratethe probability that a mutation will occur somewhere in the bacterial genome per cell divisionis relatively constant for a variety of organisms with genomes of different sizes and appears to be a significant factor in determining the fitness of a bacterial strain for survival in nature.

Most mutations are deleterious, and the risk of adverse mutations for individual bacteria must be balanced against the positive value of mutability as a mechanism for adaptation of bacterial populations to changing environmental conditions.

Complementation Tests

To determine if mutations are located in the same gene or different genes, complementation tests are performed with partially diploid bacterial strains. Two copies of the region of the bacterial chromosome harboring a mutation are present in the same bacterium, with each copy containing a different mutation. A wild-type phenotype indicates that the mutations are in different genes. This phenomenon is called complementation. If a mutant phenotype is observed, a control experiment should be performed with the mutations in the cis arrangement to exclude the possibility that the wild-type alleles cannot be expressed normally in a partially diploid bacterial strain. Complementation tests were originally called "cis-trans" tests, and the term cistron is sometimes used as a synonym for gene. Complementation tests can be performed and interpreted even if the specific biochemical functions of the gene products are unknown.

Consider using a complementation test to characterise two independently derived Lac- mutants of *E coli*. The biochemical pathway for utilisation of lactose requires ß-galactoside permease (genotypic symbol lacY) to transport lactose into the bacterial cell and b-galactosidase (genotypic symbol lacZ) to convert lactose into D-glucose and D-galactose. Mutants that lack b-galactoside permease or b-galactosidase cannot utilise lactose for growth. If the mutations in both Lac- mutants inactivated the same protein (e.g., b-galactoside) then a partial diploid strain containing the lacZ genes from both mutants in the trans arrangement would be unable to utilise lactose. In contrast, if the genotypes of the two mutants were lacZ+ lacY and lacZ lacY+, the partially diploid bacterium would produce active b-galactosidase from the lacZ+ determinant and active b-galactoside permease from the lacY+ determinant. Complementation would occur, and the partially diploid strain would utilise lactose.

RECOMBINATION DNA AND GENE CLONING

Many methods are available to make hybrid DNA molecules in vitro (recombinant DNA) and to characterise them. Such methods include isolating specific genes in hybrid replicons, determining their nucleotide sequences, and creating mutations at designated locations. A clone is a population of organisms or molecules derived by asexual reproduction from a single ancestor. Gene cloning is the process of incorporating foreign genes into hybrid DNA replicons.

Cloned genes can be expressed in appropriate host cells, and the phenotypes that they determine can be analysed. Some key concepts underlying representative methods are summarised here. The first step in gene cloning is to make fragments of the donor DNA by mechanical or enzymatic methods. Certain restriction endonucleases, designated as class II, are particularly useful for preparing defined fragments of DNA molecules. They cleave both strands of double-stranded DNA molecules at specific, palindromic sequences that usually vary from four to eight nucleotides in length, and the resulting DNA fragments are called restriction fragments.

Some restriction endonucleases cleave at coincident sites to create blunt-ended DNA fragments, and others cut at staggered positions to create DNA fragments with short, self-complementary, single-stranded 5' or 3' ends. The random probability that n adjacent nucleotides in a DNA strand will correspond to a specific restriction site is approximately 1/4n. Sites for enzymes that recognise unique 4, 6, or 8 nucleotide targets are likely to occur about once in every 256, 4096, or 65,536 nucleotides, respectively. By choosing appropriate restriction enzymes, specific DNA molecules, including bacterial chromosomes, plasmids, and phage genomes, can be digested into sets of restriction fragments that have appropriate sizes for specific applications.

A restriction map identifies the positions of target sites for specific restriction endonucleases in a DNA molecule. Restriction maps are available for many cloned DNA fragments, plasmids and phage genomes, as well as for the entire chromosome of *E coli* and several other bacteria.

The second step in gene cloning is to create hybrid replicons consisting of donor DNA fragments and a cloning vector. Cloning vectors are small plasmid or phage replicons that have one or more restriction sites into which foreign DNA can be inserted. Hybrid replicons are produced by using DNA ligase to join the restricted vector DNA with donor DNA fragments that have compatible ends, or, alternatively, synthetic oligonucleotides are used as linkers to create compatibility between donor and vector DNA molecules with different ends.

Ligating a vector to a heterogeneous set of DNA fragments from a donor genome is called shotgun cloning, and the collection of recombinant DNA molecules that contains the various fragments is called a genomic library. If a specific DNA fragment is available, it can be incorporated into a recombinant replicon by direct cloning into an appropriate vector chosen from the wide variety of vectors available.

Plasmid and phage vectors are used mainly to clone small inserts usually less than 10 kbp. Examples of more special purpose vectors include cosmids, which are plasmid vectors that can be packaged into phage capsids (lambda cosmids accept inserts up to 30-40 kbp), and phagemids, which are plasmid-phage hybrid replicons that can exist either as plasmids or as single-stranded DNA phages under different experimental conditions. Phage P1 cosmids can accept inserts up to 100 kbp, and still larger DNA molecules can be cloned in yeast artificial chromosomes (YACs) which can stably maintain inserts up to and exceeding 1 Mbp in size. Other specialised vectors detect promoters, transcription termination signals, or other regulatory elements within foreign DNA inserts or, conversely, provide promoters from which transcription of cloned genes can be initiated.

The final steps in gene cloning are to introduce hybrid replicons into appropriate recipient cells and test them for expression of donor genes of interest. Prokaryotic cells or eukaryotic cells can be used as recipients, but they differ with respect to their permissiveness for specific replicons, the transcriptional signals that they recognise, and the post-translational modifications of protein structure that they can accomplish. Recombinant DNA molecules produced in vitro can be introduced directly into recipient cells by transformation or transfection.

In addition, clones in cosmid or phage vectors can be packaged into phage coats and introduced into susceptible recipient cells by transduction. By using specialised vectors (shuttle vectors) that can replicate in multiple cell types, genes from any organism can be cloned and manipulated in a convenient bacterial system and subsequently reintroduced into cells of the original organism for analysis in their natural environment.

Many methods are available to identify bacteria that contain recombinant DNA molecules. Most cloning vectors have genes for traits that can be positively selected, such as resistance to antibiotics. Furthermore, it is often possible to introduce foreign DNA into the cloning vector at a site that inactivates a nonessential, but easily recognisable, vector function. If both of these conditions are fulfilled, bacteria that contain recombinant molecules can be selected and distinguished easily from bacteria that contain only the vector.

Bacteria in a genomic library that contain a particular cloned gene can be identified by using biochemical or immunological methods to test for the desired gene product. Alternatively, the cloned gene of interest can be detected directly by using nucleic acid hybridisation methods, provided that a specific DNA or RNA probe is available. Because insertion of foreign DNA into a cloning vector at an appropriate site does not inactivate its ability to replicate in appropriate recipient cells, hybrid replicons of interest can be amplified by replication, and the recombinant DNA molecules or their gene products can be purified and studied.

The ability to purify specific DNA molecules made it feasible to develop enzymatic and chemical methods for determining their nucleotide sequences, and current methods for introducing mutations at defined sites in cloned genes are based on knowing their restriction maps or nucleotide sequences. Recombinant DNA methods make it feasible to clone specific DNA fragments from any source into vectors that can be studied in well-characterised bacteria, in eukaryotic cells, or in vitro.

Applications of DNA cloning are expanding rapidly in all fields of biology and medicine. In medical genetics such applications range from the prenatal diagnosis of inherited human diseases to the characterisation of oncogenes and their roles in carcinogenesis. Pharmaceutical applications include large-scale production from cloned human genes of biologic products with therapeutic value, such as polypeptide hormones, interleukins, and enzymes.

Applications in public health and laboratory medicine include development of vaccines to prevent specific infections and probes to diagnose specific infections by nucleic acid hybridisation or polymerase chain reaction (PCR). The latter process uses oligonucleotide primers and DNA polymerase to amplify specific target DNA sequences during multiple cycles of synthesis in vitro, making it possible to detect rare target DNA sequences in clinical specimens with great sensitivity.

Regulation of Gene Expression

The phenotypic properties of bacteria are determined by their genotypes and growth conditions. For bacteria in pure culture, changes in growth conditions often result in predictable physiological adaptations in all members of the population. Typically, essential gene products are made in amounts that permit fastest growth in the given environment, and products required under special circumstances are made only when they are needed.

Physiological adaptations are often associated with changes in metabolic activities. The flow of metabolites through particular biochemical pathways can be controlled both by regulating the synthesis of specific enzymes and by altering the activities of existing enzymes.

Specific regulation involves a gene or group of genes involved in a particular metabolic process. Induction and repression enable bacteria to regulate production of specific gene products in response to appropriate signals. Generally catabolic enzymes are induced when the substrate for the pathway is present in the growth medium, and biosynthetic enzymes are repressed by the product of the pathway. Enzymes that participate in a single biochemical pathway often occupy adjacent positions on the bacterial chromosome and are coordinately induced or repressed. They form an operon, a group of contiguous genes that is transcribed as a single unit and translated to produce the corresponding gene products.

Organization into an operon is an important strategy for coordinately regulating the expression of genes in bacteria. Operons that can be induced or repressed are controlled by binding of specific regulatory proteins to particular nucleotide sequences that function as regulatory sites within the operon. Comparison of the amino acid sequences of many of these different regulatory proteins showed that they could be grouped together into families of regulators that may have evolved from common ancestoral genes. Members of the lysR family include regulators of such diverse phenomena as lysine, cysteine and methionine metabolism in *E coli* and iron repression in V cholerae.

Global regulation simultaneously alters expression of a group of genes and operons, collectively called a regulon, that are controlled by the same regulatory signal. Global regulation determines responses of bacteria to basic nutrients such as carbon, nitrogen or phosphate, reactions to stresses such as DNA damage or heat shock, and synthesis by pathogens of specific virulence factors during growth in their host animals.

The amount of a specific protein in a bacterial cell can vary from none to many thousands of molecules. This wide range is often determined by the combined action of several regulatory mechanisms that affect expression of the corresponding structural gene. Regulation is achieved by determining how often a gene is transcribed into functional mRNA, how efficiently the mRNA is translated into protein, how rapidly the mRNA is degraded, how rapidly the protein product turns over, and whether the activity of the protein product can be altered by allosteric effects or covalent modifications.

mRNAs as Transcriptional Units

Gene expression begins with DNA-dependent RNA polymerase (RNA polymerase) catalysing the transcription of specific mRNA from one strand of a DNA template. Binding of RNA polymerase to DNA occurs at specific sites called promoters, and transcription begins adjacent to the promoter.

Strong promoters can interact efficiently with RNA polymerase and initiate transcription at a high rate; weak promoters initiate transcription at slow rates. In either case, mRNA is synthesized from its 5' end toward its 3' end at an approximately constant rate until the RNA polymerase recognises another specific site called a terminator. RNA polymerase then dissociates from the template, and transcription of the mRNA is completed.

Individual mRNA molecules may code for one or more polypeptides. Transcription of an operon produces a polycistronic mRNA that codes for several polypeptides. Translation of polycistronic mRNAs leads to coordinate synthesis of the encoded polypeptides, but each polypeptide is synthesized as a separate molecule. A specific ribosome binding site is located just upstream from the start of each coding sequence on the mRNA molecule.

Messenger RNAs in bacteria are degraded rapidly with an average half life of several minutes, in contrast to tRNAs and rRNAs which are much more stable. Although mRNAs represent about half of the newly synthesized RNA, they represent only a small fraction of the total RNA. The short half-life of mRNAs has important consequences for gene expression. If the synthesis of a specific mRNA is prevented, production of the corresponding polypeptides declines rapidly.

Control of gene expression occurs by regulating one or more of the steps in the pathway from the DNA template to the active gene product. Simultaneous regulation at several levels permits greater control over gene expression than would be possible with a single regulatory mechanism. The most common way to regulate gene expression in bacteria is to control the production of specific mRNAs. Since the rate of elongation of an RNA molecule is approximately constant, the major factors that control mRNA synthesis are the rate of initiation and the probability that a full length transcript will be produced.

Regulation of Transcription Initiation

Some mRNAs in bacteria are synthesized at constant rates, resulting in constitutive production of the encoded polypeptides. The amounts of specific mRNAs and polypeptides produced from different constitutive genes vary greatly, however, and often reflect differences in strength of the promoters for those genes.

Transcription of many operons is regulated in response to changing environmental conditions. The promoters determine the maximum rate of transcription initiation for such operons, but regulatory proteins participate in controlling transcription. Nucleotide sequences in operons to which specific regulatory proteins bind are called regulatory sites or operators.

Operators and promoters are located close together within operons and may have overlapping DNA sequences. The binding of regulatory proteins to operators can either increase or decrease the frequency of transcription initiation. Proteins that function as negative regulators are usually called repressors. Because regulatory proteins can diffuse through the cytoplasm, the structural genes for regulatory proteins do not have to be linked to the target operons.

The ability to sense the presence or absence of specific compounds and change the rates of synthesis of appropriate gene products are central to the control of gene expression. Regulatory proteins offer one solution to this problem of stimulus-response coupling. Many regulatory proteins are bifunctional and bind not only to appropriate operators but also to specific effectors, which are small molecules such as particular sugars, amino acids, and other metabolites. Furthermore, regulatory proteins are allosteric, meaning that they can exist in different conformations which exhibit different binding affinities for their cognate operators and effectors.

A sufficient concentration of effector favors formation of the regulatory protein-effector complex, which has either high or low affinity for the operator in any specific case. In negatively regulated systems the effector functions as a corepressor if the regulatory protein-effector complex is the active repressor, and the effector functions as an inducer and causes derepression if the free regulatory protein is the active repressor. Conversely, in positively regulated systems, the effector stimulates expression of the operon if the regulatory protein-effector complex is the positive regulator, and the effector inhibits expression of the operon if the free regulatory protein is the positive regulator.

The lactose (lac) operon of *E coli* is an example of an inducible, negatively regulated operon. The lacI gene codes for a repressor that binds to the lac operator and prevents transcription from the lac promoter. The structural gene for this repressor is separate from the lac operon, and the repressor is synthesized constitutively at a low rate. When inducer binds to the lac repressor, the complex cannot bind to the operator and cannot prevent binding of the RNA polymerase to the promoter. If other conditions are favorable, the lac operon is expressed, resulting in synthesis of β-galactosidase, β-galactoside permease and β-galactoside transacetylase.

The lac operon can be induced by lactose or by structurally related compounds such as isopropyl-β-D-thiogalactoside (IPTG). IPTG is called a gratuitous inducer because it induces the lac operon, but is not a substrate for b-galactosidase. Negative regulation also occurs in many biosynthetic operons in *E coli*. In such operons a product of the biosynthetic pathway functions as the effector for the negative regulatory system.

The arabinose (ara) operon in *E coli* is both positively and negatively regulated. In the presence of arabinose the regulatory protein stimulates transcription of the ara operon. In the absence of arabinose, however, the regulatory protein represses the ara operon. Operons are often controlled by more than one mechanism. When *E coli* is grown in a medium containing glucose and an alternative carbon source such as lactose or arabinose, induction of the lac or ara operon and utilisation of the lactose or arabinose are delayed until the glucose has been consumed. This phenomenon is called diauxic growth. The failure to induce the lac or ara operon in the presence of glucose is an example of catabolite repression. The lac and ara operons are positively regulated by cyclic-3',5'-adenosine monophosphate (cAMP) and the catabolite gene activator (CAP) protein (the product of the crp gene).

The cAMP-CAP complex interacts with CAP binding sites in the regulatory regions of some operons, including the lac and ara operons, and stimulates transcription from the corresponding promoters. The level of intracellular cAMP in *E coli* is high during growth in the absence of glucose, and low during growth in the presence of glucose. Catabolite repression is due, therefore, to lack of activation of cAMP-dependent operons when the bacteria are grown in the presence of glucose or certain other rapidly metabolisable carbon sources.

Regulation of Transcription Termination

Attenuation is a mechanism for regulating operons by terminating transcription of mRNA prematurely. Attenuation is common in biosynthetic operons, including the trp, histidine (his), threonine (thr), isoleucine-valine (ilv), and phenylalanine (phe) operons. The trp operon in *E coli* is controlled both by repression and attenuation. In the presence of excess tryptophan, initiation of transcription from the trp promoter is repressed. In addition, however, those transcripts that are initiated from the trp promoter are usually terminated before any of the structural genes of the trp operon are transcribed.

The concentration of intracellular tryptophan required to maintain repression exceeds that needed for attenuation. Such dual control enables the cell to fine tune the expression of the trp operon in response to decreasing concentrations of tryptophan. The secondary structure of mRNA has an important role in the mechanism of attenuation.

All mRNAs have a leader sequence between the transcriptional start site and the beginning of the coding sequence for the first structural gene. For amino acid biosynthetic operons that are subject to attenuation, the mRNA leader sequence has two distinctive features. It encodes a short peptide containing the amino-acid produced by the regulated pathway, and it can form alternative, mutually incompatible, double-stranded RNA structures that participate in regulatory events. For example, the peptide encoded by the trp mRNA leader sequence contains two adjacent tryptophan residues, and the peptide encoded by the his mRNA leader sequence has a series of seven consecutive histidine residues.

In E coli transcription and translation are functionally coupled. Nonsense mutations that cause premature termination of translation often cause decreased transcription of more distal genes in the same operon. This phenomenon is called polarity. Ribosomes usually initiate translation of a growing mRNA molecule prior to completion of transcription, and such translation masks sites that would otherwise cause the RNA polymerase to terminate transcription. Premature termination of translation by a nonsense codon dissociates the ribosomes from the mRNA and enables RNA polymerase to interact with the unmasked transcription termination sites.

In some biological systems, including phage lambda, antitermination is used as a positive regulatory mechanism to control gene expression. Immediately after infection of *E coli* by lambda, RNA polymerase binds to two promoters in lambda DNA and initiates divergent primary transcripts which terminate at specific sites on the lambda genome. A protein encoded by one of the primary transcripts interacts with RNA polymerase and enables it to continue transcription through the primary termination sites, thereby expressing a second set of lambda genes. One of the products encoded by a secondary transcript blocks termination of another mRNA and activates expression of a third set of genes. Antitermination has a key role, therefore, in controlling the cascade of gene expression during lytic growth of phage lambda. Antitermination is also involved in the regulation of *E coli* rRNA operons.

Regulation of Translation

The ribosome binding site on mRNA is complementary to a sequence at the 3' end of 16S rRNA. Interaction between these sequences facilitates formation of the initiation complex for protein synthesis. Both the extent of homology with 16S rRNA and the spacing of the ribosome binding site from the initiation codon affect the efficiency of translation initiation. Codon usage in mRNA also influences translation efficiency. Messenger RNAs for proteins that are required in large amounts tend to use codons that are translated by the most abundant species of tRNA, and the converse is also true.

Translational control is important for regulation of synthesis of ribosomal proteins. Production of ribosomes involves a high metabolic cost for bacteria, and at high growth rates ribosomes can constitute nearly one-half of the cell weight. Most ribosomal proteins and rRNAs are found assembled into ribosomes, and the pool of free ribosomal subunits is very small. The genes for ribosomal proteins are organised into several operons. Certain of the free ribosomal proteins directly inhibit the translation of the polycistronic mRNAs that encode them, thereby ensuring that synthesis of ribosomal proteins is balanced with the requirement for their utilisation.

Regulons and Signal Transducing Proteins

A regulon is a group of genes or operons controlled by a common regulator. There are several advantages to placing different operons under control by the same regulator. It enables the sensing of a single stimulus to be coupled to expression of a large number of genes that may be needed for an appropriate response, and it eliminates the requirement for the coordinately regulated genes to be linked on the bacterial chromosome. The stimulus to which the regulon responds can be an intracellular component or an environmental signal. Individual operons may also be subject to regulation by several different mechanisms and expressed under conditions that differ from those affecting the whole regulon.

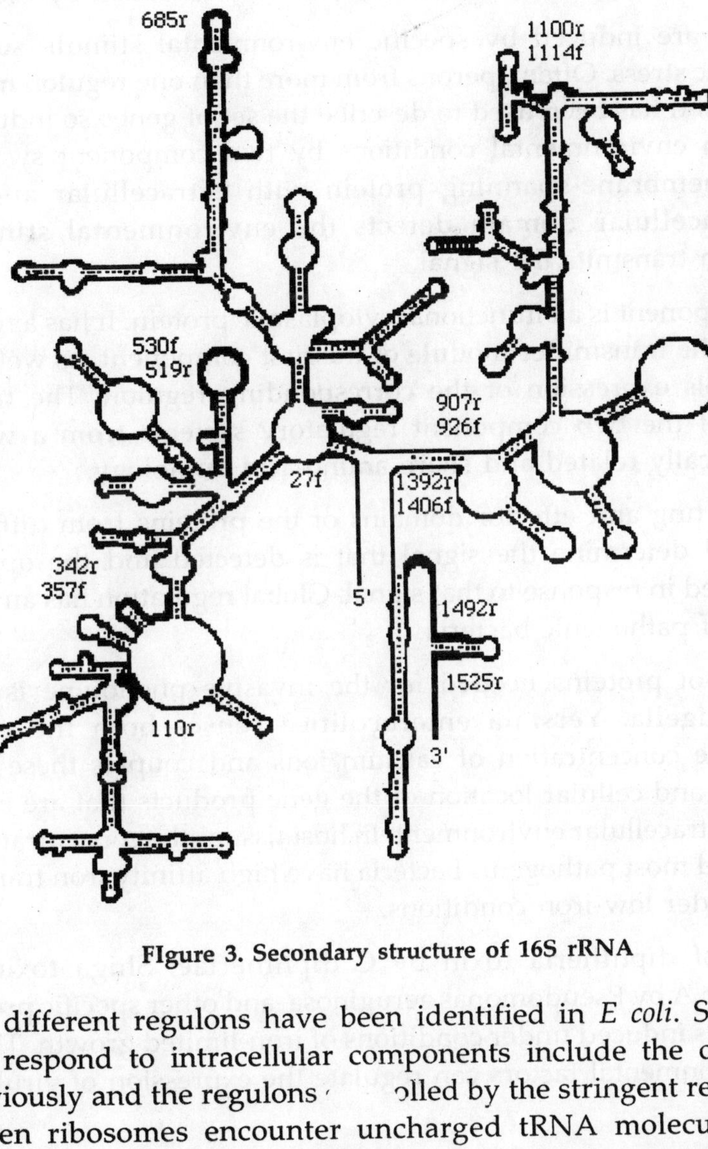

FIgure 3. Secondary structure of 16S rRNA

More than 40 different regulons have been identified in *E coli*. Specific examples of regulons that respond to intracellular components include the cAMP-CAP regulon described previously and the regulons [illegible]olled by the stringent response and the SOS response. When ribosomes encounter uncharged tRNA molecules during protein synthesis, the stringent response is activated and results in prompt cessation of rRNA synthesis. A novel nucleotide called guanosine-3'-diphosphate-5'-diphosphate (ppGpp) accumulates during amino-acid starvation.

The ppGpp produced by idling ribosomes appears to be a mediator of the stringent response, but the precise mechanism causing inhibition of rRNA synthesis is unknown. The SOS response is associated with damage to DNA and involves induction of more than 20 genes involved in several DNA repair pathways. The product of the recA gene detects inhibition of DNA synthesis and initiates events leading to proteolytic cleavage and inactivation of the repressor for the SOS pathway, encoded by the lexA gene.

Some regulons are induced by specific environmental stimuli, such as nutrient limitation or osmotic stress. Often operons from more than one regulon may be induced, and the term stimulon has been used to describe the set of genes so induced. Typically, bacteria sense such environmental conditions by two component systems. The first component is a membrane-spanning protein with extracellular and intracellular domains. Its extracellular domain detects the environmental stimulus, and its cytoplasmic domain transmits the signal.

The second component is a bifunctional cytoplasmic protein. It has a receiver domain that interacts with the transmitter module of the first component, as well as an effector domain that controls expression of the corresponding regulon. The transmitter and receiver modules of the two component regulatory systems from a wide variety of regulons are genetically related and share amino-acid homology.

The signal-detecting and effector domains of the proteins from different regulons vary, however, and determine the signal that is detected and the operons that are activated or repressed in response to that signal. Global regulation has an important role in the physiology of pathogenic bacteria.

The expression of proteins needed for the invasive phenotype is controlled by temperature in Shigella. Yersinia enterocolitica senses both the environmental temperature and the concentration of calcium ions and couples these signals to the expression of genes and cellular location of the gene products that are appropriate for an intracellular or extracellular environment. In host tissues the concentration of free iron is extremely low, and most pathogenic bacteria have high affinity iron transport systems that are induced under low-iron conditions.

The synthesis of diphtheria toxin by C diphtheriae, Shiga toxin by Shigella dysenteriae, exotoxin A by Pseudomonas aeruginosa, and other specific proteins in many pathogenic bacteria is induced under conditions of iron-limited growth. These examples illustrate how environmental factors can regulate the expression of virulence genes in pathogenic bacteria.

VIRAL GENETICS

Viruses are simple entities, lacking an energy-generating system and having very limited biosynthetic capabilities. The smallest viruses have only a few genes; the largest viruses have as many as 200. Genetically, however, viruses have many features in common with cells. Viruses are subject to mutations, the genomes of different viruses can recombine to form novel progeny, the expression of the viral genome can be regulated, and viral gene products can interact. By studying viruses, we can learn more about the mechanisms by which viruses and their host cells function.

Genetic Change in Viruses

The mechanisms by which genetic changes occur in viruses. Two principal mechanisms are involved: mutation and recombination. Alterations in the genetic material of a virus may lead to changes in the function of viral proteins. Such changes may result in the creation of new viral serotypes or viruses of altered virulence.

Mutations

Mutations arise by one of three mechanisms: (1) by the effects of physical mutagens (UV light, x-rays) on nucleic acids; (2) by the natural behavior of the bases that make up nucleic acids (resonance from keto to enol and from amino to imino forms), and (3) through the fallibility of the enzymes that replicate the nucleic acids. The first two mechanisms act similarly in all viruses; hence, the effects of physical mutagens and the natural behavior of nucleotides are relatively constant. However, viruses differ markedly in their mutation rates, which is due primarily to differences in the fidelity with which their enzymes replicate their nucleic acids. Viruses with high-fidelity transcriptases have relatively low mutation rates and vice versa.

Mutation rates and outcomes

DNA viruses have mutation rates similar to those of eukaryotic cells because, like eukaryotic DNA polymerases, their replicatory enzymes have proofreading functions. The error rate for DNA viruses has been calculated to be 10^{-8} to 10^{-11} errors per incorporated nucleotide. With this low mutation rate, replication of even the most complex DNA viruses, which have 2 X 10^5 to 3 X 10^5 nucleotide pairs per genome, will generate mutants rather rarely, perhaps once in several hundred to many thousand genome copies.

The RNA viruses, however, lack a proofreading function in their replicatory enzymes, and some have mutation rates that are many orders of magnitude higher 10^{-3} to 10^{-4} errors per incorporated nucleotide. Even the simplest RNA viruses, which have about 7,400 nucleotides per genome, will generate mutants frequently, perhaps as often as once per genome copy. Not all mutations that occur persist in the virus population. Mutations that interfere with the essential functions of attachment, penetration, uncoating, replication, assembly, and release do not permit misreplication and are rapidly lost from the population. However, because of the redundancy of the genetic code, many mutations are neutral, resulting either in no change in the viral protein or in replacement of an amino acid by a functionally similar amino acid. Only mutations that do not cripple essential viral functions can persist or become fixed in a virus population.

Phenotypic Variation by Mutations

Mutations that alter the viral phenotype but are not deleterious may be important. For example, mutation can create novel antigenic determinants. A mutation in the hemagglutinin gene of influenza A virus can give rise to a hemagglutinin molecule with an altered antigenic site (epitope). Provided the attachment function of the new hemagglutinin is intact, the mutant virus may be able to initiate an infection in an individual immune to viruses expressing the previous hemagglutinin. For example, from 1968 to 1979, mutations altered 10 percent of the amino acids in the influenza virus hemagglutinin serotype H3 molecule. This relatively modest mechanism of antigenic change through mutation, called antigenic drift, may allow a virus to outflank host defenses and cause disease in previously immune individuals.

Vaccine Strains from Mutations

Mutation has been a principal tool of virologists in developing attenuated live virus vaccines. For example, the Sabin vaccine strains of poliovirus were developed by growing polioviruses in monkey kidney cells. Mutation and selection produced variant polioviruses that were adapted for efficient replication in these cells.

Some of the mutations in these variants affected the genes coding for the poliovirus coat proteins in such a way as to produce mutants unable to attach to human neural cells but still able to infect human intestinal cells. Infection of human intestinal cells does not produce paralytic disease but does induce immunity. Poliovirus vaccine strains 1 and 2 have multiple mutations in the coat proteins and are very stable. The type 3 vaccine strain is less stable and is subject to back-mutations that restore neural virulence. This vaccine strain therefore causes paralytic disease in one out of every several million vaccinated individuals. Despite the possibility of back-mutations, the generation and selection of attenuated viral mutants remains an important mechanism for producing viral vaccines.

Recombination

Viral recombination occurs when viruses of two different parent strains coinfect the same host cell and interact during replication to generate virus progeny that have some genes from both parents. Recombination generally occurs between members of the same virus type (e.g., between two influenza viruses or between two herpes simplex viruses). Two mechanisms of recombination have been observed for viruses: independent assortment and incomplete linkage. Either mechanism can produce new viral serotypes or viruses with altered virulence.

Recombination by Independent Assortment

Independent assortment occurs when viruses that have multipartite genomes trade segments during replication. These genes are unlinked and assort at random. Recombination by independent assortment has been reported, for example, for the influenza viruses and other orthomyxoviruses and for the reoviruses. The frequency of recombination by independent assortment is 6 to 20 percent for orthomyxoviruses.

Independent assortment between an animal and a human strain of influenza virus during a mixed infection can yield an antigenically novel influenza virus strain capable of infecting humans but carrying animal-strain hemagglutinin and/or neuraminidase surface molecules. This recombinant can infect individuals that are immune to the parent human virus. This mechanism results in an immediate, major antigenic change and is called antigenic shift. Antigenic shifts in influenza virus antigens can give rise to pandemics of influenza. Such antigenic shifts have occurred relatively frequently during recent history. Because the number of different serotypes of hemagglutinin and neuraminidase are limited, a given strain reappears from time to time. For example, the H1N1 influenza virus strain was responsible for the 1918 to 1919 influenza pandemic that caused 20 million deaths. The same virus also caused pandemics in 1934 and in 1947, then disappeared after 1958 and reappeared in 1977. The reappearance of virus strains after an absence is believed to be the result of recombinational events involving the independent assortment of genes from two variant viruses.

Phenotypic Variation from Recombination

Viral recombination is important because it can generate novel progeny viruses that express new antigenic and/or virulence characteristics. For example, the novel progeny viruses may have new surface proteins that permit them to infect previously resistant individuals; they may have altered virulence characteristics; they may have novel combinations of proteins that make them infective to new cells in the original host or to new hosts; or they may carry material of cellular origin that gives them oncogenic potential.

Vaccines and Gene Therapy through Recombination

Recombination is being used experimentally by virologists to create new vaccines. Vaccinia virus, a DNA virus of the poxvirus group, was used as a live vaccine in the eradication of smallpox. Recombinant vaccinia viruses are being developed that carry vaccinia virus DNA recombined with DNA from other sources (exogenous DNA). Vaccinia virus strains carrying DNA coding for bacterial and viral antigens have been produced. It is expected that after vaccination with the recombinant vaccinia virus, the bacterial or viral antigen (immunogen) will be produced.

The presence of this immunogen will then stimulate specific antibody production by the host, resulting in protection of the host from the immunogen. Studies with these live, recombinant vaccinia viruses are currently under way to determine whether inoculation of the skin with the recombinant virus can induce a protective host antibody response to the bacterial or viral antigens. Other studies are investigating the use of live, recombinant adenoviruses containing bacterial or viral genes to infect the gastrointestinal tract and induce both mucosal and systemic immunity.

In a similar manner, recombinant viruses are also being developed that carry normal human genes. It is envisioned that such recombinant viruses could be useful for gene therapy. Target diseases for gene therapy span a wide range, including diabetes, cystic fibrosis, severe combined immunodeficiency syndrome, etc. Indeed, treatment of cystic fibrosis patients with replication deficient, recombinant adenoviruses bearing a normal copy of the cystic fibrosis transmembrane regulator gene has already been approved. If these studies give positive results, such directed generation of recombinant viruses may become an important tool in the development of vaccines and for gene therapy.

REFERENCES

Madigan M, Martinko J (ed.) 2006. Brock Biology of Microorganisms (11th ed.). Prentice Hall.

Drews G. 1999. "Ferdinand Cohn, a Founder of Modern Microbiology". ASM News 65 (8).

Burkovski A (ed). 2008. *Corynebacteria: Genomics and Molecular Biology*. Caister Academic Press.

Rehm BHA (ed). 2008. *Microbial Production of Biopolymers and Polymer Precursors: Applications and Perspectives*. Caister Academic Press.

Lerner, Brenda Wilmoth & K. Lee Lerner (eds) 2006. *Medicine, health, and bioethics : essential primary sources* (1st ed.). Thomson Gale.

Witzany, Guenther. 2008. *Bio-Communication of Bacteria and its Evolutionary Interrelations to Natural Genome Editing Competences of Viruses*.. Nature Precedings. hdl:10101/npre.2008.1738.1.

10

Microbial Nutrition and Growth

Microorganisms, like all organisms, have a basic requirement for several essential elements for growth. These are necessary for the biosynthesis of the components of cells as well as their function. Carbon has a pivotal role in the structure of all molecules found in living cells, however, nitrogen, hydrogen, oxygen, sulphur and phosphorus are all essential as well. Of less significance, in respect of the amount required, are many trace nutrients which are predominantly metal ions e.g. Zn, Cu, Mn etc.

NUTRITIONAL CLASSIFICATION OF MICROORGANISMS

Microorganisms are classified on the basis of their utilisation of nutrients as an energy source and as a carbon source. The major level of classification is into chemotrophs and phototrophs. The former derive energy from chemicals and the latter depend primarily on radiant energy. These two major groups are further classified on the basis of the carbon sources utilised.

- *Chemoautotrophs (Chemolithotrophs)*—use inorganic substrates as sources of energy and CO_2 as the main source of carbon.
- *Chemoheterotrophs (Chemorganotrophs)*—utilise organic substrates for both needs.
- *Photoautotrophs (Photolithotrophs)*—use light as energy source and CO_2 as carbon source.
- *Photoheterotrophs (photoorganotrophs)*—use light as energy source and organic carbon sources.

Not all bacteria can be so precisely categorised, for example, some phototrophic bacteria can also grow as chemotrophs.

Rhodospirillum rubrum exists as a photoheterotroph in anaerobic conditions but when oxygen is available, and in the dark, it grows as a chemoheterotroph. This important metabolic switch is also reflected in changes in internal cell structure. The vesicles which contain the photosynthetic pigments become disorganised when the organism grows as a chemoheterotroph. All fungi are all chemoheterotro-phic microorganisms.

Chemoheterotrophs show a wide diversity with respect to the sources of organic nutrients that are essential for their growth. At the two extremes are the saprotrophs (saprophytes), which utilise nonliving organic matter, and the biotrophs (parasites) which may be totally dependent on another living organism for their nutrition and cause disease (pathogens).

In between there are several types of association between this group of microorganisms and another organism. These include mutalistic or symbiotic associations whereby both partners gain benefit, organisms that live as commensals and under some circumstances cause disease, and finally obligate biotrophs that can live both a saprotrophic and biotrophic existence.

SOURCE AND SUPPLY AND NUTRIENTS

The natural environments colonised by bacteria and fungi provide a wide variety of nutrients which will support growth of these organisms. In some organisms, described as fastidious, a specific nutritional requirement can be met from only one source. These organisms are generally pathogens and cannot be cultivated in the laboratory in the absence of their host or host tissues. The mixtures of nutrients used in the Microbiology laboratory to cultivate bacteria and fungi are called media. These range from complex, undefined mixtures to simple defined mixtures where the nature and amount of a nutrient is known.

The former type of medium is used normally for the routine and regular cultivation of an organism, or when cultures are required quickly for some particular use e.g. for genetic experiments. Conversely for biochemical and physiological experiments a defined medium is favoured e.g. where the regulation of synthesis of an enzyme is being investigated. Media for the cultivation of bacteria. The nutritional diversity of bacteria requires the use of a range of different culture media. Thus autotrophs require a medium lacking organic carbon (Table 1).

In the case chemoheterotrophs one nutrient is used as the energy source. A good example of such organisms are the nitrifying bacteria which derive energy from the oxidation of NH_4 ion (*Nitrosomonas*) and NO_2 ion (*Nitrobacter*). A defined medium for a chemoheterotroph has a very similar composition, the notable addition being the organic carbon source (Table 2).

A more complex medium is given in Table 3. This comprises extracts of natural materials, in this case an extract from animal tissues.

Table 1. Chemically defined medium for a chemoautotrophic bacterium

Component	*Function*	*Amount (g)*
$(NH_4)_2 SO_4$	N and energy source	0.5
$NaHCO_3$	C Source	0.5
Na_2HPO_4	Buffer and essential ions	13.5
NaH_2PO_4	Buffer and essential ions	0.7
$MgSO_4 7H_2O$	Essential ions	0.1
$FeC1_3.6H_2O$	Essential ions	0.014
$CaC1_2.2H_2O$	Essential ions	0.18
Water		1 litre

Table 2. Defined medium for chemoheterotrophic bacteria

Component	*Function*	*Amount (g)*
Glucose	C and energy source	1
$NH_4H_2PO_4$	N source, buffer and essential ions	5
K_2HPO_4	Buffer and essential ions	1
NaCl	Essential ions	5
$MgSO_4.7H_2O$	Essential ions	0.2
Water	1 litre	

Table 3. A complex medium (Nutrient broth)

Component	*Function*	*Amount (g)*
Beef extract	Water soluble extract of animal tissues, source of C, N, vitamins & salts	3
Peptone	Protein digest, organic N and some vitamins	5
NaCl	Ions & osmotic requirements	8
Water		1 litre

MEDIA FOR THE CULTIVATION

Fungi

Fungi are also cultivated on defined or complex media. Examples of these are given in Tables 4. Note the inclusion of organic carbon in both cases, reflecting the chemoheterotrophic nature of fungi, however, in fungal media extracts of plant materials are used in complex media. Comparison of bacterial and fungal media shows an important difference in the ratio of carbon to nitrogen; bacterial media have a low C:N ratio and fungi a high C:N.

Table 4 A defined medium for the cultivation of fungi

Component	*Function*	*Amount (g)*
Glucose	C & energy source	20
$NaNO_3$	N source	6
$MgSO_4.7H_2O$	Essential ions	1.5
KCl	Essential ions	1.5
$ZnSO_4$	Essential ions	Trace
Fe $SO_4.7H_2O$	Essential ions	Trace
Water		1 litre

Special Media

Different types of organisms require the use of particular media. For example *anaerobic bacteria* grow only on a medium lacking oxygen. Reducing agents, such as sodium thioglycollate, are commonly used, but most effective is a piece of apparatus called an anaerobe jar. This is a vessel in which the atmosphere can be made totally oxygen free. Selective media are used to enhance growth of particular types of microorganisms.

These are particularly useful for the isolation of particular types of organism(s) from a very mixed population e.g. the use of a medium containing brilliant green (a dye) which inhibits gram +ve bacteria, for the isolation of gram -ve types. Differential media are used to distinguish particular types of organisms in a mixed population. MacConkey medium, which contains bile salts and crystal violet dye, is used to differentiate Enterobacteriaceae (*Salmonella, Shigella* and *Escherichia*) from other gram -ve bacteria.

An enrichment medium is used to selectively isolate organisms with a particular nutritional requirement of function. If organisms that can hydrolyse chitin are required samples of natural material e.g. soil are spread on a medium containing chitin. Only chitin degrading organisms will grow.

Agar

All media can be used in liquid or solidified form. The latter include agar as an ingredient. This is a complex polysaccharide extracted from marine algae. It is normally added at a concentration of 1.5%. Agar has important properties being liquid at 100°C but solidifying at 40°C. This allows the incorporation of bacteria into the medium as an alternative to spreading on the medium surface.

MICROBIAL GROWTH

Microorganisms represent a significant percentage of the total biomass on the earth. As a result of high rates of their growth and metabolic activity and their adaptability, they are found almost everywhere and occupy an important trophic position in the proper functioning of ecosystems.

They are ubiquitous in distribution. Any part of the biosphere supporting the complex, multicellular higher organisms shall contain microorganisms, whereass the converse is not true.

Growth may be defined as the orderly increase in all cellular constituents and results from the biosynthetic and energy generating processes. The growth may refer to the increase in mass of a single cell or to the increase in size of a population of cells. Growth implies that all chemical components of the cell increase with the same speed and after a certain time this leads to increase in cell number, which causes increase in size or number of the individuals. Growth is normally performed batch-wise or continuously.

The growth of a culture is related to the composition of the medium. If all the essential components are available, the growth is balanced. If, however, one or several essential components are missing the growth is terminated due to unbalanced growth, which often leads to death of the culture.

Methods of Determining Microorganisms Growth

A variety of direct as well as indirect methods are available to determine growth and growth rates of microorganisms. The choice of the method will however, depend on whether it is bacteria or fungi and also certain inherent characteristics of the microorganisms such as clumping.

The direct methods involve determining the increase in cell number, dry weight or the increase in any other cellular component as a function of time while indirect methods include measurement of optical density, turbidity, etc.

The total number of cells in a population can be determined either microscopically or electronically. In the microscopic methods number of cells in an appropriately diluted sample can be determined of using a haemocytometer.

One such is the electronic Coulter counter which measures the increase in resistance to the flow of current through a microorifice that separates two electrodes every times a microbial cell passes through the orifice.

The microscopic method has the advantage in that cells are viewed and counted while in the electronic method, particles other than cells are also counted since the instrument cannot differentiate between a cell and inert particle. However, the former method is laborious and subject to errors the latter is quick and the degree of precision is very high. Both methods however, do not distinguish between cells which are viable from the non-viable cells

The viable cell number can be determined only by the dilution plating technique. In this, a diluted sample is plated on an appropriate medium and after an incubation period, the number of colonies are determined.

The difference between the total cell number and the viable cell number will give the number of cells in the population that are nonviable other direct methods of estimating growth include estimation of cell nitrogen or increase in dry weight, or any other cellular component. However, these methods are not as easy as the determining by cell number

The direct methods of determining growth by cell number are applicable for use with organisms which divide by binary fission such the bacteria or by budding as in yeast. In fungi or in organisms that grow in clumps, methods that determine cell mass, (dry weight or wet weight), packed cell volume etc., will have to be used.

In these methods, a known amount of medium (in multiples) is inoculated with a known number of spores or cells and incubated. At lated one or more flasks are withdrawn and the cell mass is collected by cent ifugation. If centrifuged in a graduated conical glass tube, the direct packed volume of the cell pellet can be determined. For determining the dry weight. The culture is centrifuged in preweighed tubes and the pellet dried to constant weight at 80-85°C.

The most popular indirect method of determining growth in bacteria or yeast is by the use of colorimeters or turbidometers in which density of the cell suspension can be determined. By taking adequate the precautions, the method can be made reliable and useful.

The type of curves that one gets when growth is measured hi a liquid medium by different methods. The changes that occur in the cell population after inoculation into fresh growth medium are more correctl indicated by the dry weight or optical density measurements. In the lag period, as said earlier, biosynthetic activity leading to increase in mass occurs but this cannot be iden ified if the cell number is determined.

MICROBIAL GROWTH CONTROL

In the 19th century, surgery was risky and dangerous, and patients undergoing even the most routine operations were at very high risk of infection. This was so because surgery was not performed under aseptic conditions. The operating room, the surgeon's hands, and the surgical instruments were laden with microbes, which caused high levels of infection and mortality.

Surgeons in the mid-1800s often operated wearing their street clothes, without washing their hands. They frequently used ordinary sewing thread to suture wounds, and stuck the needles in the lapels of their frock coats in between patients.

Surgical dressings were often made up of surplus cotton or jute from the floors of cotton mills. It was against this background that French scientist Louis Pasteur demonstrated that invisible microbes caused disease. It was clear that Lister's techniques were effective in increasing the rates of surviving surgery, but his theories were controversial because many 19th century surgeons were unwilling to accept something they could not see. Also, perhaps another reason that surgeons were slow to pick up on Lister's methods was the fact that during surgery they were required to breathe an irritating aerosol of phenol.

The control of microbial growth is necessary in many practical situations, and significant advances in agriculture, medicine, and food science have been made through study of this area of microbiology.

"Control of microbial growth", as used here, means to inhibit or prevent growth of microorganisms. This control is affected in two basic ways: (1) by killing microorganisms or (2) by inhibiting the growth of microorganisms. Control of growth usually involves the use of physical or chemical agents which either kill or prevent the growth of microorganisms. Agents which kill cells are called cidal agents; agents which inhibit the growth of cells are referred to as static agents. Thus, the term bactericidal refers to killing bacteria, and bacteriostatic refers to inhibiting the growth of bacterial cells. A bactericide kills bacteria, a fungicide kills fungi, and so on. In microbiology, sterilisation refers to the complete destruction or elimination of all viable organisms in or on a substance being sterilised. There are no degrees of sterilisation: an object or substance is either sterile or not. Sterilisation procedures involve the use of heat, radiation or chemicals, or physical removal of cells.

Methods of Sterilisation

Heat: Most important and widely used. For sterilisation one must consider the type of heat, and most importantly, the time of application and temperature to ensure destruction of all microorganisms. Endospores of bacteria are considered the most thermoduric of all cells so their destruction guarantees sterility.

Incineration: Burns organisms and physically destroys them. Used for needles, inoculating wires, glassware, etc. and objects not destroyed in the incineration process.

Boiling: 100° for 30 minutes. Kills everything except some endospores. To kill endospores, and therefore sterilise a solution, very long (>6 hours) boiling, or intermittent boiling is required.

Autoclaving (steam under pressure or pressure cooker). Autoclaving is the most effective and most efficient means of sterilisation. All autoclaves operate on a time/temperature relationship. These two variables are extremely important. Higher temperatures ensure more rapid killing. The usual standard temperature/pressure employed is 121ºC/15 psi for 15 minutes. Longer times are needed for larger loads, large volumes of liquid, and more dense materials. Autoclaving is ideal for sterilising biohasardous waste, surgical dressings, glassware, many types of microbiologic media, liquids, and many other things. However, certain items, such as plastics and certain medical instruments (e.g. fiber-optic endoscopes), cannot withstand autoclaving and should be sterilised with chemical or gas sterilants. When proper conditions and time are employed, no living organisms will survive a trip through an autoclave.

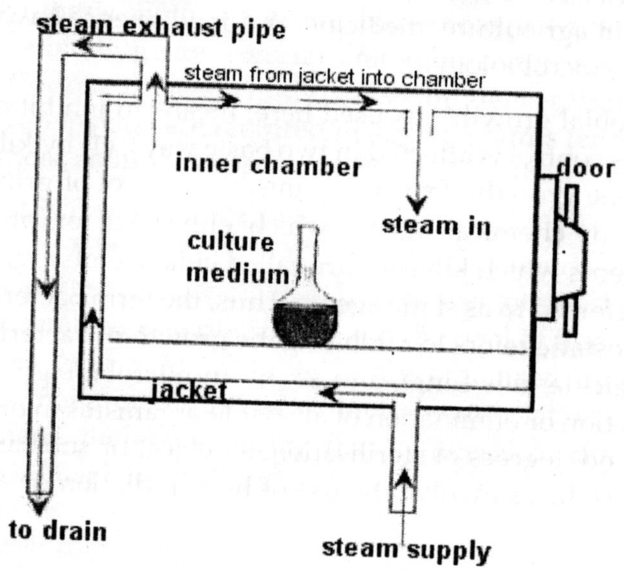

Figure 1. Schematic diagram of a laboratory autoclave in use to sterilise microbiological culture medium.

Why is an autoclave such an effective steriliser? The autoclave is a large pressure cooker; it operates by using steam under pressure as the sterilising agent. High pressures enable steam to reach high temperatures, thus increasing its heat content and killing power. Most of the heating power of steam comes from its latent heat of vaporisation. This is the amount of heat required to convert boiling water to steam. This amount of heat is large compared to that required to make water hot. For example, it takes 80 calories to make 1 liter of water boil, but 540 calories to convert that boiling water to steam. Therefore, steam at 100º C has almost seven times more heat than boiling water.

Moist heat is thought to kill microorganisms by causing denaturation of essential proteins. Death rate is directly proportional to the concentration of microorganisms at any given time. The time required to kill a known population of microorganisms in a specific suspension at a particular temperature is referred to as thermal death time (TDT). Increasing the temperature decreases TDT, and lowering the temperature increases TDT. Processes conducted at high temperatures for short periods of time are preferred over lower temperatures for longer times.

Environmental conditions also influence TDT. Increased heat causes increased toxicity of metabolic products and toxins. TDT decreases with pronounced acidic or basic pHs. However, fats and oils slow heat penetration and increase TDT. It must be remembered that thermal death times are not precise values; they measure the effectiveness and rapidity of a sterilisation process. Autoclaving 121ºC/15 psi for 15 minutes exceeds the thermal death time for most organisms except some extraordinary sporeformers.

Dry heat (hot air oven): basically the cooking oven. The rules of relating time and temperature apply, but dry heat is not as effective as moist heat (i.e., higher temperatures are needed for longer periods of time). For example 160°/2hours or 170°/1hour is necessary for sterilisation. The dry heat oven is used for glassware, metal, and objects that won't melt.

Irradiation: usually destroys or distorts nucleic acids. Ultraviolet light is commonly used to sterilise the surfaces of objects, although x-rays, gamma radiation and electron beam radiation are also used.

Ultraviolet lamps are used to sterilise workspaces and tools used in microbiology laboratories and health care facilities. UV light at germicidal wavelengths (two peaks, 185 nm and 265 nm) causes adjacent thymine molecules on DNA to dimerize, thereby inhibiting DNA replication (even though the organism may not be killed outright, it will not be able to reproduce). However, since microorganisms can be shielded from ultraviolet light in fissures, cracks and shaded areas, UV lamps should only be used as a supplement to other sterilisation techniques.

Gamma radiation and electron beam radiation are forms of ionising radiation used primarily in the health care industry. Gamma rays, emitted from cobalt-60, are similar in many ways to microwaves and x-rays. Gamma rays delivered during sterilisation break chemical bonds by interacting with the electrons of atomic constituents. Gamma rays are highly effective in killing microorganisms and do not leave residues or have sufficient energy to impart radioactivity.

Electron beam (e-beam) radiation, a form of ionising energy, is generally characterised by low penetration and high-dose rates. E-beam irradiation is similar to gamma radiation in that it alters various chemical and molecular bonds on contact. Beams produced for e-beam sterilisation are concentrated, highly-charged streams of electrons generated by the acceleration and conversion of electricity. e-beam and gamma radiation are for sterilisation of items ranging from syringes to cardiothoracic devices.

Filtration involves the physical removal (exclusion) of all cells in a liquid or gas. It is especially important for sterilisation of solutions which would be denatured by heat (e.g. antibiotics, injectable drugs, amino acids, vitamins, etc.). Portable units can be used in the field for water purification and industrial units can be used to "pasteurise" beverages. Essentially, solutions or gases are passed through a filter of sufficient pore diameter (generally 0.22 micron) to remove the smallest known bacterial cells.

Chemical and Gas

Chemicals used for sterilisation include the gases ethylene oxide and formaldehyde, and liquids such as glutaraldehyde. Ozone, hydrogen peroxide and peracetic acid are also examples of chemical sterilisation techniques are based on oxidative capabilities of the chemical.

Ethylene oxide (ETO) is the most commonly used form of chemical sterilisation. Due to its low boiling point of 10.4ºC at atmospheric pressure, EtO) behaves as a gas at room temperature. EtO chemically reacts with amino acids, proteins, and DNA to prevent microbial reproduction.

The sterilisation process is carried out in a specialised gas chamber. After sterilisation, products are transferred to an aeration cell, where they remain until the gas disperses and the product is safe to handle. ETO is used for cellulose and plastics irradiation, usually in hermetically sealed packages. Ethylene oxide can be used with a wide range of plastics (e.g. petri dishes, pipettes, syringes, medical devices, etc.) and other materials without affecting their integrity.

Ozone sterilisation has been recently approved for use in the U.S. It uses oxygen that is subjected to an intense electrical field that separates oxygen molecules into atomic oxygen, which then combines with other oxygen molecules to form ozone.

Ozone is used as a disinfectant for water and food. It is used in both gas and liquid forms as an antimicrobial agent in the treatment, storage and processing of foods, including meat, poultry and eggs. Many municipalities use ozone technology to purify their water and sewage. Los Angeles has one of the largest municipal ozone water treatment plants in the world. Ozone is used to disinfect swimming pools, and some companies selling bottled water use ozonated water to sterilize containers.

Low Temperature Gas Plasma (LTGP) is used as an alternative to ethylene oxide. It uses a small amount of liquid hydrogen peroxide (H_2O_2), which is energised with radio frequency waves into gas plasma. This leads to the generation of free radicals and other chemical species, which destroy organisms.

Non Sterilising Methods to Control Microbial Growth

Many physical and chemical technologies are employed by our civilisation to control the growth of (certain) microbes, although sterility may not the desired end-point. Rather, preventing spoilage of food or curing infectious disease might be the desired outcome.

Applications of Heat

The lethal temperature varies in microorganisms. The time required to kill depends on the number of organisms, species, nature of the product being heated, pH, and temperature. Autoclaving, which kills all microorganisms with heat, is commonly employed in canning, bottling, and other sterile packaging procedures. This is an ultimate form of preservation against microbes. But, there are some other uses of heat to control growth of microbes although it may not kill all organisms present.

Boiling: 100° for 30 minutes. Kills everything except some endospores. It also inactivates viruses. For the purposes of purifying drinking water, 100° for five minutes is a "standard" in the mountains" though there have been some reports that Giardia cysts can survive this process. Longer boiling might be recommended for Mississippi River water the closer to the Gulf.

Pasteurisation is the use of mild heat to reduce the number of microorganisms in a product or food. In the case of pasteurisation of milk, the time and temperature depend on killing potential pathogens that are transmitted in milk, i.e., staphylococci, streptococci, *Brucella abortus* and *Mycobacterium tuberculosis*. But pasteurisation kills many spoilage organisms, as well, and therefore increases the shelf life of milk especially at refrigeration temperatures (2°C).

Milk is usually pasteurised by heating, typically at 63°C for 30 minutes (batch method) or at 71°C for 15 seconds (flash method), to kill bacteria and extend the milk's usable life. The process kills pathogens [illegible]aves relatively benign microorganisms that can sour improperly stored milk.

During the process of ultrapasteurisation, also known as ultra high-temperature (UHT) pasteurisation, milk is heated to temperatures of 140 °C. In the direct method, the milk is brought into contact with steam at 140°C for one or two seconds. A thin film of milk falls through a chamber of high-pressure steam, heating the milk instantaneously. The milk is flash cooled by application of a slight vacuum, which serves the dual purpose of removing excess water in the milk from condensing steam. In the indirect method of ultrapasteurisation, milk is heated in a plate heat exchanger. It takes several seconds for the temperature of the milk to reach 140°C, and it is during this time that the milk is scalded, invariably leading to a burned taste. If ultrapasteurisation is coupled with aseptic packaging, the result is a long shelf life and a product that does not need refrigeration.

Low temperature (refrigeration and freezing): Most organisms grow very little or not at all at 0°C. Perishable foods are stored at low temperatues to slow rate of growth and consequent spoilage (e.g. milk). Low temperatures are not bactericidal. Psychrotrophs, rather than true psychrophiles, are the usual cause of food spoilage in refrigerated foods. Although a few microbes will grow in supercooled solutions as low as minus 20°C, most foods are preserved against microbial growth in the household freezer.

Drying (removal of H_2O): Most microorganisms cannot grow at reduced water activity ($A_w < 0.90$). Drying is often used to preserve foods (e.g. fruits, grains, etc.). Methods involve removal of water from product by heat, evaporation, freeze-drying, and addition of salt or sugar.

Irradiation (UV, x-ray, gamma radiation): destroys microorganisms as described under "sterilisation". Many spoilage organisms are readily killed by irradiation.

In some parts of Europe, fruits and vegetables are irradiated to increase their shelf life up to 500 percent. The practice has not been accepted in the U.S. UV light can be used to pasteurise fruit juices by flowing the juice over a high intensity ultraviolet light source. UV systems for water treatment are available for personal, residential and commercial applications and may be used to control bacteria, viruses and protozoan cysts.

Control of Microbial Growth by Chemical Agents

Antimicrobial agents are chemicals that kill or inhibit the growth microorganisms. Antimicrobial agents include chemical preservatives and antiseptics, as well as drugs used in the treatment of infectious diseases of plants and animals. Antimicrobial agents may be of natural or synthetic origin, and they may have a static or cidal effect on microorganisms.

Types of Antimicrobial Agents

Antiseptics: Microbicidal agents harmless enough to be applied to the skin and mucous membrane; should not be taken internally. Examples include alcohols, mercurials, silver nitrate, iodine solution, alcohols, detergents.

Disinfectants: agents that kill microorganisms, but not necessarily their spores, but are not safe for application to living tissues; they are used on inanimate objects such as tables, floors, utensils, etc. Examples include, hypochlorites, chlorine compounds, lye, copper sulfate, quaternary ammonium compounds, formaldehyde and phenolic compounds.

Chemotherapeutic agents (synthetic antibiotics): antimicrobial agents of synthetic origin useful in the treatment of microbial or viral disease. Examples are sulfonilamides, isoniazid, ethambutol, AZT, nalidixic acid and chloramphenicol. Note that the microbiologist's definition of a chemotherapeutic agent requires that the agent be used for antimicrobial purpose and excludes synthetic agents used for therapy against diseases that are not of microbial origin. Hence, pharmacology distinguishes the microbiologist's chemotherapeutic agent as a "synthetic antibiotic".

Antibiotics: antimicrobial agents produced by microorganisms that kill or inhibit other microorganisms. This is the microbiologist's definition. A more broadened definition of an antibiotic includes any chemical of natural origin which has the effect to kill or inhibit the growth of other types cells. Since most clinically-useful antibiotics are produced by microorganisms and are used to kill or inhibit infectious Bacteria, we will follow the classic definition. Note also, pharmacologists refer to any antimicrobial chemical used in the treatment of infectious disease as as antibiotic.

Antibiotics are low molecular-weight (non-protein) molecules produced as secondary metabolites, mainly by microorganisms that live in the soil. Most of these microorganisms form some type of a spore or other dormant cell, and there is thought to be some relationship (besides temporal) between antibiotic production and the processes of sporulation. Among the molds, the notable antibiotic producers are *Penicillium* and *Cephalosporium*, which are the main source of the beta-lactam antibiotics (penicillin and its relatives).

In the Bacteria, the Actinomycetes, notably *Streptomyces* species, produce a variety of types of antibiotics including the aminoglycosides (e.g. streptomycin), macrolides (e.g. erythromycin), and the tetracyclines. Endospore-forming *Bacillus* species produce polypeptide antibiotics such as polymyxin and bacitracin. Semisynthetic antibiotics are molecules produced my a microbe that are subsequently modified by an organic chemist to enhance their antimicrobial properties or to render them unique for a pharmaceutical patent.

Antimicrobial Agents Used in the Treatment of Infectious Disease

The modern era of antimicrobial chemotherapy began following Fleming's discovery in 1929 of the powerful bactericidal substance penicillin, and Domagk's discovery in 1935 of synthetic chemicals (sulfonamides) with broad antimicrobial activity. In the early 1940's, spurred partially by the need for antibacterial agents in WW II, penicillin was isolated, purified and injected into experimental animals, where it was found to not only cure infections but also to possess incredibly low toxicity for the animals. This fact ushered into being the age of antibiotic chemotherapy and an intense search for similar antimicrobial agents of low toxicity to animals that might prove useful in the treatment of infectious disease. The rapid isolation of streptomycin, chloramphenicol and tetracycline soon followed, and by the 1950's, these and several other antibiotics were in clinical usage.

The most important property of a clinically-useful antimicrobial agent, especially from the patient's point of view, is its selective toxicity, i.e., the agent acts in some way that inhibits or kills bacterial pathogens but has little or no toxic effect on the animal taking the drug This implies that the biochemical processes in the bacteria are in some way different from those in the animal cells, and that the advantage of this difference can be taken in chemotherapy.

Antimicrobial Agents

Cell wall synthesis inhibitors: Cell wall synthesis inhibitors generally inhibit some step in the synthesis of bacterial peptidoglycan. Generally they exert their selective toxicity against eubacteria because human cells lack cell walls.

Beta lactam antibiotics Chemically, these antibiotics contain a 4-membered beta lactam ring. They are the products of two groups of fungi, *Penicillium* and *Cephalosporium* molds, and are correspondingly represented by the penicillins and cephalosporins. The beta lactam antibiotics inhibit the last step in peptidoglycan synthesis, the final cross-linking between between peptide side chains, mediated by bacterial carboxypeptidase and transpeptidase enzymes. Beta lactam antibiotics are normally bactericidal and require that cells be actively growing in order to exert their toxicity.

Natural penicillins, such as Penicillin G or Penicillin V, are produced by fermentation of *Penicillium chrysogenum*. They are effective against streptococcus, gonococcus and staphylococcus, except where resistance has developed. They are considered narrow spectrum since they are not effective against Gram-negative rods. Semisynthetic penicillins first appeared in 1959. A mold produces the main part of the molecule (6-aminopenicillanic acid) which can be modified chemically by the addition of side chains. Many of these compounds have been developed to have distinct benefits or advantages over penicillin G, such as increased spectrum of activity (e.g. effectiveness against Gram-negative rods), resistance to penicillinase or effectiveness when administered orally. Amoxycillin and Ampicillin have broadened spectra against Gram-negatives and are effective orally; Methicillin is penicillinase-resistant.

Clavulanic acid is a chemical sometimes added to a semisynthetic penicillin preparation. Thus, amoxycillin plus clavulanate is clavamox or augmentin. The clavulanate is not an antimicrobial agent. It inhibits beta lactamase enzymes and has given extended life to penicillinase-sensitive beta lactams. Although nontoxic, penicillins occasionally cause death when administered to persons who are allergic to them. In the

U.S. there are 300-500 deaths annually due to penicillin allergy. In allergic individuals the beta lactam molecule attaches to a serum protein which initiates an IgE-mediated inflammatory response.

Cephalolsporins are beta lactam antibiotics with a similar mode of action to penicillins that are produced by species of *Cephalosporium*. The have a low toxicity and a somewhat broader spectrum than natural penicillins. They are often used as penicillin substitutes, against Gram-negative bacteria, and in surgical prophylaxis. They are subject to degradation by some bacterial beta-lactamases, but they tend to be resistant to beta-lactamases from *S. aureus*.

Bacitracin is a polypeptide antibiotic produced by *Bacillus* species. It prevents cell wall growth by inhibiting the release of the muropeptide subunits of peptidoglycan from the lipid carrier molecule that carries the subunit to the outside of the membrane. Teichoic acid synthesis, which requires the same carrier, is also inhibited. Bacitracin has a high toxicity which precludes its systemic use. It is present in many topical antibiotic preparations, and since it is not absorbed by the gut, it is given to "sterilise" the bowel prior to surgery.

Cell membrane inhibitors: disorganise the structure or inhibit the function of bacterial membranes. The integrity of the cytoplasmic and outer membranes is vital to bacteria, and compounds that disorganise the membranes rapidly kill the cells. However, due to the similarities in phospholipids in bacterial and eucaryotic membranes, this action is rarely specific enough to permit these compounds to be used systemically. The only antibacterial antibiotic of clinical importance that acts by this mechanism is Polymyxin, produced by *Bacillus polymyxa*. Polymyxin is effective mainly against Gram-negative bacteria and is usually limited to topical usage. Polymyxins bind to membrane phospholipids and thereby interfere with membrane function. Polymyxin is occasionally given for urinary tract infections caused by *Pseudomonas* that are gentamicin, carbenicillin and tobramycin resistant. The balance between effectiveness and damage to the kidney and other organs is dangerously close, and the drug should only be given under close supervision in the hospital.

Protein synthesis inhibitors: Many therapeutically useful antibiotics owe their action to inhibition of some step in the complex process of translation. Their attack is always at one of the events occurring on the ribosome rather than the stage of amino acid activation or attachment to a particular tRNA. Most have an affinity or specificity for 70S (as opposed to 80S) ribosomes, and they achieve their selective toxicity in this manner. The most important antibiotics with this mode of action are the tetracyclines, chloramphenicol, the macrolides (e.g. erythromycin) and the aminoglycosides (e.g. streptomycin). The aminoglycosides are products of *Streptomyces* species and are represented by streptomycin, kanamycin, tobramycin and gentamicin. These antibiotics exert their activity by binding to bacterial ribosomes and preventing the initiation of protein synthesis. Aminoglycosides have been used against a wide variety of bacterial infections caused by Gram-positive and Gram-negative bacteria. Streptomycin has been used extensively as a primary drug in the treatment of tuberculosis.

Effects on Nucleic Acids: Some chemotherapeutic agents affect the synthesis of DNA or RNA, or can bind to DNA or RNA so that their messages cannot be read. Either case, of course, can block the growth of cells. The majority of these drugs are unselective, however, and affect animal cells and bacterial cells alike and therefore have no therapeutic application. Two classes of nucleic acid synthesis inhibitors which have selective activity against procaryotes and some medical utility are quinolones and rifamycins.

Competitive Inhibitors: The competitive inhibitors are mostly all synthetic chemotherapeutic agents. Most are "growth factor analogs", chemicals which are structurally similar to a bacterial growth factor but which do not fulfill its metabolic function in the cell. Some are bacteriostatic and some are bactericidal.

Sulfonamides were introduced as chemotherapeutic agents by Domagk in 1935, who showed that one of these compounds (prontosil) had the effect of curing mice with infections caused by beta-hemolytic streptococci. Chemical modifications of the compound sulfanilamide gave compounds with even higher and broader antibacterial activity. The resulting sulfonamides have broadly similar antibacterial activity, but differ widely in their pharmacological actions. Bacteria which are almost always sensitive to the sulfonamides include *Streptococcus pneumoniae*, beta-hemolytic streptococci and *E. coli*. The sulfonamides have been extremely useful in the treatment of uncomplicated UTI caused by *E. coli*, and in the treatment of meningococcal meningitis. The most useful sulfonamides are sulfanilamide, Gantrisin and Trimethoprim.

REFERENCES

Moses, Vivian et al. 1999. *Biotechnology: The Science and the Business*. CRC Press. p. 563.

Schmidt A, Ragazzi E, Coppellotti O, Roghi G. 2006. "A microworld in Triassic amber". *Nature* 444 (7121): 835.

Soni, S.K. 2007. *Microbes: A Source of Energy for 21st Century*. New India Publishing.

11

Microbial Epidemiology

Epidemiology is a descriptive science and includes the determination of rates, that is, the quantification of disease occurrence within a specific population. The most commonly studied rate is the attack rate: the number of cases of the disease divided by the population among whom the cases have occurred. Epidemiology can accurately describe a disease and many factors concerning its occurrence before its cause is identified. For example, Snow described many aspects of the epidemiology of cholera in the late 1840s, fully 30 years before Koch described the bacillus and Semmelweis described puerperal fever in detail in 1861 and recommended appropriate control and prevention measures a number of years before the streptococcal agent was fully described. One goal of epidemiologic studies is to define the parameters of a disease, including risk factors, in order to develop the most effective measures for control.

Proper interpretation of disease-specific epidemiologic data requires information concerning past as well as present occurrence of the disease. An increase in the number of reported cases of a disease that is normal and expected, representing a seasonal pattern of change in host susceptibility, does not constitute an epidemic. Therefore, the regular collection, collation, analysis, and reporting of data concerning the occurrence of a disease is important to properly interpret short-term changes in occurrence.

A sensitive and specific surveillance program is important for the proper interpretation of disease occurrence data. Almost every country has a national disease surveillance program that regularly collects data on selected diseases. The quality of these programs varies, but, generally, useful data are collected that are important in developing control and prevention measures. There is an international agreement that the occurrence of three diseasescholera, plague, and yellow feverwill be reported to the World Health Organisation in Geneva, Switzerland. In the United States, the Centers for Disease Control and Prevention (CDC), U.S. Public Health Service, and the state health officers of all 50 states have agreed to report the occurrence of 51 diseases weekly and of another 10 diseases annually from the states to the CDC. Many states have regulations or laws that mandate reporting of these diseases and often of other diseases of specific interest to the state health department.

The methods of case reporting vary within each state. Passive reporting is one of the main methods. In such a case, physicians or personnel in clinics or hospitals report occurrences of relevant diseases by telephone, postcard, or a reporting form, usually at weekly intervals. In some instances, the report may be initiated by the public health or clinical laboratory where the etiologic agent is identified. Some diseases, such as human rabies, must be reported by telephone as soon as diagnosed. In an active surveillance program, the health authority regularly initiates the request for reporting. The local health department may call all or some health care providers at regular intervals to inquire about the occurrence of a disease or diseases. The active system may be used during an epidemic or if accurate data concerning all cases of a disease are desired.

The health care provider usually makes the initial passive report to a local authority, such as a city or county health department. This unit collates its data and sends a report to the next highest health department level, usually the state health department.

The number of cases of each reportable disease are presented weekly, via computer linkage, by the state health department to the CDC. Data are analysed at each level to develop needed information to assist public health authorities in disease control and prevention. For some diseases, such as hepatitis, the CDC requests preparation of a separate case reporting form containing more specific details.

In addition, the CDC prepares and distributes routine reports summarising and interpreting the analyses and providing information on epidemics and other appropriate public health matters. Most states and some county health departments also prepare and distribute their own surveillance reports. The CDC publishes *Morbidity and Mortality Weekly Report,* which is available for a small fee from the Massachusetts Medical Society. The CDC also prepares more detailed surveillance reports for specific diseases, as well as an annual summary report, all of which can also be obtained through the Massachusetts Medical Society.

Infection is the replication of organisms in the tissue of a host; when defined in terms of infection, disease is overt clinical manifestation. In an inapparent (subclinical) infection, an immune response can occur without overt clinical disease. A carrier (colonised individual) is a person in whom organisms are present and may be multiplying, but who shows no clinical response to their presence. The carrier state may be permanent, with the organism always present; intermittent, with the organism present for various periods; or temporary, with carriage for only a brief period. Dissemination is the movement of an infectious agent from a source directly into the environment; when infection results from dissemination, the source, if an individual, is referred to as a dangerous disseminator.

Infectiousness is the transmission of organisms from a source, or reservoir, to a susceptible individual. A human may be infective during the preclinical, clinical, postclinical, or recovery phase of an illness. The incubation period is the interval in the preclinical period between the time at which the causative agent first infects the host and the onset of clinical symptoms; during this time the agent is replicating. Transmission is most likely during the incubation period for some diseases such as measles; in other diseases such as shigellosis, transmission occurs during the clinical period. The individual may be infective during the convalescent phase, as in diphtheria, or may become an asymptomatic carrier and remain infective for a prolonged period, as do approximately 5% of persons with typhoid fever.

The spectrum of occurrence of disease in a defined population includes sporadic (occasional occurrence); endemic (regular, continuing occurrence); epidemic (significantly increased occurrence); and pandemic (epidemic occurrence in multiple countries).

CHAIN OF INFECTION

The chain of infection includes the three factors that lead to infection: the etiologic agent, the method of transmission, and the host (Fig. 1). These links should be characterised before control and prevention measures are proposed. Environmental factors that may influence disease occurrence must be evaluated.

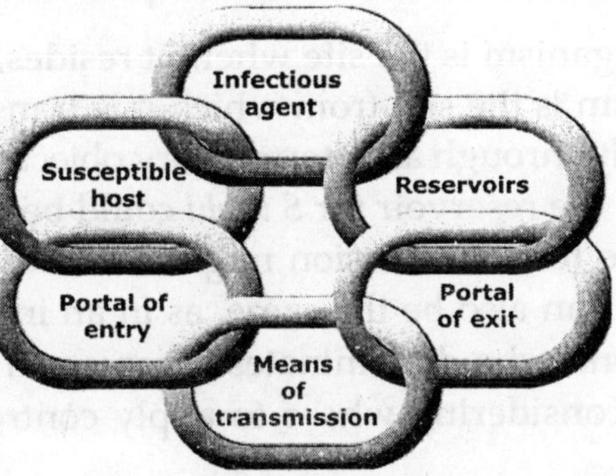

Figure 1. Chain of infection

Infectious Agent

The infectious agent may be any microorganism that can cause infection. The pathogenicity of an agent is its ability to cause disease; pathogenicity is further characterised by describing the organism's virulence and invasiveness. Virulence refers to the severity of infection, which can be expressed by describing the morbidity (incidence of disease) and mortality (death rate) of the infection. An example of a highly virulent organism is *Yersinia pestis,* the agent of plague, which almost always causes severe disease in the susceptible host.

The invasiveness of an organism refers to its ability to invade tissue. *Vibrio cholerae* organisms are noninvasive, causing symptoms by releasing into the intestinal canal an exotoxin that acts on the tissues. In contrast, *Shigella* organisms in the intestinal canal are invasive and migrate into the tissue.

No microorganism is assuredly avirulent. An organism may have very low virulence, but if the host is highly susceptible, as when therapeutically immunosuppressed, infection with that organism may cause disease. For example, the poliomyelitis virus used in oral polio vaccine is highly attenuated and thus has low virulence, but in some highly susceptible individuals it may cause paralytic disease.

Other factors should be considered in describing the agent. The infecting dose (the number of organisms necessary to cause disease) varies according to the organism, method of transmission, site of entrance of the organism into the host, host defenses, and host species. Another agent factor is specificity; some agents (for example, *Salmonella typhimurium)* can infect a broad range of hosts; others have a narrow range of hosts. *S typhi,* for example, infects only humans. Other agent factors include antigenic composition, which can vary within a species (as in influenza virus or *Streptococcus* species); antibiotic sensitivity; resistance transfer plasmids; and enzyme production.

The reservoir of an organism is the site where it resides, metabolises, and multiplies. The source of the organism is the site from which it is transmitted to a susceptible host, either directly or indirectly through an intermediary object. The reservoir and source can be different; for example, the reservoir for *S typhi* could be the gallbladder of an infected individual, but the source for transmission might be food contaminated by the carrier. The reservoir and source can also be the same, as in an individual who is a permanent nasal carrier of *S aureus* and who disseminates organisms from this site. The distinction can be important when considering where to apply control measures.

METHOD OF TRANSMISSION

The method of transmission is the means by which the agent goes from the source to the host. The four major methods of transmission are by contact, by common vehicle, by air or via a vector.

In contact transmission the agent is spread directly, indirectly, or by airborne droplets. Direct contact transmission takes place when organisms are transmitted directly from the source to the susceptible host without involving an intermediate object; this is also referred to as person-to-person transmission. An example is the transmission of hepatitis A virus from one individual to another by hand contact. Indirect transmission occurs when the organisms are transmitted from a source, either animate or inanimate, to a host by means of an inanimate object. An example is transmission of *Pseudomonas* organisms from one individual to another by means of a shaving brush. Droplet spread refers to organisms that travel through the air very short distances, that is, less than 3 feet from a source to a host. Therefore, the organisms are not airborne in the true sense. An example of a disease that may be spread by droplets is measles.

Common-vehicle transmission refers to agents transmitted by a common inanimate vehicle, with multiple cases resulting from such exposure. This category includes diseases in which food or water as well as drugs and parenteral fluids are the vehicles of infection. Examples include food-borne salmonellosis, waterborne shigellosis, and bacteremia resulting from use of intravenous fluids contaminated with a gram-negative organism.

The third method of transmission, airborne transmission, refers to infection spread by droplet nuclei or dust. To be truly airborne, the particles should travel more than 3 feet through the air from the source to the host. Droplet nuclei are the residue from the evaporation of fluid from droplets, are light enough to be transmitted more than 3 feet from the source, and may remain airborne for prolonged periods. Tuberculosis is primarily an airborne disease; the source may be a coughing patient who creates aerosols of droplet nuclei that contain tubercle bacilli. Infectious agents may be contained in dust particles, which may become resuspended and transmitted to hosts. An example occurred in an outbreak of salmonellosis in a newborn nursery in which *Salmonella*-contaminated dust in a vacuum cleaner bag was resuspended when the equipment was used repeatedly, resulting in infections among the newborns.

The fourth method of transmission is vector borne transmission, in which arthropods are the vectors. Vector transmission may be external or internal. External, or mechanical, transmission occurs when organisms are carried mechanically on the vector (for example, *Salmonella* organisms that contaminate the legs of flies). Internal transmission occurs when the organisms are carried within the vector. If the pathogen is not changed by its carriage within the vector, the carriage is called harborage (as when a flea ingests plague bacilli from an infected individual or animal and contaminates a susceptible host when it feeds again; the organism is not changed while in the flea). The other form of internal transmission is called biologic. In this form, the organism is changed biologically during its passage through the vector (for example, malaria parasites in the mosquito vector).

An infectious agent may be transmitted by more than one route. For example, *Salmonella* may be transmitted by a common vehicle (food) or by contact spread (human carrier). *Francisella tularensis* may be transmitted by any of the four routes.

Host

The third link in the chain of infection is the host. The organism may enter the host through the skin, mucous membranes, lungs, gastrointestinal tract, or genitourinary tract, and it may enter fetuses through the placenta. The resulting disease often reflects the point of entrance, but not always: meningococci that enter the host through the mucous membranes may nonetheless cause meningitis. Development of disease in a host reflects agent characteristics and is influenced by host defense mechanisms, which may be nonspecific or specific.

Nonspecific defense mechanisms include the skin, mucous membranes, secretions, excretions, enzymes, the inflammatory response, genetic factors, hormones, nutrition, behavioral patterns, and the presence of other diseases. Specific defense mechanisms or immunity may be natural, resulting from exposure to the infectious agent, or artificial, resulting from active or passive immunisation.

The environment can affect any link in the chain of infection. Temperature can assist or inhibit multiplication of organisms at their reservoir; air velocity can assist the airborne movement of droplet nuclei; low humidity can damage mucous membranes; and ultraviolet radiation can kill the microorganisms. In any investigation of disease, it is important to evaluate the effect of environmental factors. At times, environmental control measures are instituted more on emotional grounds than on the basis of epidemiologic fact. It should be apparent that the occurrence of disease results from the interaction of many factors. Some of these factors are outlined here.

EPIDEMIOLOGIC METHODS

The three major epidemiologic techniques are descriptive, analytic, and experimental. Although all three can be used in investigating the occurrence of disease, the method used most is descriptive epidemiology. Once the basic epidemiology of a disease has been described, specific analytic methods can be used to study the disease further, and a specific experimental approach can be developed to test a hypothesis.

Descriptive Epidemiology

In descriptive epidemiology, data that describe the occurrence of the disease are collected by various methods from all relevant sources. The data are then collated by time, place, and person. Four time trends are considered in describing the epidemiologic data. The secular trend describes the occurrence of disease over a prolonged period, usually years; it is influenced by the degree of immunity in the population and possibly nonspecific measures such as improved socioeconomic and nutritional levels among the population. For example, the secular trend of tetanus in the United States since 1920 shows a gradual and steady decline.

The second time trend is the periodic trend. A temporary modification in the overall secular trend, the periodic trend may indicate a change in the antigenic characteristics of the disease agent. For example, the change in antigenic structure of the prevalent influenza A virus every 2 to 3 years results in periodic increases in the occurrence of clinical influenza caused by lack of natural immunity among the population. Additionally, a lowering of the overall immunity of a population or a segment thereof (known as herd immunity) can result in an increase in the occurrence of the disease. This can be seen with some immunisable diseases when periodic decreases occur in the level of immunisation in a defined population. This may then result in an increase in the number of cases, with a subsequent rise in the overall level of herd immunity. The number of new cases then decreases until the herd's immunity is low enough to allow transmission to occur again and new cases then appear.

The third time trend is the seasonal trend. This trend reflects seasonal changes in disease occurrence following changes in environmental conditions that enhance the ability of the agent to replicate or be transmitted. For example, food-borne disease outbreaks occur more frequently in the summer, when temperatures favor multiplication of bacteria. This trend becomes evident when the occurrence of salmonellosis is examined on a monthly basis.

The fourth time trend is the epidemic occurrence of disease. An epidemic is a sudden increase in occurrence due to prevalent factors that support transmission.

A description of epidemiologic data by place must consider three different sites: where the individual was when disease occurred; where the individual was when he or she became infected from the source; and where the source became infected with the etiologic agent. Therefore, in an outbreak of food poisoning, the host may become clinically ill at home from food eaten in a restaurant. The vehicle may have been undercooked chicken, which became infected on a poultry farm. These differences are important to consider in attempting to prevent additional cases.

The third focus of descriptive epidemiology is the infected person. All pertinent characteristics should be noted: age, sex, occupation, personal habits, socioeconomic status, immunisation history, presence of underlying disease, and other data.

Once the descriptive epidemiologic data have been analysed, the features of the epidemic should be clear enough that additional areas for investigation are apparent.

Analytic Epidemiology

The second epidemiologic method is analytic epidemiology, which analyses disease determinants for possible causal relations. The two main analytic methods are the case-control (or case-comparison) method and the cohort method. The case-control method starts with the effect (disease) and retrospectively investigates the cause that led to the effect. The case group consists of individuals with the disease; a comparison group has members similar to those of the case group except for absence of the disease. These two groups are then compared to determine differences that would explain the occurrence of the disease. An example of a case-control study is selecting individuals with meningococcal meningitis and a comparison group matched for age, sex, socioeconomic status, and residence, but without the disease, to see what factors may have influenced the occurrence in the group that developed disease.

The second analytic approach is the cohort method, which prospectively studies two populations: one that has had contact with the suspected causal factor under study and a similar group that has had no contact with the factor. When both groups are observed, the effect of the factor should become apparent. An example of a cohort approach is to observe two similar groups of people, one composed of individuals who received blood transfusions and the other of persons who did not. The occurrence of hepatitis prospectively in both groups permits one to make an association between blood transfusions and hepatitis; that is, if the transfused blood was contaminated with hepatitis B virus, the recipient cohort should have a higher incidence of hepatitis than the nontransfused cohort.

The case-control approach is relatively easy to conduct, can be completed in a shorter period than the cohort approach, and is inexpensive and reproducible; however, bias may be introduced in selecting the two groups, it may be difficult to exclude subclinical cases from the comparison group, and a patient's recall of past events may be faulty. The advantages of a cohort study are the accuracy of collected data and the ability to make a direct estimate of the disease risk resulting from factor contact; however, cohort studies take longer and are more expensive to conduct.

Another analytic method is the cross-sectional study, in which a population is surveyed over a limited period to determine the relationship between a disease and variables present at the same time that may influence its occurrence.

Experimental Epidemiology

The third epidemiologic method is the experimental approach. A hypothesis is developed and an experimental model is constructed in which one or more selected factors are manipulated. The effect of the manipulation will either confirm or disprove the hypothesis. An example is the evaluation of the effect of a new drug on a disease. A group of people with the disease is identified, and some members are randomly selected to receive the drug. If the only difference between the two is use of the drug, the clinical differences between the groups should reflect the effectiveness of the drug.

Epidemic Investigation

An epidemic investigation describes the factors relevant to an outbreak of disease; once the circumstances related to the occurrence of disease are defined, appropriate control and prevention measures can be identified. In an epidemic investigation, data are collected, collated according to time, place, and person, and analysed and inferences are drawn.

In the investigation, the first action should be to confirm the existence of the epidemic by noting from past surveillance data the number of cases suspected and comparing this with the number of cases initially reported. Additionally, the investigator should discuss the occurrence of the disease with physicians or others who have seen or reported cases after examining patients and reviewing laboratory and hospital records. These diagnoses should then be verified. A case definition should be developed to differentiate patients who represent actual cases, those who represent suspected or presumptive cases, and those who should be omitted from further study. Additional cases may be sought or additional patient data obtained, and a rough case count made.

This initial phase consists basically of collecting data, which then must be organised according to time, place, and person. The population at risk should be identified and a hypothesis developed concerning the occurrence of the disease. If appropriate, specimens should be collected and transported to the laboratory. More specific studies may be indicated. Additional data from these studies should be analysed and the hypothesis confirmed or altered. After analysis, control and prevention measures should be developed and, as far as possible, implemented. A report containing this information should be prepared and distributed to those involved in investigating the outbreak and in implementing control and/or prevention measures. Continued surveillance activities may be appropriate to evaluate the effectiveness of the control and prevention measures.

In the United States, the CDC assists state health departments by providing epidemiologic and laboratory support services on request. Its assistance supports disease investigations and diagnostic laboratory activities and includes various training programs conducted in the states and at the CDC. A close working relationship exists between the CDC and state health departments. Additionally, physicians frequently consult with CDC personnel on a variety of health-related problems and attend public health training programs.

The use of epidemiology to characterise a disease before its etiology has been identified is exemplified by the initial studies of acquired immune deficiency syndrome (AIDS). The first cases came to the attention of the CDC late in 1981 when an increase was observed in requests for pentamidine for treatment of *Pneumocystis carinii* pneumonia. This initiated specific surveillance activities and epidemiologic studies that provided important information about this newly diagnosed disease.

Initial symptoms include fever, loss of appetite, weight loss, extreme fatigue, and enlargement of lymph nodes. A severe immune deficiency then develops, which appears to be associated with opportunistic infections. These infections include *P carinii* pneumonia, diagnosed in 52 percent of cases; Kaposi sarcoma in 26 percent of cases; and both *P carinii* pneumonia and Kaposi sarcoma in 7 percent of cases. The remaining 15 percent of AIDS patients have other parasitic, fungal, bacterial, or viral infections associated with immunodeficiencies. Among the first 2,640 cases reported to the CDC, there were 1,092 deaths, a case-fatality rate of 41 percent. Approximately 95 percent of the cases were male; 70 percent were 20 to 49 years of age at the time of diagnosis. Approximately 40 percent of the cases were reported from New York City, 12 percent from San Francisco, 8 percent from Los Angeles, and the remainder from 32 other states. Cases were reported from at least 16 other countries. Among the 90 percent of patients who were categorised according to possible risk factors, those at highest risk were homosexuals or bisexuals (70 percent), intravenous drug abusers (17 percent), Haitian entrants into the United States (9.5 percent), and persons with hemophilia (1 percent).

Analysis of these initial data, collected before the etiologic agent of AIDS was identified, supported the hypothesis that transmission occurred primarily by sexual contact, receipt of contaminated blood or blood products, or contact with contaminated intravenous needles. Spread through casual contact did not seem likely. The epidemiologic data indicated that AIDS was an infectious disease. It has now been determined that AIDS results from infection with a retrovirus of the human T cell leukemia/lymphoma virus family, which has been designated human immunodeficiency virus type I (HIV-l). The initial hypotheses have been proven as shown by analysis of data subsequently collected.

ASPECTS OF VIRAL EPIDEMIOLOGY

Within the field of medical virology, pathogenesis concerns the processes by which viruses infect individuals and cause disease, whereas epidemiology examines the transfer and persistence of viruses in human populations. Epidemiology and evolution are linked because epidemiologic mechanisms of transfer largely determine the natural selection component of viral evolution. Since viruses multiply only within cells, the epidemiology of viral diseases does not involve multiplication in food, water, or soil. However, some viruses that infect man may multiply and persist in other animals, such as arthropods, rodents and bats.

Portals of Entry and Exit

The human body presents three large epithelial surfaces to the environmentthe skin, the respiratory mucosa, and the alimentary tract, and two lesser surfacesthe genital tract and the conjunctiva. (Fig. 2) To gain entry to the body, viruses must either (1) infect cells in one of these surfaces, (2) otherwise breach the surface (by trauma, the bite of an arthropod or animal, or injection, transfusion or transplantation), or (3) be transmitted congenitally. Viruses escape from the body via the same surfaces, often but not necessarily by the route used as a portal of entry.

Infection via the Skin

Intact skin has a tough outer layer of cornified cells. This barrier protects the body from infection, but is frequently breached by trauma or by inoculation (e.g., by a needle or an insect bite.

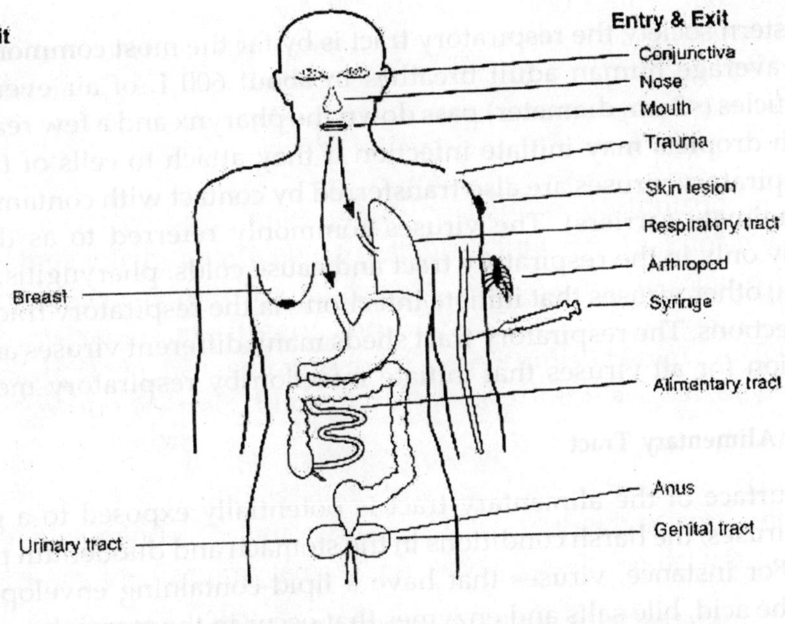

Figure 2: Body Surfaces as Sites of Vial Infection and Shedding

Virus inoculation by injection or transfusion is now common as a result both of medical procedures and of social practices such as sharing needles by intravenous drug users. In Western society, hepatitis B and hepatitis C viruses are usually transmitted in this way; less often, cytomegalovirus, Epstein-Barr virus (EBV), and the human immunodeficiency viruses (HIV) may be transferred in this manner. Hepatitis B and hepatitis C viruses may also be transferred by minor "surgical" procedures like tattooing, dentistry, ear piercing, and even (in the past) arm-to-arm vaccination.

Infection by arthropod bite is important for the large number of viruses that multiply in both arthropods and vertebrates (the arboviruses, which include most togaviruses and flaviviruses, the Orbivirus genus of the reoviruses, and all bunyaviruses except the hantaviruses). In such infections the virus is usually injected directly into small blood vessels.

In contrast to the many viruses that enter the body through the skin, only a few are shed from it in an infectious form. Herpes zoster lesions usually shed few virus particles, but they are epidemiologically important in that adults shedding virus may transmit chickenpox to susceptible children. Some viruses that infect humans by the respiratory tract may be shed from superficial lesions of the oral mucosa, (e.g., measles, and in the past, smallpox viruses) or from infected salivary glands, (e.g., mumps virus).

Infection via the Respiratory Tract

In modern Western society the respiratory tract is by far the most common route of viral infection. The average human adult breathes in about 600 L of air every hour; small suspended particles (<2 μm diameter) pass down the pharynx and a few reach the alveoli. Viruses in such droplets may initiate infection if they attach to cells of the respiratory tract. Many respiratory viruses are also transferred by contact with contaminated fingers or fomites (inanimate carriers). The viruses commonly referred to as the respiratory viruses multiply only in the respiratory tract and cause colds, pharyngitis, bronchiolitis, and pneumonia; other viruses that initiate infection via the respiratory tract can produce generalised infections. The respiratory tract sheds many different viruses and is the main route of excretion for all viruses that initiate infection by respiratory means.

Infection via the Alimentary Tract

Although the surface of the alimentary tract is potentially exposed to a great number and variety of viruses, the harsh conditions in the stomach and duodenum protect it from many viruses. For instance, viruses that have a lipid-containing envelope are usually inactivated by the acid, bile salts and enzymes that occur in the stomach and duodenum. Infection via the gut, therefore, is due to viruses that resist these chemicals. These viruses multiply in the cells of the small intestine and are excreted in the feces. Such viruses usually resist environmental conditions and may cause water- and food-borne epidemics. Recently, the significance of trauma to the mucosa of the lower rectum as a result of anal intercourse has been highlighted by the frequency of sexually transmitted viruses, notably HIV in homosexual men.

Infection via the Genital Tract

During the last decade the list of sexually transmitted viruses of the female genital tract has been enlarged by the demonstration of heterosexual transmission of HIV, the human T-cell lymphotropic virus type 1 (HTLV-1) and possibly hepatitis C virus. The importance of genital transmission of particular papillomaviruses in the causation of cervical carcinoma is receiving much attention. Genital ulcers due to herpes simplex type 2 (HSV-2), a sexually transmitted virus, are important in themselves and increase the likelihood of heterosexual transmission of HIV. A few viruses are shed in the urine of humans or, in the case of arenavirus and hantavirus infections, of rodents. The viruses from rodent urine may then cause human disease as a result of the inhalation of dust containing virus particles (hemorrhagic fever in the case of arenaviruses and hemorrhagic fever with renal syndrome or hantavirus pulmonary syndrome from hantavirus infections).

Vertical Transmission

Vertical transmission refers to the transfer of virus from parent to offspring, and may occur via the ovum, across the placenta, during birth, or via the mother's milk. Viruses that cross the placenta include rubella virus and cytomegaloviruses, which may cause congenital defects or severe neonatal disease, and HIV. The classic examples of vertical transmission of viruses in animals are lymphocytic choriomeningitis virus in mice, transmitted via the cytoplasm of the egg or the placenta, and the retroviruses that cause avian leukosis and sarcoma and murine leukemia. The retroviruses are transferred either as an integrated DNA copy of the viral RNA genome, or more rarely, in birds, as infectious virions via the egg. HTLV-1, the retrovirus that causes adult T-cell leukemia/lymphoma, appears to be transmitted horizontally, although integrated provirus is found in the lymphocytes of affected individuals.

Vertical transmission of cytomegalovirus may occur through the mother's milk, and both cytomegalovirus and herpes simplex virus type 1 can be transmitted from parents to infants by salivary contamination. Then, because of its long latency and the periodic recurrence of lesions, the same virus may be transferred to the next generation. In small, isolated human populations, infections with zoster-chickenpox may be maintained by a similar cycle, zoster in the grandmother causing chickenpox in the grandchild by horizontal transmission. Perinatal transmission of hepatitis B virus is important in much of Africa and Asia because it is common and often produces a persistent infection that may lead to cirrhosis of the liver or primary hepatocellular carcinoma.

Viral Zoonoses

A wide range of viruses that can cause human diseases survive in nature as infections of other animals; humans are only occasionally infected, and infection of humans is usually unimportant for viral survival. These infections are called zoonoses; many are caused by arboviruses (viruses that are transmitted by arthropod vectors) and some are due to direct infection. However, some arbovirus infections, notably dengue and yellow fever, can be maintained indefinitely by human-to-human mosquito transmission, although both have animal reservoir hosts also.

Epidemiologic Features of Viral Infections

The control of infections requires knowledge not only the mode of transmission but also of the incubation period, period of communicability, and seasonal incidence. Not all infections cause disease; inapparent infection (which may nevertheless be responsible for new cases) is the rule with many viruses, especially enteroviruses and some of the herpesviruses. Only in a few diseases, such as measles, does virtually every infection of a susceptible individual cause obvious clinical disease.

Humoral immunity affects the behavior of viral infections as much in human populations as in individuals. The frequency of immunity in a population is sometimes called the herd immunity. Most generalised virus diseases are associated with lifelong immunity; therefore, in the absence of an animal reservoir or of recurrent infectivity, these diseases survive only in large populations and die out in small isolated communities. For example, even in the absence of vaccination, measles and poliomyelitis do not occur as endemic infections in remote populations of Eskimos or the populations of small islands.

In superficial infections of the respiratory and alimentary tracts, humoral antibodies are less important than secretory antibodies (IgA). However, IgA is produced for a much shorter period, so that reinfections with viruses such as respiratory syncytial virus are relatively common. Further, the effect of antibody in preventing respiratory and enteric infections is often circumvented by the great number of non-cross-reacting antigenic types of most viruses that cause superficial infections of these surfaces.

REFERENCES

Beran GW. 1994. *Handbook of Zoonoses*, 2nd Ed., Section B. Viral. CRC Press, Boca Raton, Florida,

Evans AS (ed). 1989. *Viral Infections of Humans: Epidemiology and Control.* 3rd Ed. Plenum Medical, New York,

Fox JP, Hall CE, Elveback LR. 1970. *Epidemiology, Man and Disease.* Macmillan, New York,

Hennekens CH, Buring JE. 1987. *Epidemiology in Medicine.* Little, Brown, Boston,

Langmuir AD. 1963. "The surveillance of communicable diseases of national importance". *N Engl J Med* 268: 182.

Lilienfeld DE, Stolley PD. 1994. *Foundations of Epidemiology.* 3rd Ed. Oxford University Press, New York.

MacMahon B, Pugh TF. 1970. *Epidemiology Principles and Methods.* Little, Brown, Boston.

Mandell GL, Douglas RG, Jr, Bennett JE. 1990 . *Principles and Practice of Infectious Diseases*, 3rd Ed. Churchill Livingstone, New York.

White DO, Fenner F. 1994. *Medical Virology*, 4th Ed. Academic Press, San Diego,.

12

Bacterial Pathogens

Historically, bacteria have been the cause of some of the most deadly diseases and widespread epidemics of human civilisation. Smallpox and malaria, diseases caused by other microbes, have killed more humans than bacterial diseases, but diseases such as tuberculosis, typhus, plague, diphtheria, typhoid, cholera, dysentery and pneumonia have taken a large toll of humanity.

At the beginning of the twentieth century, pneumonia, tuberculosis and diarrhea were the three leading causes of death. Water purification, immunisation (vaccination) and antibiotic treatment have reduced the morbidity and the mortality of bacterial disease in the twenty-first century, at least in the developed world where these are acceptable cultural practices.

Albeit, some bacterial diseases have been conquered (for the present), but many new bacterial pathogens have been recognised in the past 30 years, and many "old" bacterial pathogens, such as *Staphylococcus aureus* and *Mycobacterium tuberculosis,* have emerged with new forms of virulence and new patterns of resistance to antimicrobial agents. Great vigilance is warranted, and research and study are needed to control both old and new bacterial pathogens.

MAJOR GROUPS OF BACTERIAL PATHOGENS

This article deals with the major groups of bacterial pathogens. All groups are defined by at least one bacteriological criterion such as Gram stain, metabolism, morphology, spore formation, etc. However, there is often some genetic or phylogenetic relationship between members of a group.

Although we organise bacterial pathogens into natural groups for discussion based on bacteriological criteria, rather than on the basis of affected organ, mode of transmission, or type of disease, two summary tables are provided at the end of this reading that identify bacterial pathogens of humans on the basis of specific bacterium, type of disease, and usual mode of transmission. When one searches for clusters of pathogens in the Bacterial Domain of the Tree of Life, they are found primarily among the Gram-positive bacteria and the Gram-negative proteobacteria. Most of the bacterial pathogens of humans are classified as Gram-positive or Gram-negative, but some notable exceptions include the mycoplasmas, chlamydiae, spirochetes and the mycobacteria.

Table 1. Examples of bacterial pathogens and diseases recognised or reemerged since 1977

Bacterium	*Disease*
Legionella pneumophila	Legionnaires' pneumonia
Listeria monocytogenes	listeriosis
Campylobacter jejuni	gastroenteritis distributed world-wide
Staphylococcus aureus	toxic shock syndrome
E. coli O157:H7	hemorrhagic colitis; hemolytic uremic syndrome
Borrelia burgdorferi	Lyme Disease and complications
Helicobacter pylori	gastric and duodenal ulcers
Ehrlichia chaffeensis	human ehrlichiosis
Clostridium difficile	antibiotic induced diarrhea; pseudomembranous colitis
Vibrio cholerae O139	epidemic cholera
Salmonella enterica	salmonellosis
Bartonella henselae	cat scratch fever
Streptococcus pyogenes	necrotising fasciitis (GAS); streptococcal toxic shock syndrome
S. aureus (e.g. MRSA)	nosocomial and community associated infections
Chlamydia pneumoniae	atherosclerosis
Clostridium botulinum	sudden infant death syndrome (SIDS)
Vibrio vulnificus	wound infection, septicemia, gastrointestinal disease
Parachlamydia	pneumonia
Corynebacterium amycolatum	hospital-acquired endocarditis
Klebsiella pneumoniae	blood stream infections
(*E. faecalis* and *E. faecium*)	nosocomial infections
Acinetobacter baumannii	nosocomial infections

Spirochetes

The spirochetes are a phylogenetically distinct group of bacteria which have a unique cell morphology and mode of motility. Spirochetes are very thin, flexible, spiral-shaped bacteria that move by means of structures called axial filaments or endoflagella. The flagellar filaments are contained within a sheath between the cell wall peptidoglycan and an outer membrane.

The filaments flex or rotate within their sheath, which causes the cells to bend, flex and rotate during movement. Most spirochetes are free living, or live in associations with animals (e.g. in the oral cavity or GI tract). A few are pathogens of animals, occasionally transmitted to humans (e.g. leptospirosis). The two major pathogens of humans are *Treponema pallidum,* the agent of syphilis, a sexually transmitted disease, and *Borrelia burgdorferi,* cause of Lyme Disease, transmitted by the bite of the deer tick.

Spirilla and other Curved Bacteria

Spirilla are Gram-negative bacteria with a helical or spiral shape. Their metabolism is respiratory and never fermentative. Unlike spirochetes, they have a rigid cell wall and are motile by means of ordinary polar flagella. Two important pathogens of humans are found among the spiral forms. *Campylobacter jejuni* is the cause of bacterial diarrhea, especially in children. The bacterium is transmitted via contaminated food, usually undercooked poultry or shellfish, or untreated drinking water. *Helicobacter pylori* is able to colonise the gastric mucosal cells of humans, i.e., the lining of the stomach, and it has been well established as the cause of peptic ulcers and there is strong evidence for its involvement in adenocarcinoma.

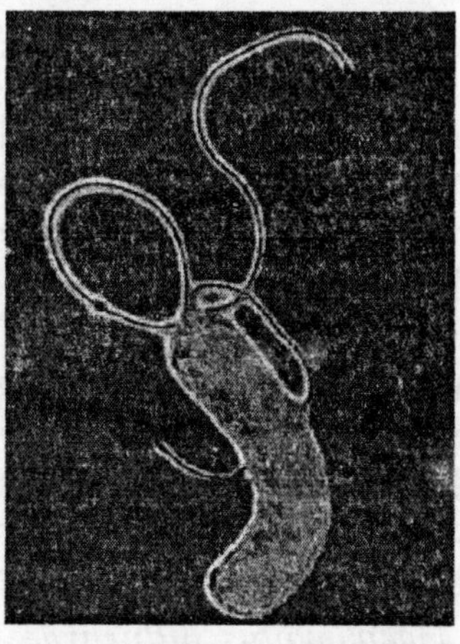

Figure 1. ***Helicobacter Pylori***

Vibrios

The term vibrio refers to a Gram-negative bacterium which has the cell shape of a curved rod or a comma. Members of the genus *Vibrio* consists of common bacteria in aquatic environments, especially marine environments. They have structural and metabolic properties that overlap with both the enterics and the pseudomonads.

Vibrios are facultative anaerobes, like enterics, but they have polar flagella, are oxidase-positive, and degrade sugars in the same manner as the pseudomonads. In aquatic habitats, they overlap with the pseudomonads in their ecology, although pseudomonads favor fresh water and vibrios prefer salt water.

Some marine vibrios are bioluminescent (they emit light) and some are symbionts of fish, squid and other marine life. *Vibrio cholerae* causes epidemic or Asiatic cholera which, untreated, is one of the most rapidly fatal infectious diseases known.

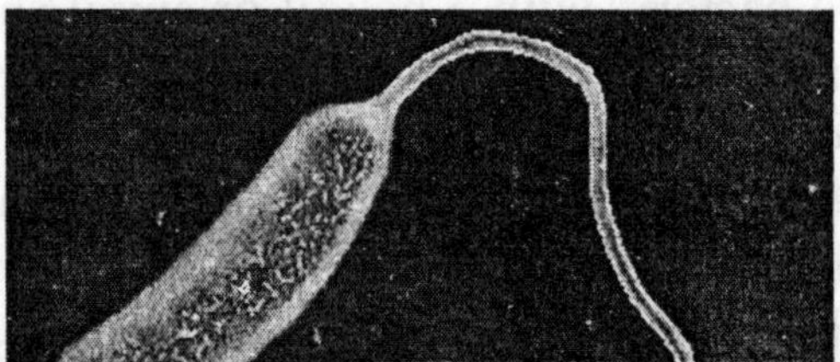

Figure 2. Vibrio cholerae

The pathology is related to diarrheal diseases caused by the enteric bacteria, except it is relentless, and a patient can die rapidly from dehydration. The cholera toxin, which is the classic model of a bacterial enterotoxin, is also produced by some strains of *E. coli.*

The Gram-negative Aerobic Rods and Cocci

This group consists of Gram-negative bacteria phenotypically related to members of the genus *Pseudomonas.* Their metabolism is respiratory and never fermentative. Important human pathogens include *Pseudomonas aeruginosa, Neisseria gonorrhoeae, Neisseria meningitidis, Bordetella pertussis, Haemophilus influenzae, Legionella pneumophila, Brucella, Francisella,* and a few others.

Many bacteria in this physiological group are free-living in soil and water, and they play an important role in decomposition, biodegradation, and the C and N cycles. Also, many bacteria that are pathogens of plants are found in this group, including *Pseudomonas, Xanthomonas and Agrobacterium.*

Pseudomonas aeruginosa is the quintessential opportunistic pathogen of humans. It is a leading cause of hospital-acquired infections (nosocomial infections), and it is difficult to eradicate due to its resistance to most antimicrobial agents. There is probably no tissue that cannot become infected by *Pseudomonas* if the host defenses are weakened, and it is difficult to treat due to inherent and acquired resistance to antimicrobial agents. It is usually involved in soft tissue infections, urinary tract infections and pneumonia.

Whooping cough (or pertussis) is caused by *Bordetella pertussis*. The disease is particularly serious in infants and young children and has a high mortality rate. Whooping cough is controlled by vaccination with the acellular pertussis vaccine, which is usually given in association with diphtheria, tetanus and sometimes *H. influenzae* type b (Hib), as part of the childhood immunisation program in the U.S.

Legionnaires' pneumonia is caused by *Legionella pneumophila*. This pneumonia, and the bacterium, were not discovered until 1976, when there was an outbreak of disease at a Legionnaire's meeting in Philadelphia. It took several months to find, culture and grow the bacterium.

The incident was a wake-up call to public health officials that there were probably a lot of disease-producing bacteria in the environment that they know nothing about. *Neisseria gonorrhoeae* causes the sexually-transmitted disease gonorrhea, and *Neisseria meningitidis* is the agent of meningococcal meningitis.

Haemophilus influenzae is also a cause of meningitis, but the incidence of the disease has declined rapidly with the use of the Hib vaccine which began in 1994. *Haemophilus* is sometimes involved in infections of the upper respiratory tract, particularly the sinuses.

Brucellosis is a chronic debilitating infection in humans associated with reproductive failure in domestic animals. Person-to-person transmission of brucellae is extremely rare. *Brucella abortus* is the species usually involved in human disease. The primary reservoir of the organism is in cattle, although bison are sometimes wrongfully accused.

Enterics

Enteric bacteria are Gram-negative rods with facultative anaerobic metabolism that live in the intestinal tracts of animals in health and disease. This group consists of *Escherichia coli* and its relatives, the members of the family *Enterobacteriaceae*. Enteric bacteria are related phenotypically to several other genera of bacteria such as *Pseudomonas* and Vibrios.

Generally, a distinction can be made on the ability to ferment glucose; enteric bacteria all ferment glucose to acid end products while similar Gram-negative bacteria (e.g.

pseudomonads) cannot ferment glucose. Because they are consistent members of the normal flora humans, and because of their medical importance, an extremely large number of enteric bacteria have been isolated and characterised.

- *Escherichia coli* is, of course, the type species of the enterics. *E. coli* is such a regular inhabitant of the intestine of humans that it is used by public health authorities as an indicator of fecal pollution of drinking water supplies, swimming beaches, foods, etc. *E. coli* is the most studied of all organisms in biology because of its natural occurrence and the ease and speed of growing the bacterium in the laboratory. It has been used in hundreds of thousands of experiments in cell biology, physiology, and genetics, and was among the first cells for which the entire chromosomal DNA base sequence (genome) was determined.

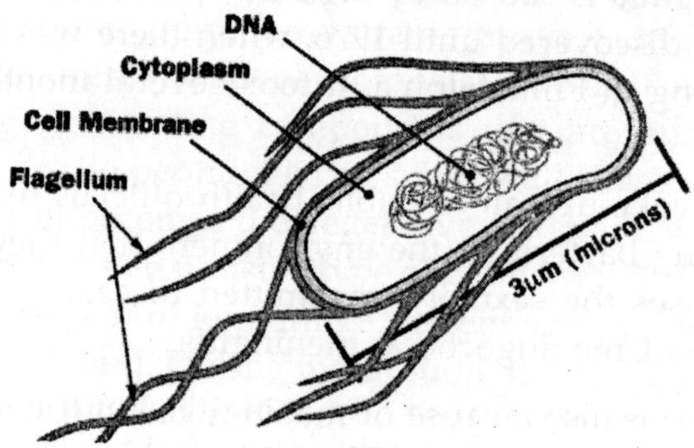

Figure 3. Escherichia coli

In spite of the knowledge gained about the molecular biology, genetics and physiology of *E. coli,* surprisingly little is known about its ecology, for example, why it consistently associates with humans, how it helps its host, how it harms its host, etc.

A few strains of *E. coli* are pathogenic (one is now notorious, strain 0157:H7, that has been found to contaminate raw hamburger, vegetables, unpasteurised milk and drinking water) . *Escherichia coli* causes intestinal tract infections (usually acute and uncomplicated, except in the very young) or uncomplicated urinary tract infections and neonatal meningitis.

The enteric group includes two other important some other intestinal pathogens of humans: *Salmonella* and *Shigella. Shigella dysenteriae* causes bacillary dysentery: *Salmonella enterica,* causes food poisoning and gastroenteritis. *Salmonella typhi,* which infects via the intestinal route, causes typhoid fever.

Some bacteria that don't have an intestinal habitat resemble *E. coli* in enough ways to warrant inclusion in the enteric group. This includes *Proteus,* a common saprophyte of decaying organic matter and *Yersinia pestis,* which causes bubonic plague. Also classified as an enteric is *Erwinia,* a pathogen of plants that causes fireblight in pear and apple trees and soft rot of carrots and potatoes.

Pyogenic Cocci

The pyogenic cocci are spherical bacteria that cause various suppurative (pus-producing) infections in animals. Included are the Gram-positive cocci *Staphylococcus aureus, Streptococcus pyogenes* and *Streptococcus pneumoniae,* and the Gram-negative cocci, *Neisseria gonorrhoeae* and *N. meningitidis.* In terms of their phylogeny, physiology and genetics, these genera of bacteria are unrelated to one another. They share a common ecology, however, as parasites of humans.

The Gram-positive cocci are the leading pathogens of humans. It is estimated that they produce at least a third of all the bacterial infections of humans, including strep throat, pneumonia, otitis media, meningitis, food poisoning, various skin diseases and severe types of septic shock. The Gram-negative cocci, notably the neisseriae, cause gonorrhea and meningococcal meningitis.

Two species of *Staphylococcus* live in association with humans: *Staphylococcus epidermidis* which lives normally on the skin and mucous membranes, and *Staphylococcus aureus,* which may occur normally at various locales, but in particular on the nasal membranes (nares). *S. epidermidis* is rarely a pathogen and probably benefits its host by producing acids on the skin that retard the growth of dermatophytic fungi.

S. aureus always has the potential to cause disease and so is considered a pathogen. Different strains of *S. aureus* differ in the range of diseases they can cause, including boils and pimples, wound infections, pneumonia, osteomyelitis, septicemia, food intoxication, and toxic shock syndrome.

S. aureus is the leading cause of nosocomial infections by Gram-positive bacteria. Also, it is notoriously resistant to penicillin and many other antibiotics. Recently, a strain of *S. aureus* has been reported that is resistant to all known antibiotics in clinical usage, which is a grim reminder that the clock is ticking on the lifetime of the usefulness of current antibiotics in treatment of infectious disease.

Staphylococcus aureus is a successful bacterial pathogen because it has a very wide range of virulence determinants (structural, biochemical or genetic features that allow the bacterium to cause disease), and it occurs as normal flora of humans (on skin, nasal membranes and the GI tract), which ensures that it is readily transmitted from one individual to another.

Streptococcus pyogenes, more specifically the beta-hemolytic group A streptococci, like *S. aureus*, cause an array of suppurative diseases and toxinoses, in addition to some autoimmune or allergic diseases. *S. pyogenes* is occasionally found as normal flora in the upper respiratory tract (<15% of individuals), but it is the main streptococcal pathogen for man, most often causing tonsillitis or strep throat.

Streptococci also invade the skin to cause localised infections and lesions, and produce toxins that cause scarlet fever and toxic shock. Sometimes, as a result of an acute streptococcal infection, anomalous immune responses are started that lead to diseases like rheumatic fever and glomerulonephritis, which are called post-streptococcal sequelae. Unlike the staphylococci, the streptococci have not developed widespread resistance to penicillin and the other beta lactam antibiotics, so that the beta lactams remain drugs of choice for the treatment of acute streptococcal infections.

Streptococcus pneumoniae is the most frequent cause of bacterial pneumonia in humans. It is also a frequent cause of otitis media (infection of the middle ear) and meningitis. The bacterium colonises the nasopharynx and from there gains access to the lung or to the eustachian tube. If the bacteria descend into the lung they can impede engulfment by alveolar macrophages if they possess a capsule which somehow prevents the engulfment process. Thus, encapsulated strains are able to invade the lung and are virulent (cause disease), and noncapsulated strains, which are readily removed by phagocytes, are nonvirulent.

The *Neisseriae* cause gonorrhea and meningitis. *Neisseriaceae* is a family of Gram-negative bacteria with characteristics of enterics and pseudomonads. The neisseriae are small, Gram-negative cocci usually seen in pairs with flattened adjacent sides.

Most neisseriae are normal flora or harmless commensals of mammals living on mucous membranes. In humans they are common residents of the throat and upper respiratory tract. Two species are primary pathogens of man, *Neisseria gonorrhoeae* and *Neisseria meningitidis*.

Neisseria gonorrhoeae is the second leading bacterial cause of sexually-transmitted disease in the U.S., causing over 300,000 cases of gonorrhea annually. Sometimes, in females, the disease may be unrecognised or asymptomatic such that an infected mother can give birth and unknowingly transmit the bacterium to the infant during its passage through the birth canal.

The bacterium is able to colonise and infect the newborn eye resulting neonatal ophthalmia, which may produce blindness. For this reason (as well as to control Chlamydia which may also be present), an antimicrobial agent is usually added to the newborn eye at the time of birth.

Neisseria meningitidis is an important cause of bacterial meningitis, an inflammation of the meninges of the brain and spinal cord. Other bacteria that cause meningitis include *Haemophilus influenzae, Staphylococcus aureus* and *Escherichia coli.* Meningococcal meningitis differs from other causes in that it is often responsible for epidemics of meningitis. It occurs most often in children aged 6 to 11 months, but it also occurs in older children and in adults. Meningococcal meningitis can be a rapidly fatal disease, and untreated meningitis has a mortality rate near 50 percent. However, early intervention with antibiotics is highly effective, and with treatment most individuals recover without permanent damage to the nervous system.

Endospore-forming Bacteria

Endospore-forming bacteria produce a unique resting cell called an endospore. They are Gram-positive and usually rod-shaped, but there are exceptions. The two medically-important genera are *Bacillus,* the members of which are aerobic sporeformers in the soils, and *Clostridium,* whose species are anaerobic sporeformers of soils, sediments and the intestinal tracts of animals.

Some sporeformers are pathogens of animals, usually due to the production of powerful toxins. *Bacillus anthracis* causes anthrax, a disease of domestic animals (cattle, sheep, etc.), which may be transmitted to humans. *Bacillus cereus* causes food poisoning. *Clostridium botulimum* causes botulism, a form of food poisoning, and *Clostridium tetani* is the agent of tetanus.

Clostridium perfringens causes food poisoning, anaerobic wound infections and gas gangrene, and *Clostridium difficile* causes a severe form of colitis called pseudomembranous colitis. Whenever the spore-formers act as pathogens, it is not uncommon or surprising that their spores are somehow involved in transmission or survival of the organism between hosts.

Listeria monocytogenes is a Gram-positive rod-shaped bacterium related to *Bacillus* and *Clostridium,* but it does not form endospores. *Listeria monocytogenes* is the agent of listeriosis, a serious infection caused by eating food contaminated with the bacteria. Listeriosis has recently been recognised as an important public health problem in the United States. The disease affects primarily pregnant women, newborns, and adults with weakened immune systems.

Actinomycetes and related Bacteria

The actinomycetes are not thought of as pathogenic bacteria, but two of their relatives are among the most important pathogens of humans, these being the agents of tuberculosis and diphtheria. Actinomycetes are a large group of Gram-positive bacteria that usually grow by filament formation, or at least show a tendency towards branching and filament formation. Many of the organisms can form resting structures called spores, but they are not the same as endospores.

Branched forms superficially resemble molds and are a striking example of convergent evolution of a procaryote and a eucaryote together in the soil habitat. Actinomycetes such as *Streptomyces* have a world-wide distribution in soils. They are important in aerobic decomposition of organic compounds and have an important role in biodegradation and the carbon cycle.

Actinomycetes are the main producers of antibiotics in industrial settings, being the source of most tetracyclines, macrolides (e.g. erythromycin), and aminoglycosides (e.g. streptomycin, gentamicin, etc.). Two genera of bacteria that are related to the actinomycetes, *Corynebacterium* and *Mycobacterium,* contain important pathogens of humans: Otherwise, many nonpathogenic mycobacteria and corynebacteria live in normal associations with animals.

*Mycobacterium tuberculosis*_is the etiologic agent of tuberculosis (TB) in humans. Tuberculosis is the leading cause of death in the world from a single infectious disease. *Mycobacterium tuberculosis*_infects 1.7 billion people/year which is equal to 33% of the entire world population.

The bacterium is responsible for over 3 million deaths/year. After a century of decline in the United States, cases of tuberculosis have increased slightly, and multiple drug-resistant strains have emerged, This increase in cases is attributable to changes in the social structure in cities, the HIV epidemic, and patient failure to comply with treatment programs. A related organism, *Mycobacterium leprae,* causes leprosy.

The genus *Corynebacterium* consists of a diverse group of bacteria including animal and plant pathogens, as well as saprophytes. Some corynebacteria are part of the normal flora of humans, finding a suitable niche in virtually every anatomic site. The best known and most widely studied species is *Corynebacterium diphtheriae,* the causal agent of diphtheria.

The *Corynebacterium diphtheriae* traces closely the development of medical microbiology, immunology and molecular biology. Many contributions to these fields, as well as to our understanding of host-bacterial interactions, have been made studying diphtheria and the diphtheria toxin. Rickettsias and chlamydiae are two unrelated groups of bacteria that are obligate intracellular parasites of eucaryotic cells. Rickettsias cannot grow outside of a host cell because they have leaky membranes and are unable to obtain nutrients in an extracellular habitat. Chlamydiae are unable to produce ATP in amounts required to sustain metabolism outside of a host cell and are, in a sense, energy-parasites.

Rickettsias occur in nature in the gut lining of arthropods (ticks, fleas, lice, etc.). They are transmitted to vertebrates by an arthropod bite and produce diseases such as typhus fever, Rocky Mountain Spotted Fever, Q fever and ehrlichiosis.

Chlamydiae are tiny bacteria that infect birds and mammals. They may colonise and infect tissues of the eye and urogenital tract in humans. *Chlamydia trachomatis* causes several important diseases in humans: chlamydia, the most prevalent sexually transmitted disease in the U.S., trachoma, a leading cause of blindness worldwide, and lymphogranuloma venereum. *Chlamydia pneumoniae* is a cause of pneumonia and has been recently linked to atherosclerosis. Mycoplasmas are a group of bacteria that lack a cell wall. The cells are bounded by a single triple-layered membrane. They may be free-living in soil and sewage, parasitic inhabitants of the mouth and urinary tract of humans, or pathogens in animals and plants. In humans, *Mycoplasma pneumoniae* causes primary atypical pneumonia, also called walking pneumonia.

PATHOGENIC BACTERIA AND INFECTIOUS DISEASES

Pathogenic bacteria are bacteria that cause infectious diseases. Although the vast majority of bacteria are harmless or beneficial, quite a few bacteria are pathogenic. The most common bacterial disease is tuberculosis, caused by the bacterium Mycobacterium tuberculosis, which kills about 2 million people a year, mostly in sub-Saharan Africa. Pathogenic bacteria contribute to other globally important diseases, such as pneumonia, which can be caused by bacteria such as Streptococcus and Pseudomonas, and foodborne illnesses, which can be caused by bacteria such as Shigella, Campylobacter and Salmonella. Each pathogenic species has a characteristic spectrum of interactions with its human hosts.

Conditionally Pathogenic

Some organisms, such as Staphylococcus or Streptococcus, can cause skin infections, pneumonia, meningitis and even overwhelming sepsis, a systemic inflammatory response producing shock, massive vasodilation and death. Yet these organisms are also part of the normal human flora and usually exist on the skin or in the nose without causing any disease at all.

Intracellular

Other organisms invariably cause disease in humans, such as the Rickettsia, which are obligate intracellular parasites able to grow and reproduce only within the cells of other organisms. One species of Rickettsia causes typhus, while another causes Rocky Mountain spotted fever. Chlamydia, another phylum of obligate intracellular parasites, contains species that can cause pneumonia, or urinary tract infection and may be involved in coronary heart disease.

Opportunistic

Some species, such as Pseudomonas aeruginosa, Burkholderia cenocepacia, and Mycobacterium avium, are opportunistic pathogens and cause disease mainly in people suffering from immunosuppression or cystic fibrosis.

TREATMENT

Bacterial infections may be treated with antibiotics, which are classified as bacteriocidal if they kill bacteria, or bacteriostatic if they just prevent bacterial growth. There are many types of antibiotics and each class inhibits a process that is different in the pathogen from that found in the host. For example, the antibiotics, chloramphenicol and tetracyclin inhibit the bacterial ribosome, but not the structurally-different eukaryotic ribosome, and so exhibit selective toxicity. Antibiotics are used both in treating human disease and in intensive farming to promote animal growth. Both uses may be contributing to the rapid development of antibiotic resistance in bacterial populations. Infections can be prevented by antiseptic measures such as sterilizating the skin prior to piercing it with the needle of a syringe, and by proper care of indwelling catheters. Surgical and dental instruments are also sterilized to prevent contamination and infection by bacteria. Disinfectants such as bleach are used to kill bacteria or other pathogens on surfaces to prevent contamination and further reduce the risk of infection.

REFERENCES

Fish D. "Optimal antimicrobial therapy for sepsis". *Am J Health Syst Pharm* 59 Suppl 1: S13 – 9. PMID 11885408.

Belland R, Ouellette S, Gieffers J, Byrne G (2004). "Chlamydia pneumoniae and atherosclerosis". *Cell Microbiol* 6 (2): 117 – 27. doi:10.1046/j.1462-5822.2003.00352.x.

Saiman, L. "Microbiology of early CF lung disease". *Paediatr Respir Rev.* Volume = 5 Suppl A: S367–369. PMID 14980298

Yonath A, Bashan A (2004). "Ribosomal crystallography: initiation, peptide bond formation, and amino acid polymerization are hampered by antibiotics". *Annu Rev Microbiol* 58: 233 – 51.

13

Viral Pathogens

Pathogenesis is the process by which an infection leads to disease. Pathogenic mechanisms of viral disease include (1) implantation of virus at the portal of entry, (2) local replication, (3) spread to target organs (disease sites), and (4) spread to sites of shedding of virus into the environment. Factors that affect pathogenic mechanisms are (1) accessibility of virus to tissue, (2) cell susceptibility to virus multiplication, and (3) virus susceptibility to host defenses. Natural selection favors the dominance of low-virulence virus strains.

Direct cell damage and death from viral infection may result from (1) diversion of the cell's energy, (2) shutoff of cell macromolecular synthesis, (3) competition of viral mRNA for cellular ribosomes, (4) competition of viral promoters and transcriptional enhancers for cellular transcriptional factors such as RNA polymerases, and inhibition of the interferon defense mechanisms. Indirect cell damage can result from integration of the viral genome, induction of mutations in the host genome, inflammation, and the host immune response.

TISSUE TROPISM

Viral affinity for specific body tissues (tropism) is determined by (1) cell receptors for virus, (2) cell transcription factors that recognise viral promoters and enhancer sequences, (3) ability of the cell to support virus replication, (4) physical barriers, (5) local temperature, pH, and oxygen tension ensymes and non-specific factors in body secretions, and (6) digestive enzymes and bile in the gastrointestinal tract that may inactivate some viruses.

Virions implant onto living cells mainly via the respiratory, gastrointestinal, skin-penetrating, and genital routes although other routes can be used. The final outcome of infection may be determined by the dose and location of the virus as well as its infectivity and virulence.

Most virus types spread among cells extracellularly, but some may also spread intracellularly. Establishment of local infection may lead to localised disease and localised shedding of virus.

Viremic: The most common route of systemic spread from the portal of entry is the circulation, which the virus reaches via the lymphatics. Virus may enter the target organs from the capillaries by (1) multiplying in endothelial cells or fixed macrophages, (2) diffusing through gaps, and (3) being carried in a migrating leukocyte.

Neural: Dissemination via nerves usually occurs with rabies virus and sometimes with herpesvirus and poliovirus infections.

The incubation period is the time between exposure to virus and onset of disease. During this usually asymptomatic period, implantation, local multiplication, and spread (for disseminated infections) occur.

Depending on the balance between virus and host defenses, virus multiplication in the target organ may be sufficient to cause disease and death.

Although the respiratory tract, alimentary tract, urogenital tract and blood are the most frequent sites of shedding, diverse viruses may be shed at virtually every site.

CONGENITAL INFECTIONS

Infection of the fetus as a target "organ" is special because the virus must traverse additional physical barriers, the early fetal immune and interferon defense systems may be immature, transfer of the maternal defenses are partially blocked by the placenta, the developing first-trimester fetal organs are vulnerable to infection, and hormonal changes are taking place.

Pathogenesis is the process by which virus infection leads to disease. Pathogenic mechanisms include implantation of the virus at a body site (the portal of entry), replication at that site, and then spread to and multiplication within sites (target organs) where disease or shedding of virus into the environment occurs. Most viral infections are subclinical, suggesting that body defenses against viruses arrest most infections before disease symptoms become manifest.

Knowledge of subclinical infections comes from serologic studies showing that sizeable portions of the population have specific antibodies to viruses even though the individuals have no history of disease. These inapparent infections have great epidemiologic importance: they constitute major sources for dissemination of virus through the population, and they confer immunity.

Many factors affect pathogenic mechanisms. An early determinant is the extent to which body tissues and organs are accessible to the virus. Accessibility is influenced by physical barriers (such as mucus and tissue barriers), by the distance to be traversed within the body, and by natural defense mechanisms. If the virus reaches an organ, infection occurs only if cells capable of supporting virus replication are present.

Cellular susceptibility requires a cell surface attachment site (receptor) for the virions and also an intracellular environment that permits virus replication and release. Even if virus initiates infection in a susceptible organ, replication of sufficient virus to cause disease may be prevented by host defenses.

Other factors that determine whether infection and disease occur are the many virulence characteristics of the infecting virus. To cause disease, the infecting virus must be able to overcome the inhibitory effects of physical barriers, distance, host defenses, and differing cellular susceptibilities to infection.

The inhibitory effects are genetically controlled and therefore may vary among individuals and races. Virulence characteristics enable the virus to initiate infection, spread in the body, and replicate to large enough numbers to impair the target organ. These factors include the ability to replicate under certain circumstances during inflammation, during the febrile response, in migratory cells, and in the presence of natural body inhibitors and interferon. Extremely virulent strains often occur within virus populations.

Occasionally, these strains become dominant as a result of unusual selective pressures. The viral proteins and genes responsible for specific virulence functions are only just beginning to be identified. Fortunately for the survival of humans and animals (and hence for the infecting virus), most natural selective pressures favor the dominance of less virulent strains. Because these strains do not cause severe disease or death, their replication and transmission are not impaired by an incapacitated host.

Mild or inapparent infections can result from absence of one or more virulence factors. For example, a virus that has all the virulence characteristics except the ability to multiply at elevated temperatures is arrested at the febrile stage of infection and causes a milder disease than its totally virulent counterpart.

Live virus vaccines are composed of viruses deficient in one or more virulence factors; they cause only inapparent infections and yet are able to replicate sufficiently to induce immunity. The occurrence of spontaneous or induced mutations in viral genetic material may alter the pathogenesis of the induced disease, e.g. HIV. These mutations can be of particular importance with the development of drug resistant strains of virus.

Disease does not always follow successful virus replication in the target organ. Disease occurs only if the virus replicates sufficiently to damage essential cells directly, to cause the release of toxic substances from infected tissues, to damage cellular genes or to damage organ function indirectly as a result of the host immune response to the presence of virus antigens.

As a group, viruses use all conceivable portals of entry, mechanisms of spread, target organs, and sites of excretion. This abundance of possibilities is not surprising considering the astronomic numbers of viruses and their variants.

Direct cell damage and death may result from disruption of cellular macromolecular synthesis by the infecting virus. Also, viruses cannot synthesize their genetic and structural components, and so they rely almost exclusively on the host cell for these functions. Their parasitic replication therefore robs the host cell of energy and macromolecular components, severely impairing the host's ability to function and often resulting in cell death and disease.

Pathogenesis at the cellular level can be viewed as a process that occurs in progressive stages leading to cellular disease. An essential aspect of viral pathogenesis at the cellular level is the competition between the synthetic needs of the virus and those of the host cell. Since viruses must use the cell's machinery to synthesize their own nucleic acids and proteins, they have evolved various mechanisms to subvert the cell's normal functions to those required for production of viral macromolecules and eventually viral progeny.

The function of some of the viral genetic elements associated with virulence may be related to providing conditions in which the synthetic needs of the virus compete effectively for a limited supply of cellular macromolecule components and synthetic machinery, such as ribosomes.

Most viruses have an affinity for specific tissues; that is, they display tissue specificity or tropism. This specificity is determined by selective susceptibility of cells, physical barriers, local temperature and pH, and host defenses. Many examples of viral tissue tropism are known. Polioviruses selectively infect and destroy certain nerve cells, which have a higher concentration of surface receptors for polioviruses than do virus-resistant cells. Rhinoviruses multiply exclusively in the upper respiratory tract because they are adapted to multiply best at low temperature and pH and high oxygen tension. Enteroviruses can multiply in the intestine, partly because they resist inactivation by digestive enzymes, bile, and acid.

The cell receptors for some viruses have been identified. Rabies virus uses the acetylcholine receptor present on neurons as a receptor, and hepatitis B virus binds to polymerised albumin receptors found on liver cells. Similarly, Epstein-Barr virus uses complement CD21 receptors on B lymphocytes, and human immunodeficiency virus uses the CD4 molecules present on T lymphocytes as specific receptors.

Viral tropism is also dictated in part by the presence of specific cell transcription factors that require enhancer sequences within the viral genome. Recently, enhancer sequences have been shown to participate in the pathogenesis of certain viral infections. Enhancer sequences within the long terminal repeat (LTR) regions of Moloney murine leukemia retrovirus are active in certain host tissues. In addition, JV papovavirus appears to have an enhancer sequence that is active specifically in oligodendroglia cells, and hepatitis B virus enhancer activity is most active in hepatocytes.

SEQUENCE OF VIRUS SPREAD IN THE HOST

The range of structural and biochemical effects that viruses have on the hosts cell is extensive. These are called cytopathic effects. Most virus infections eventually result in the death of the host cell. The causes of death include cell lysis, alterations to the cell's surface membrane and apoptosis. Often cell death is caused by cessation of its normal activities due to suppression by virus-specific proteins, not all of which are components of the virus particle.

Some viruses cause no apparent changes to the infected cell. Cells in which the virus is latent and inactive show few signs of infection and often function normally. This causes persistent infections and the virus is often dormant for many months or years. This is often the case with herpes viruses. Viruses, such as Epstein-Barr virus often cause cells to proliferate without causing malignancy, but viruses, such as papillomaviruses are an established cause of cancer.

Implantation at Portal of Entry

Viruses are carried to the body by all possible routes (air, food, bites, and any contaminated object). Similarly, all possible sites of implantation (all body surfaces and internal sites reached by mechanical penetration) may be used. The frequency of implantation is greatest where virus contacts living cells directly.

With some viruses, implantation in the fetus may occur at the time of fertilisation through infected germ cells, as well as later in gestation via the placenta, or at birth. Even at the earliest stage of pathogenesis (implantation), certain variables may influence the final outcome of the infection. For example, the dose, infectivity, and virulence of virus implanted and the location of implantation may determine whether the infection will be inapparent (subclinical) or will cause mild, severe, or lethal disease.

Local Replication and Local Spread

Successful implantation may be followed by local replication and local spread of virus. Virus that replicates within the initially infected cell may spread to adjacent cells extracellularly or intracellularly. Extracellular spread occurs by release of virus into the extracellular fluid and subsequent infection of the adjacent cell. Intracellular spread occurs by fusion of infected cells with adjacent, uninfected cells or by way of cytoplasmic bridges between cells.

Most viruses spread extracellularly, but herpesviruses, paramyxoviruses, and poxviruses may spread through both intracellular and extra cellular routes. Intracellular spread provides virus with a partially protected environment because the antibody defense does not penetrate cell membranes. Spread to cells beyond adjacent cells may occur through the liquid spaces within the local site (e.g., lymphatics) or by diffusion through surface fluids such as the mucous layer of the respiratory tract. Also, infected migratory cells such as lymphocytes and macrophages may spread the virus within local tissue.

Establishment of infection at the portal of entry may be followed by continued local virus multiplication, leading to localised virus shedding and localised disease. In this way, local sites of implantation also are target organs and sites of shedding in many infections. Respiratory tract infections that fall into this category include influenza, the common cold, and parainfluenza virus infections.

Alimentary tract infections caused by several gastroenteritis viruses (e.g., rotaviruses and picornaviruses) also may fall into this category. Localised skin infections of this type include warts, cowpox, and molluscum contagiosum. Localised infections may spread over body surfaces to infect distant surfaces.

Other viruses may spread internally to distant target organs and sites of excretion (disseminated infection). A third category of viruses may cause both local and disseminated disease, as in herpes simplex and measles.

DISSEMINATION FROM THE PORTAL OF ENTRY

Dissemination in the Bloodstream

At the portal of entry, multiplying virus contacts pathways to the blood and peripheral nerves, the principal routes of widespread dissemination through the body. The most common route of systemic spread of virus involves the circulation.

Viruses such as those causing poliomyelitis, smallpox, and measles disseminate through the blood after an initial period of replication at the portal of entry (the alimentary and respiratory tracts), where the infection often causes no significant symptoms or signs of illness because the virus kills cells that are expendable and easily replaced.

Virus progeny diffuse through the afferent lymphatics to the lymphoid tissue and then through the efferent lymphatics to infect cells in close contact with the bloodstream (e.g., endothelial cells, especially those of the lymphoreticular organs). This initial spread may result in a brief primary viremia. Subsequent release of virus directly into the bloodstream induces a secondary viremia, which usually lasts several days and puts the virus in contact with the capillary system of all body tissues.

Virus may enter the target organ from the capillaries by replicating within a capillary endothelial cell or fixed macrophage and then being released on the target organ side of the capillary. Virus may also diffuse through small gaps in the capillary endothelium or penetrate the capillary wall through an infected, migrating leukocyte.

The virus may then replicate and spread within the target organ or site of excretion by the same mechanisms as for local dissemination at the portal of entry. Disease occurs if the virus replicates in a sufficient number of essential cells and destroys them. For example, in poliomyelitis the central nervous system is the target organ, whereas the alimentary tract is both the portal of entry and the site of shedding. In some situations, the target organ and site of shedding may be the same.

Dissemination in Nerves

Dissemination through the nerves is less common than bloodstream dissemination, but is the means of spread in a number of important diseases. This mechanism occurs in rabies virus, herpesvirus, and, occasionally, poliomyelitis virus infections. Rabies virus implanted by a bite from a rabid animal replicates subcutaneously and within muscular tissue to reach nerve endings. Evidence indicates that the virus spreads centrally in the neurites (axons and dendrites) and perineural cells, where virus is shielded from antibody. This nerve route leads rabies virus to the central nervous system, where disease originates. Rabies virus then spreads centrifugally through the nerves to reach the salivary glands, the site of shedding.

Incubation Period

During most virus infections, no signs or symptoms of disease occur through the stage of virus dissemination. Thus, the incubation period extends from the time of implantation through the phase of dissemination, ending when virus replication in the target organs causes disease. Occasionally, mild fever and malaise occur during viremia, but they often are transient and have little diagnostic value.

The incubation period tends to be brief (1 to 3 days) in infections in which virus travels only a short distance to reach the target organ (i.e., in infections in which disease is due to virus replication at the portal of entry). Conversely, incubation periods in generalised infections are longer because of the stepwise fashion by which the virus moves through the body before reaching the target organs.

Other factors also may influence the incubation period. Generalised infections produced by togaviruses may have an unexpectedly short incubation period because of direct intravascular injection (insect bite) of a rapidly multiplying virus. The mechanisms governing the long incubation period (months to years) of persistent infections are poorly understood.

The persistently infected cell is often not lysed, or lysis is delayed. In addition, disease may result from a late immune reaction to viral antigen (e.g., arenaviruses in rodents), from unknown mechanisms in slow viral infections during which no immune response has been detected (as in the scrapie-kuru group), or mutation in the host genetic material resulting in cellular transformation and cancer.

Multiplication in Target Organs

Virus replication in the target organ resembles replication at other body sites except that (1) the target organ in systemic infections is usually reached late during the stepwise progression of virus through the body, and (2) clinical disease originates there. At each step of virus progression through the body, the local recovery mechanisms are activated. Thus, when the target organ is infected, the previously infected sites may have reached various stages of recovery.

Circulating interferon and immune responses probably account for the termination of viremia, but these responses may be too late to prevent seeding of virus into the target organ and into sites of shedding. Nevertheless, these systemic defenses can diffuse in various degrees into target organs and thereby help retard virus replication and disease. Depending on the balance between virus and host defenses, virus multiplication in the target organ may be sufficient to produce dysfunction manifested by disease or death.

Additional constitutional disease such as fever and malaise may result from diffusion of toxic products of virus replication and cell necrosis, as well as from release of lymphokines and other inflammatory mediators. Release of leukotriene C4 during respiratory infection may cause bronchospasm. Viral antigens also may participate in immune reactions, leading to disease manifestations. In addition, impairment of leukocytes and immuno-suppression by some viruses may cause secondary bacterial infection.

Shedding of Virus

Because of the diversity of viruses, virtually every possible site of shedding is utilised; however, the most frequent sites are the respiratory and alimentary tracts. Blood and lymph are sites of shedding for the arboviruses, since biting insects become infected by this route. HIV is shed in blood and semen. Milk is a site of shedding for viruses such as some RNA tumor viruses (retroviruses) and cytomegalovirus (a herpesvirus).

Several viruses (e.g., cytomegaloviruses) are shed simultaneously from the urinary tract and other sites more commonly associated with shedding. The genital tract is a common site of shedding for herpesvirus type 2 and may be the route through which the virus is transmitted to sexual partners or the fetus. Saliva is the primary source of shedding for rabies virus. Cytomegalovirus is also shed from these last two sites. Finally, viruses such as tumor viruses that are integrated into the DNA of host cells can be shed through germ cells.

Congenital Infections

Infection of the fetus is a special case of infection in a target organ. The factors that determine whether a target organ is infected also apply to the fetus, but the fetus presents additional variables. The immune and interferon systems of the very young fetus are immature. This immaturity, coupled with the partial placental barrier to transfer of maternal immunity and interferon, deprive the very young fetus of important defense mechanisms.

Another variable is the high vulnerability to disruption of the rapidly developing fetal organs, especially during the first trimester of pregnancy. Furthermore, susceptibility to virus replication may be modulated by the undifferentiated state of the fetal cells and by hormonal changes during pregnancy. Although virus multiplication in the fetus may lead to congenital anomalies or fetal death, the mother may have only a mild or inapparent infection.

To cause congenital anomalies, virus must reach the fetus and multiply in it, thereby causing maldeveloped organs. Generally, virus reaches the fetus during maternal viremia by infecting or passing through the placenta to the fetal circulation and then to fetal target organs. Sufficient virus multiplication may disrupt development of fetal organs, especially during their rapid development. Although many viruses occasionally cause congenital anomalies, cytomegalovirus and rubella virus are the most common offenders. Virus shedding by the congenitally infected newborn infant may occur as a result of persistence of the virus infection at sites of shedding.

VIRUSES AND HUMAN DISEASE

Examples of common human diseases caused by viruses include the common cold, influenza, chickenpox and cold sores. Many serious diseases such as ebola, AIDS, avian influenza and SARS are caused by viruses. The relative ability of viruses to cause disease is described in terms of virulence. Other diseases are under investigation as to whether they too have a virus as the causative agent, such as the possible connection between human herpes virus six (HHV6) and neurological diseases such as multiple sclerosis and chronic fatigue syndrome. There is current controversy over whether the borna virus, previously thought to cause neurological diseases in horses, could be responsible for psychiatric illnesses in humans.

Viruses have different mechanisms by which they produce disease in an organism, which largely depends on the viral species. Mechanisms at the cellular level primarily include cell lysis, the breaking open and subsequent death of the cell. In multicellular organisms, if enough cells die the whole organism will start to suffer the effects. Although viruses cause disruption of healthy homeostasis, resulting in disease, they may exist relatively harmlessly within an organism. An example would include the ability of the herpes simplex virus, which cause cold sores, to remain in a dormant state within the human body. This is called latency and is a characteristic of the all herpes viruses including the Epstein-Barr virus, which causes glandular fever, and the varicella zoster virus, which causes chicken pox. Latent chickenpox infections return in later life as the disease called shingles.

Some viruses can cause life-long or chronic infections, where the viruses continue to replicate in the body despite the hosts' defence mechanisms. This is common in hepatitis B virus and hepatitis C virus infections. People chronically infected are known as carriers, as they serve as reservoirs of infectious virus. In populations with a high proportion of carriers, the disease is said to be endemic.

Host Defence Mechanisms

The body's first line of defence against viruses is the innate immune system. This comprises cells and other mechanisms that defend the host from infection in a non-specific manner. This means that the cells of the innate system recognise, and respond to, pathogens in a generic way, but unlike the adaptive immune system, it does not confer long-lasting or protective immunity to the host.

RNA interference is an important innate defence against viruses. Many viruses have a replication strategy that involves double-stranded RNA (dsRNA). When such a virus infects a cell, it releases its RNA molecule or molecules, which immediately bind to a protein complex called dicer that cuts the RNA into smaller pieces. A biochemical pathway called the RISC complex is activated which degrades the viral mRNA and the cell survives the infection. Rotaviruses avoid this mechanism by not uncoating fully inside the cell and by releasing newly produced mRNA through pores in the particle's inner capsid. The genomic dsRNA remains protected inside the core of the virion.

When the adaptive immune system of a vertebrate encounters a virus, it produces specific antibodies which bind to the virus and render it non-infectious. This is called humoral immunity. Two types of antibodies are important. The first called IgM is highly effective at neutralizing viruses but is only produced by the cells of the immune system for a few weeks. The second, called, IgG is produced indefinitely. The presence of IgM in the blood of the host is used to test for acute infection, whereas IgG indicates an infection sometime in the past. IgG antibody is measured when tests for immunity are carried out.

A second defence of vertebrates against viruses is called cell-mediated immunity and involves immune cells known as T cells. The body's cells constantly display short fragments of their proteins on the cell's surface, and if a T cell recognises a suspicious viral fragment there, the host cell is destroyed by T killer cells and the virus-specific T-cells proliferate. Cells such as the macrophage are specialists at this antigen presentation. The production of interferon is an important host defence mechanism. This is a hormone produced by the body when viruses are present. Its role in immunity is complex, but it eventually stops the viruses from reproducing by killing the infected cell and its close neighbours

Not all virus infections produce a protective immune response in this way. HIV evades the immune system by constantly changing the amino acid sequence of the proteins on the surface of the virion. These persistent viruses evade immune control by sequestration, blockade of antigen presentation, cytokine resistance, evasion of natural killer cell activities, escape from apoptosis, and antigenic shift. Other viruses, called neurotropic viruses, are disseminated by neural spread where the immune system may be unable to reach them.

Prevention and Treatment

Because viruses use the machinery of a host cell to reproduce and reside within them, they are difficult to eliminate without killing the host cell. The most effective medical approaches to viral diseases so far are vaccinations to provide resistance to infection, and antiviral drugs.

Vaccines

Vaccination is a cheap and effective way of preventing infections by viruses. Vaccines were used to prevent viral infections long before the discovery of the actual viruses. Their use has resulted in a dramatic decline in morbidity (illness) and mortality (death) associated with viral infections such as polio, measles, mumps and rubella. Smallpox infections have been eradicated. Currently vaccines are available to prevent over thirteen viral infections of humans, and more are used to prevent viral infections of animals. Vaccines can consist of live-attenuated or killed viruses, or viral proteins (antigens). Live vaccines contain weakened forms of the virus that causes the disease. Such viruses are called attenuated. Live vaccines can be dangerous when given to people with a weak immunity, because in these people, the weakened virus can cause the original disease.

Biotechnology and genetic engineering techniques are used to produce subunit vaccines. These vaccines use only the capsid proteins of the virus. Hepatitis B vaccine is an example of this type of vaccine. Subunit vaccines are safe for immunocompromised patients because they cannot cause the disease. However, the yellow fever virus vaccine, a live-attenuated strain called 17D, is probably the safest and most effective vaccine ever generated.

Antiviral Drugs

Antiviral drugs are a class of medication used specifically for treating viral infections. Like antibiotics for bacteria, specific antivirals are used for specific viruses. Unlike antibiotics, however, antiviral drugs do not kill the virus, they only stall their development.

Antiviral drugs are one class of antimicrobials, a larger group which also includes antibiotic, antifungal and antiparasitic drugs. They are relatively harmless to the host, and therefore can be used to treat infections. They should be distinguished from viricides, which actively destroy virus particles outside the body.

Most of the antivirals now available are designed to help deal with HIV, herpes viruses (best known for causing cold sores and genital herpes, but actually causing a wide range of diseases), the hepatitis B and C viruses, which can cause liver cancer, and influenza A and B viruses. Researchers are now working to extend the range of antivirals to other families of pathogens. Designing safe and effective antiviral drugs is difficult, because viruses use the host's cells to replicate. This makes it difficult to find targets for the drug that would interfere with the virus without harming the host organism's cells.

The emergence of antivirals is the product of a greatly expanded knowledge of the genetic and molecular function of organisms, allowing biomedical researchers to understand the structure and function of viruses, major advances in the techniques for finding new drugs, and the intense pressure placed on the medical profession to deal with the human immunodeficiency virus (HIV), the cause of the deadly acquired immunodeficiency syndrome (AIDS) pandemic.

Almost all anti-microbials, includin; anti-virals, are subject to drug resistance as the pathogens mutate over time, becomin; less susceptible to the treatment.

The general idea behind modern antiviral drug design is to identify viral proteins, or parts of proteins, that can be disabled. These "targets" should generally be as unlike any proteins or parts of proteins in humans as possible, to reduce the likelihood of side effects. The targets should also be common across many strains of a virus, or even among different species of virus in the same family, so a single drug will have broad effectiveness. For example, a researcher might target a critical enzyme synthesized by the virus, but not the patient, that is common across strains, and see what can be done to interfere with its operation.

Once targets are identified, candidate drugs can be selected, either from drugs already known to have appropriate effects, or by actually designing the candidate at the molecular level with a computer-aided design program.

The target proteins can be manufactured in the lab for testing with candidate treatments by inserting the gene that synthesizes the target protein into bacteria or other kinds of cells. The cells are then cultured for mass production of the protein, which can then be exposed to various treatment candidates and evaluated with "rapid screening" technologies.

INFECTION IN OTHER SPECIES

Viruses infect all cellular life and, although viruses occur universally, each cellular species has its own specific range that often only infect that species. Viruses are important pathogens of livestock. Diseases such as Foot and Mouth Disease and bluetongue, are caused by viruses. Companion animals such as cats, dogs and horses, if not vaccinated, are susceptible to serious viral infections. Canine parvovirus is caused by a small DNA virus and infections are often fatal in pups. Like all invertebrates, the honey bee is susceptible to many viral infections. Fortunately, most viruses co-exist harmlessly in their host and cause no signs or symptoms of disease.

Plants

There are many types of plant virus, but often they only cause a loss of yield, and it is not economically viable to try to control them. Plant viruses are often spread from plant to plant by organisms, known as vectors. These are normally insects, but some fungi, nematode worms and single-celled organisms have been shown to be vectors. When control of plant virus infections is considered economical, (for perennial fruits for example), efforts are concentrated on killing the vectors and removing alternate hosts such as weeds. Plant viruses are harmless to humans and other animals because they can only reproduce in living plant cells.

Plants have elaborate and effective defence mechanisms against viruses. One of the most effective is the presence of so-called resistance (R) genes. Each R gene confers resistance to a particular virus by triggering localised areas of cell death around the infected cell, which can often be seen with the unaided eye as large spots. This stops the infection from spreading. RNA interference is also an effective defence in plants. When they are infected, plants often produce natural disinfectants which kill viruses, such as salicylic acid, nitric oxide and reactive oxygen molecules.

Bacteria

Bacteriophages are an extremely common and diverse group of viruses. For example, bacteriophages are the most common form of biological entity in aquatic environments, with up to ten times more of these viruses in the oceans than bacteria, reaching levels of 250,000,000 bacteriophages per millilitre of seawater. These viruses infect specific bacteria by binding to surface receptor molecules and then entering the cell. Within a short amount of time, in some cases just minutes, bacterial polymerase starts translating viral mRNA into protein. These proteins go on to become either new virions within the cell, helper proteins which help assembly of new virions, or proteins involved in cell lysis. Viral enzymes aid in the breakdown of the cell membrane, and, in the case of the T4 phage, in just over twenty minutes after injection over three hundred phages could be released. The major way bacteria defend themselves from bacteriophages is by producing enzymes which destroy foreign DNA. These enzymes, called restriction endonucleases, cut up the viral DNA that bacteriophages inject into bacterial cells. Bacteria also contain a system that uses CRISPR sequences to retain fragments of the genomes of viruses that the bacteria have come into contact with in the past, which allows them to block the virus's replication through a form of RNA interference. This genetic system provides bacteria with acquired immunity to infection.

Archaea

Some viruses replicate within archaea: these are double-stranded DNA viruses that appear to be unrelated to any other form of virus and have a variety of unusual shapes, with some resembling bottles, hooked rods, or teardrops. These viruses have been studied in most detail in the thermophilic archaea, particularly the orders Sulfolobales and Thermoproteales. Defences against these viruses may involve RNA interference from repetitive DNA sequences within archaean genomes that are related to the genes of the viruses.

REFERENCES

Almela MJ, González ME, Carrasco L (May 1991). "Inhibitors of poliovirus uncoating efficiently block the early membrane permeabilization induced by virus particles". *J. Virol.* 65 (5): 2572–7.

Bai J, Rossi J, Akkina R (March 2001). "Multivalent anti-CCR ribozymes for stem cell-based HIV type 1 gene therapy". *AIDS Res. Hum. Retroviruses* 17 (5): 385–99.

Beringer, Paul; Troy, David A.; Remington, Joseph P. 2006. *Remington, the science and practice of pharmacy.* Hagerstwon, MD: Lippincott Williams & Wilkins. pp. 1419.

Sodeik B, Griffiths G, Ericsson M, Moss B, Dom. 1994. "Assembly of vaccinia virus: effects of rifampin on the intracellular distribution of viral protein p65". *J. Virol.* 68 (2): 1103–14.

Stein DA, Skilling DE, Iversen PL, Smith. 2001. "Inhibition of Vesivirus infections in mammalian tissue culture with antisense morpholino oligomers". *Antisense Nucleic Acid Drug Dev.* 11 (5): 317–25. doi:10.1089/108729001753231696.

14
Basic Mycology

Mycology is the branch of biology concerned with the study of fungi, including their genetic and biochemical properties, their taxonomy, and their use to humans as a source for tinder, medicinals (e.g., penicillin), food (e.g., beer, wine, cheese, edible mushrooms), entheogens, as well as their dangers, such as poisoning or infection. From mycology arose the field of phytopathology, the study of plant diseases, and the two disciplines remain closely related because the vast majority of plant pathogens are fungi. A biologist who studies mycology is called a mycologist.Historically, mycology was a branch of botany (fungi are evolutionarily more closely related to animals than to plants but this was not recognized until a few decades ago). Pioneer mycologists included Elias Magnus Fries, Christian Hendrik Persoon, Anton de Bary and Lewis David von Schweinitz. Today, the most comprehensively studied and understood fungi are the yeasts and eukaryotic model organisms Saccharomyces cerevisiae and Schizosa-ccharomyces pombe.

Many fungi produce toxins, antibiotics, and other secondary metabolites. For example, the cosmopolitan (worldwide) genus Fusarium and their toxins associated with fatal outbreaks of alimentary toxic aleukia in humans were extensively studied by Abraham Joffe. Fungi are fundamental for life on earth in their roles as symbionts, e.g. in the form of mycorrhizae, insect symbionts and lichens, potency in breaking down complex organic biomolecules such as lignin, the more durable component of wood, and by playing a role in xenobiotics, a critical step in the global carbon cycle. Fungi and other organisms traditionally recognized as fungi, such as oomycetes and myxomycetes (slime molds), often are economically and socially important as some cause diseases of animals (such as histoplasmosis) as well as plants (such as Dutch elm disease and Rice blast).

STRUCTURE OF FUNGI

Fungi are eukaryotic microorganisms. Fungi can occur as yeasts, molds, or as a combination of both forms. Some fungi are capable of causing superficial, cutaneous, subcutaneous, systemic or allergic diseases. Yeasts are microscopic fungi consisting of solitary cells that reproduce by budding. Molds, in contrast, occur in long filaments known as hyphae, which grow by apical extension. Hyphae can be sparsely septate to regularly septate and possess a variable number of nuclei. Regardless of their shape or size, fungi are all heterotrophic and digest their food externally by releasing hydrolytic enzymes into their immediate surroundings (absorptive nutrition). Other characteristics of fungi are the ability to synthesize lysine by the L-a-adipic acid biosynthetic pathway and possession of a chitinous cell wall, plasma membranes containing the sterol ergosterol, 80S rRNA, and microtubules composed of tubulin.

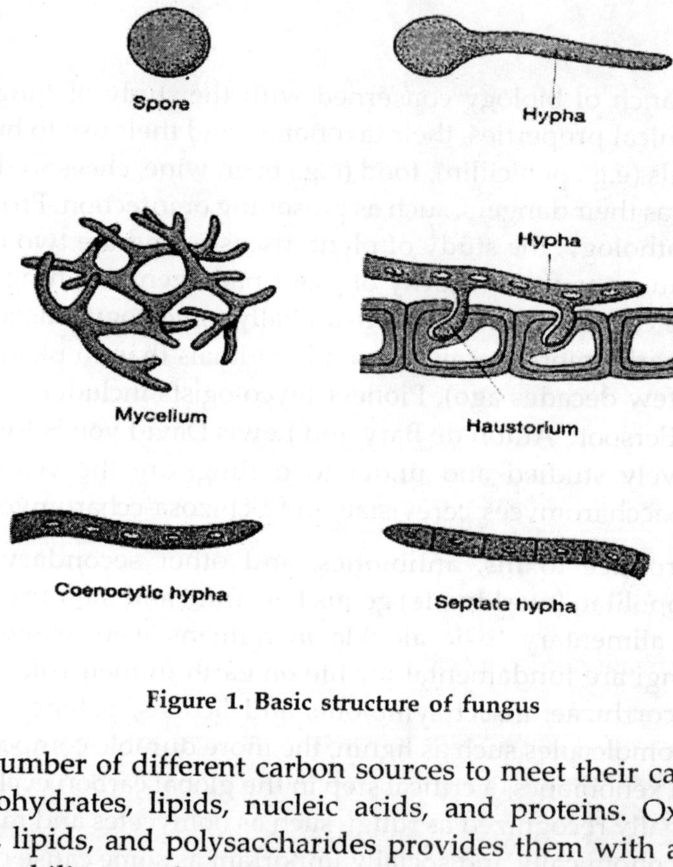

Figure 1. Basic structure of fungus

Fungi can use a number of different carbon sources to meet their carbon needs for the synthesis of carbohydrates, lipids, nucleic acids, and proteins. Oxidation of sugars, alcohols, proteins, lipids, and polysaccharides provides them with a source of energy. Differences in their abilty to utilize different carbon sources, such as simple sugars, sugar acids, and sugar alcohols, are used, along with morphology, to differentiate the various yeasts. Fungi require a source of nitrogen for synthesis of amino acids for proteins, purines and pyrimidines for nucleic acids, glucosamine for chitin, and various vitamins. Depending on the fungus, nitrogen may be obtained in the form of nitrate, nitrite, ammonium, or organic nitrogen; no fungus can fix nitrogen. Most fungi use nitrate, which is reduced first to nitrite (with the aid of nitrate reductase) and then to ammonia.

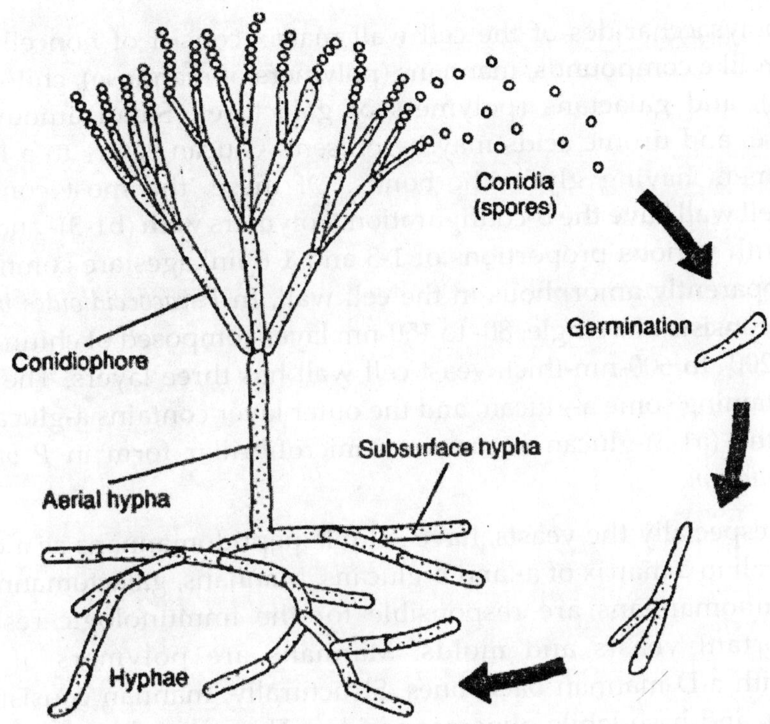

Figure 2. The microscopic structures of a septate fungus showing asexually producedconidia that leave the fungus and germinate to produce a new mycelium

Nonfungal organisms, including bacteria, synthesize the amino acid lysine by the *meso*-a,e-diaminopimelic acid pathway (DAP pathway), whereas fungi synthesize lysine by only the L-a-adipic acid pathway (AAA pathway). Use of the DAP pathway is one of the reasons microorganisms previously considered to be fungi, such as the myxomycetes, oomycetes, and hyphochytrids, are no longer classified as fungi. The DAP and AAA biosynthetic pathways for lysine synthesis represent dichotomous evolution.

Cell Wall

The rigid cell wall of fungi is a stratified structure consisting of chitinous microfibrils embedded in a matrix of small polysaccharides, proteins, lipids, inorganic salts, and pigments that provides skeletal support and shape to the enclosed protoplast. Chitin is a (b1-4)-linked polymer of *N*-acetyl-D-glucosamine (GlcNAc). It is produced in the cytosol by the transfer of GlcNAc from uridine diphosphate GlcNAc into chains of chitin by chitin synthetase, which is located in the cytosol in organelles called chitosomes. The chitin microfibrils are transported to the plasmalemma and subsequently integrated into the new cell wall.

The major polysaccharides of the cell wall matrix consist of noncellulosic glucans such as glycogen-like compounds, mannans (polymers of mannose), chitosan (polymers of glucosamine), and galactans (polymers of galactose). Small amounts of fucose, rhamnose, xylose, and uronic acids may be present. Glucan refers to a large group of D-glucose polymers having glycosidic bonds. Of these, the most common glucans composing the cell wall have the b-configuration. Polymers with (b1-3)- and (b1-6)-linked glucosyl units with various proportions of 1-3 and 1-6 linkages are common. Insoluble b-glucans are apparently amorphous in the cell wall. In *Paracoccidioides brasiliensis*, the hyphal cell wall consists of a single, 80- to 150-nm layer composed of chitin and b-glucan. In contrast, the 200- to 600-nm-thick yeast cell wall has three layers. The inner surface is chitinous, containing some a-glucan, and the outer layer contains a-glucan. It has been suggested that the (a1-3)-glucan occurs in a microfibrillar form in *P brasiliensis* and *Histoplasma capsulatum*.

Many fungi, especially the yeasts, have soluble peptidomannans as a component of their outer cell wall in a matrix of a- and b-glucans. Mannans, galactomannans, and, less frequently, rhamnomannans are responsible for the immunologic response to the medically important yeasts and molds. Mannans are polymers of mannose or heteroglucans with a-D-mannan backbones. Structurally, mannan consists of an inner core, outer chain, and base-labile oligomannosides. The outer-chain region determines its antigenic specificity. Determination of mannan concentrations in serum from patients with disseminated candidiasis has proven a useful diagnostic technique.

Cryptococcus neoformans produces a capsular polysaccharide composed of at least three distinct polymers: glucuronoxylomannan, galactoxylomannan, and mannoprotein. On the basis of the proportion of xylose and glucuronic acid residues, the degree to which mannose has side-chain substituents, and the percentage of O-acetyl attachments of the capsular polysaccharides, isolates of *C neoformans* can be separated into four antigenic groups designated A, B, C, and D. The capsule is antiphagocytic, serves as a virulence factor, persists in body fluids, and allows the yeast to avoid detection by the host immune system.

In addition to chitin, glucan, and mannan, cell walls may contain lipid, protein, chitosan, acid phosphatase, a-amylase, protease, melanin, and inorganic ions such as phosphorus, calcium, and magnesium. The outer cell wall of dermatophytes contains glycopeptides that may evoke both immediate and delayed cutaneous hypersensitivity. In the yeast *Candida albicans*, for example, the cell wall contains approximately 30 to 60 percent glucan, 25 to 50 percent mannan (mannoprotein), 1 to 2 percent chitin (located primarily at the bud scars in the parent yeast cell wall), 2 to 14 percent lipid, and 5 to

15 percent protein. The proportions of these components vary greatly from fungus to fungus.

Plasma Membrane

Fungal plasma membranes are similar to mammalian plasma membranes, differing in having the nonpolar sterol ergosterol, rather than cholesterol, as the principal sterol. The plasma membrane regulates the passage of materials into and out of the cell by being selectively permeable. Membrane sterols provide structure, modulation of membrane fluidity, and possibly control of some physiologic events.

The plasma membrane contains primarily lipids and protein, along with small quantities of carbohydrates. The major lipids are the amphipathic phospholipids and sphingolipids that form the lipid bilayer. The hydrophilic heads are toward the surface, and the hydrophobic tails are buried in the interior of the membrane. Proteins are interspersed in the bilayer, with peripheral proteins being weakly bound to the membrane. In contrast, integral proteins are tightly bound. The lipoprotein structure of the membrane provides an effective barrier to many types of molecules. Molecules cross the membrane by either diffusion or active transport. The site of interaction for most antifungal agents is the ergosterol in the membrane or its biosynthetic pathway. Polyene antifungal agents such as amphotericin B bind to ergosterol to form complexes that permit the rapid leakage of the cellular potassium, other ions, and small molecules. The loss of potassium results in the inhibition of glycolysis and respiration.

Several antifungal agents interfere with ergosterol synthesis. The first step in the synthesis of both ergosterol and cholesterol is demethylation of lanosterol. The necessary enzymes are associated with fungal microsomes, which contain an electron transport system analogous to the one in liver microsomes. Cytochrome P450 catalyzes the 1 4-a-demethylation of lanosterol, an essential step in the synthesis of ergosterol. The imidazole and triazole antifungal agents interfere with cytochrome P450-dependent 1 4-a-demethylase, which inhibits the formation of ergosterol. This results in plasma membrane permeability changes and inhibition of growth. Ergosterol may also be involved in regulating chitin synthesis. Inhibition of ergosterol synthesis by antifungal agents can result in a general activation of chitin synthetase zymogen, leading to excessive chitin production and abnormal growth.

Microtubules

Fungi possess microtubules composed of the protein tubulin. This protein consists of a dimer composed of two protein subunits. Microtubules are long, hollow cylinders approximately 25 nm in diameter that occur in the cytoplasm as a component of larger structures. These structures are involved in the movement of organelles, chromosomes, nuclei, and Golgi vesicles containing cell wall precursors.

Microtubules are the principal components of the spindle fibers, which assist in the movement of chromosomes during mitosis and meiosis. When cells are exposed to antimicrotubule agents, the movement of nuclei, mitochondria, vacuoles, and apical vesicles is disrupted. Griseofulvin, which is used to treat dermatophyte infections, binds with microtubule-associated proteins involved in the assembly of the tubulin dimers. By interfering with tubulin polymerization, griseofulvin stops mitosis at metaphase. The destruction of cytoplasmic microtubules interferes with the transport of secretory materials to the cell periphery, which may inhibit cell wall synthesis.

The fungal nucleus is bounded by a double nuclear envelope and contains chromatin and a nucleolus. Fungal nuclei are variable in size, shape, and number. The DNA and associated proteins occur as long filaments of chromatin, which condenses during nuclear division. The number of chromosomes varies with the particular fungus. Within the cell, 80 to 99 percent of the genetic material occurs in chromosomes as chromatin, and approximately 1 to 20 percent in the mitochondria. In some isolates of *Saccharomyces cerevisiae,* up to 5 percent of their DNA can be found in nuclear plasmids. When the DNA helix unwinds, one strand serves as the template for the synthesis of rRNA, tRNA, and mRNA. mRNA passes into the cytoplasm and attaches to one of the ribosomes, which are complexes of RNA and protein that serve as sites for the synthesis of protein.

YEASTS

Yeasts are fungi that grow as solitary cells that reproduce by budding. Yeast taxa are distinguished on the basis of the presence or absence of capsules, the size and shape of the yeast cells, the mechanism of daughter cell formation (conidiogenesis), the formation of pseudohyphae and true hyphae, and the presence of sexual spores, in conjunction with physiologic data. Morphology is used primarily to distinguish yeasts at the genus level, whereas the ability to assimilate and ferment various carbon sources and to utilize nitrate as a source of nitrogen are used in conjunction with morphology to identify species.

Yeasts such as *C albicans* and *Cryptococcus neoformans* produce budded cells known as blastoconidia. The formation of blastoconidia involves three basic steps: bud emergence, bud growth, and conidium separation. During bud emergence, the outer cell wall of the parent cell thins. Concurrently, new inner cell wall material and plasma membrane are synthesized at the site where new growth is occurring. New cell wall material is formed locally by activation of the polysaccharide synthetase zymogen. The process of bud emergence is regulated by the synthesis of these cellular components as well as by the turgor pressure in the parent cell. Mitosis occurs, as the bud grows, and both the developing conidium and the parent cell will contain a single nucleus. A ring of chitin forms between the developing blastoconidium and its parent yeast cell. This ring grows in to form a septum. Separation of the two cells leaves a bud scar on the parent cell wall. The bud scar contains much more chitin than does the rest of the parent cell wall. When the production of blastoconidia continues without separation of the conidia from each other, a pseudohypha, consisting of a filament of attached blastoconidia, is formed. In addition to budding yeast cells and pseudohyphae, yeasts such as C *albicans* may form true hyphae.

Candida Albicans

Candida albicans may form a budding yeast, pseudohyphae, germ tubes, true hyphae, and chlamydospores. A number of investigators are interested in germ tube formation because it represents a transition between a yeast and a mold. Generally, either low temperature or pH favors the development of a budding yeast. Other substances such as biotin, cysteine, serum transferrin, and zinc stimulate dimorphism in this yeast.

Approximately 20 percent of the C *albicans* yeast cell wall is mannan, whereas the mycelial cell wall contains a substantially smaller amount of this sugar. *Candida albicans* has three serotypes, designated A, B, and C. These are distinguished from each other on the basis of their mannans. The antigenic determinant for serotype A is its mannoheptaose side chain. In serotype B, it is the mannohexaose side chain. Serotype B tends to be more resistant to 5-fluorocytosine than is serotype A. Glucans with (b1-3)- and (b1-6)-linked groups compose about 50 to 70 percent of the yeast cell wall. It has been suggested that these glucans may impede the access of amphotericin B to the plasma membrane.

Molds

Molds are characterized by the development of hyphae, which result in the colony characteristics seen in the laboratory. Hyphae elongate by a process known as apical elongation, which requires a careful balance between cell wall lysis and new cell wall synthesis. Because molds are often differentiated on the basis of conidiogenesis, structures such as conidiophores and conidiogenous cells must be carefully evaluated. Some molds produce special sac-like cells called sporangia, the entire protoplasm of which becomes cleaved into spores called sporangiospores. Sporangia are typically formed on special hyphae called sporangiophores.

Dimorphism

A number of medically important fungi express themselves phenotypically as two different morphologic forms, which correlate with the saprophytic and parasitic modes of growth. Such fungi are called dimorphic fungi. Some researchers restrict the term to pathogens that grow as a mold at room temperature in the laboratory and as a budding yeast or as spherules either in tissue or at 37°C. In contrast, others use dimorphic for any fungus that can exist as two different phenotypes, regardless of whether it is pathogenic. We prefer to use the term "dimorphic" to describe fungi that typically grow as a mold in vitro and as either yeast cells or spherules *in vivo.* Examples of medically important dimorphic fungi include *Blastomyces dermatitidis* (hyphae and yeast cells) and *Coccidioides immitis* (hyphae and spherules).

A number of external factors contribute to the expresşion of dimorphism. Increased incubation temperature is the single most important factor. Increased carbon dioxide concentration, which probably affects the oxidation-reduction potential, enhances the conversion of the mycelial form to the tissue form in *C immitis* and *Sporothrix schenckii.* pH affects the development of the yeast form in some fungi, and cysteine or other sulfhydryl-containing compounds affects it in others. Some fungi require a combination of these factors to induced dimorphism.

Blastomyces dermatitidis

The conversion of the mycelial form of *Blastomyces dermatitidis* to the large, globose, thick-walled, broadly based budding yeast form requires only increased temperature. Hyphal cells enlarge and undergo a series of changes resulting in the transformation of these cells into yeast cells. The cells enlarge, separate, and then begin to reproduce by budding. The yeast cell wall contains approximately 95 percent (a1-3)-glucan and 5 percent (b1-3)-glucan. In contrast, the mycelial cell wall contains 60 percent (b1-3)-glucan and 40 percent (a1 -3)-glucan.

Coccidioides immitis

Coccidioides immitis is a unique dimorphic fungus because it produces spherules containing endospores in tissue, and hyphae at 25°C. Increased temperature, nutrition, and increased carbon dioxide are important for the production of sporulating spherules. A uninucleate arthroconidium begins to swell and undergo mitosis to produce additional nuclei. Once mitosis stops, initiation of spherule septation occurs. The spherule is segmented into peripheral compartments with a persistent central cavity. Uninucleate endospores occurring in packets enclosed by a thin membranous layer differentiate within the compartments. As the endospores enlarge and mature, the wall of the spherule ruptures to release the endospores. Pairs of closely appressed endospores that have not completely separated from each other may resemble the budding yeast cells of *B dermatitidis.*

Histoplasma capsulatum

Dimorphism in *Histoplasma capsulatum* involves three stages. In the first stage, induced by an increase in temperature, respiration ceases and the level of cytochromes decreases.

During the second stage of the mycelial-to-yeast conversion, cysteine or other sulfhydryl-containing compounds are required. Shunt pathways are initiated that restore the appropriate cytochrome levels, which supply the needed ATP. Cysteine is also required for the yeast form to grow. The final stage is characterized by normal cytochrome levels and respiration as the yeast grows and reproduces. The conversion of terminal or intercalary hyphal cells to a yeast form requires 3 to 14 days. In tissue, *H capsulatum* proliferates within giant cells.

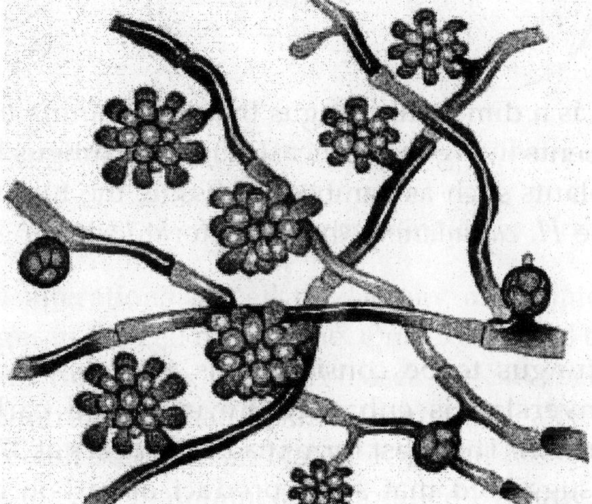

Figure 3. Histoplasma capsulatum

Yeast cells of *H capsulatum* have been divided into two chemotypes. Chemotype 1, which correlates with serotypes 1, 2, and 3, contains large amounts of (b1-3)-glucans and small amounts of chitin. Chemotype 2 contains (a1-3)-glucans, a little (b1-3)-glucan, and more chitin. Chemotype 2 correlates with serotypes 1 and 4. It is difficult to assess the importance of the chemotypes, because only a few isolates have been studied.

Paracoccidioides brasiliensis

A great deal of work has been done with the mycelial-to-yeast conversion of *Paracoccidioides brasiliensis*. In tissue the yeast is characterized by multiple budding. Series of smaller yeast daughter cells attached by narrow tubular necks are formed around a large central cell. Hyphal cells first swell and then separate from each other. The separated cells begin to bud, resulting in a yeast growth. As with *H capsulatum*, growth stops briefly as a result of increased temperature. At 37°C, synthesis of (b1-3)-glucan decreases and the hyphal cell wall softens. The hyphal cells then separate, and (a1-3)-glucans are formed as a layer on the outer cell wall surface of the yeast cells.

The yeast cell wall of *P brasiliensis* has three layers and is approximately 200 to 600 nm thick. The inner surface has chitin and some (b1-3)-glucan. The outer cell wall layer consists of (a1-3)-glucans. In contrast, the mycelial cell wall consists of one layer that is 80 to 150 nm thick, composed of chitin and (b1-3)-glucans. The (a1-3)-glucan of *P brasiliensis* is an important virulence factor. Only the yeast form has this glucan. When the a-linked glucan is absent, pathogenicity is attenuated; regeneration of a-glucan results in increased virulence.

Penicillium marneffei

Penicillium marneffei is a dimorphic fungus that is becoming extremely important as a pathogen in AIDS patients living in Southeast Asia. The fungus has been recovered from soil associated with plants such as bamboo. In tissue, the fungus forms yeast cells that divide by fission. Like *H. capsulatum*, they proliferate within giant cells.

Sporothrix schenckii

The last dimorphic fungus to be considered is *Sporothrix schenckii*. In this species, mycelial-to-yeast conversion is enhanced by increased carbon dioxide, increased temperature, and nutrition. The yeast form readily appears at 37°C and 5 percent carbon dioxide. It has been suggested that some product of carbon dioxide fixation may be required for development of the yeast form. Unlike the other dimorphic fungi capable of producing a yeast form, *S schenckii* initially produces yeast cells by direct budding from hyphae. In addition, the cell wall chemistry of the hyphal and yeast forms is similar. Glucans having (b1-3)-, (b1-4)-, and (b1-6)-linkages are present in addition to chitin. Rhamnomannan is the major antigenic determinant. The yeast cell wall is thought to contain more peptidorhamnomannan than the hyphal cell wall.

PROPAGULES

Spores can be produced either asexually or sexually. Asexual spores are always formed in a sporangium following mitosis and cytoplasmic cleavage. The number of sporangiospores and their arrangement in the sporangium are used to differentiate the various zygomycetes. Sexual spores occur following meiosis. Ascospores are formed in a saclike cell (called an ascus) by free-cell formation, basidiospores form on basidia, and zygospores form within zygosporangia. Oospores are sexual spores that are produced by one group of fungi that will not be considered because they are medically unimportant. Sexual spores are rarely seen in clinical isolates because most fungi are heterothallic (i.e., sexually self-sterile). Typically, only one of the two mating types is isolated from a particular clinical specimen. When homothallic isolates are recovered in the clinical laboratory, they often produce sexual spores because they are sexually self-fertile.

Conidia are always asexual in origin and develop in any manner that does not involve cytoplasmic cleavage. The ontogeny of conidia (conidiogenesis) and their arrangement, color, and septation are used to differentiate the various genera of molds. Some fungi have melanin in the cell wall of the conidia, the hyphae, or both. Such fungi are considered to be dematiaceous. Many of the name changes that have been recently proposed reflect a better understanding of conidiogenesis.

CLASSIFICATION

In mycology, fungi are classified on the basis of their ability to reproduce sexually, asexually, or by a combination of both. Asexual reproductive structures, which are referred to as anamorphs, are the basis for one of the sets of criteria. Because the criteria are based upon asexual morphologic forms, this system does not reflect phylogenetic relationships. It exists so that we can communicate in a simple and consistent manner by using names based upon similar morphologic structures. The second set of criteria is based upon sexual reproductive structures, which are referred to as teleomorphs. Ascospores, basidiospores, oospores, and zygospores, as well as any specialized structures associated with their development, are the basis of the second set of criteria. These criteria reflect phylogenetic relationships because they are based upon structures that form following meiosis. The term holomorph is used to describe the whole fungus, which consists of its teleomorph and anamorphs.

For example, the dimorphic fungus *Blastomyces dermatitidis* produces two anamorphs, one consisting of hyphae and one-celled conidia at 25°C and one consisting of budding yeast cells at 37°C. The name *B dermatitidis* summarizes these two anamorphs. When two sexually compatible isolates of *B dermatitidis* are mated under the appropriate conditions, a sexual fruiting body, called a gymnothecium, containing ascospores will develop. The name that is used for this sexual form or teleomorph is *Ajellomyces dermatitidis*. When one wishes to refer to the whole fungus, the name for the teleomorph is used because it reflects phylogenetic relationships. It is important to note that the name *B dermatitidis* can be used whenever one wishes to refer to the hyphal or yeast forms of this fungus.

REPRODUCTION

Sexual reproduction in the fungi typically involves fusion of two haploid nuclei (karyogamy), followed by meiotic division of the resulting diploid nucleus. In some cases, sexual spores are produced only by fusion of two nuclei of different mating types, which necessitates prior conjugation of different thalli. This condition of sexual reproduction is known as heterothallism, and the nuclear fusion is referred to as heterokaryosis. Normally plasmogamy (union of two hyphal protoplasts which brings the nuclei close together in the same cell) is followed almost immediately by karyogamy. In certain members of the Basidiomycotina, however, these two processes are separated in time and space, with plasmogamy resulting in a pair of nuclei (dikaryon) contained within a single cell. Karyogamy may be delayed until considerably later in the life history of the fungus. Meanwhile, growth and cell division of the binucleate cell occur. The development of a dikaryotic mycelium results from simultaneous division of the two closely associated nuclei and separation of the sister nuclei into two daughter cells. An alternative mechanism of sexual reproduction in the fungi is homothallism, in which a nucleus within the same thallus can fuse with another nucleus of that thallus (i.e., homokaryosis). An understanding of these nuclear cycles is fundamental to investigations of fungal genetics.

Some fungi are classified as strictly asexually reproducing forms. These include the large group of asexual (imperfect) yeasts (e.g., *Candida* species) and conidial fungi (e.g., *Coccidioides immitis*). Most members of this group have permanently lost their ability to produce meiospores. A few undergo rare sexual reproduction, and perhaps for some species we have yet to discover their sexual (perfect) stage. The most common methods of asexual reproduction, in addition to simple budding in yeasts, are blastic development of conidia from specialized hyphae (conidiogenous cells), fragmentation of hyphae into conidia, and conversion of hyphal elements into conidia or chlamydospores (thick-walled resting spores).

Despite the absence of meiosis during the life cycle of these imperfect fungi, recombination of hereditary properties and genetic variation still occur by a mechanism called parasexuality. The major events of this process include the production of diploid nuclei in a heterokaryotic, haploid mycelium that results from plasmogamy and karyogamy; multiplication of the diploid along with haploid nuclei in the heterokaryotic mycelium; sorting out of a diploid homokaryon; segregation and recombination by crossing over at mitosis; and haploidization of the diploid nuclei. Sexual and parasexual cycles are not mutually exclusive. Some fungi that reproduce sexually also exhibit parasexuality.

An extensive foundation of knowledge on the basic biology of fungi is at hand, including fungi that cause superficial, deep-seated, and systemic infections of humans and other animals. Much less is known, however, of the intricacies of interactions between these largely opportunistic pathogens and their hosts. Many areas of research in medical mycology are still in their infancy and offer formidable challenges and potential rewards. The current application of methods of recombinant DNA technology to problems of fungus-host interactions, especially the identification of pathogenicity genes, holds promise for significant contributions to our knowledge of medically important fungi.

DISEASE OF MECHANISMS OF FUNGI

Fungi are ubiquitous in nature and exist as free-living saprobes that derive no obvious benefits from parasitizing humans or animals. Since they are widespread in nature and are often cultured from diseased body surfaces, it may be difficult to assess whether a fungus found during disease is a pathogen or a transient environmental contaminant. Before a specific fungus can be confirmed as the cause of a disease, the same fungus must be isolated from serial specimens and fungal elements morphologically consistent with the isolate must be observed in tissues taken from the lesion.

In general, fungal infections and the diseases they cause are accidental. A few fungi have developed a commensal relationship with humans and are part of the indigenous microbial flora (e.g., various species of *Candida*, especially *Candida albicans*, and *Malassezia furfur*). Although a great deal of information is available concerning the molecular basis of bacterial pathogenesis, little is known about mechanisms of fungal pathogenesis. Infection is defined as entry into body tissues followed by multiplication of the organism.

The infection may be clinically inapparent or may result in disease due to a cellular injury from competitive metabolism, elaboration of toxic metabolites, replication of the fungus, or an immune response. Immune responses may be transient or prolonged and may be cell-mediated, humoral (with production of specific antibody to components of the infecting organism), or both. Successful infection may result in disease, defined as a deviation from or interruption of the normal structure or function of body parts, organs, or systems (or combinations thereof) that is marked by a characteristic set of symptoms and signs and whose etiology, pathology, and prognosis are known or unknown.

Entry

Fungi infect the body through several portals of entry. The first exposure to fungi that most humans experience occurs during birth, when they encounter the yeast *C. albicans* while passing through the vaginal canal. During this process the fungus colonizes the buccal cavity and portions of the upper and lower gastrointestinal tract of the newborn, where it maintains a life-long residence as a commensal.

Another fungus, *Malassezia furfur*, is common in areas of skin rich in sebaceous glands. How it colonizes the skin is not known, but both *M furfur* and *C albicans* are the only fungi that exist as commensals of humans and are considered part of the indigenous flora. Only under certain unusual circumstances have they caused disease. Other fungi that have been implicated in human diseases come from exogenous sources, where they exist as saprobes on decaying vegetation or as plant parasites.

Fungi rarely cause disease in healthy, immuno-competent hosts, even though we are constantly exposed to infectious propagules. It is only when fungi accidentally penetrate barriers such as intact skin and mucous membrane linings, or when immunologic defects or other debilitating conditions exist in the host, that conditions favorable for fungal colonization and growth occur.

When *C albicans*, for example, is implicated in disease processes, it may indicate that the patient has a coexisting immune, endocrine, or other debilitating disorder. In most cases, the underlying disorder must be corrected to effectively manage the fungal disease.

Adaptation and Propagation

Although most fungal diseases are the result of accidental encounters with the agent, many fungi have developed mechanisms that facilitate their multiplication within the host. For example, the dermatophytes that colonize skin, hairs, and nails elaborate enzymes that digest keratin. *Candida albicans* as a commensal organism exists in a unicellular yeastlike morphology, but when it invades tissues it becomes filamentous; conversely, the systemic fungi *Histoplasma capsulatum*, *Blastomyces dermatitidis*, and *Paracoccidoides brasiliensis* exist as molds in nature and change to a unicellular morphology when they cause disease. Other properties, such as capsule production by *C neoformans* and the adherence properties of *Candida* species to host tissues, also contribute to their pathogenicity. In general, the fungi that cause systemic disease must be able to grow and multiply at 37°C.

Dissemination

Disseminated fungal diseases usually indicate a breach in host defenses. Such a breach may be caused by endocrinopathies or immune disorders, or it may be induced iatrogenically. Effective management of the fungal infection requires a concerted effort to uncover and correct the underlying defects.

Host Factors

The high degree of innate resistance of humans to fungal invasion is based primarily on the various protective mechanisms that prevent fungi from entering host tissues. Fungal growth is discouraged by the intact skin and factors such as naturally occurring long-chain unsaturated fatty acids, pH competition with the normal bacterial flora, epithelial turnover rate, and the desiccated nature of the stratum corneum. Other body surfaces, such as the respiratory tree, gastrointestinal tract, and vaginal vault, are lined with mucous membranes (epithelium) bathed in fluids that contain antimicrobial substances, and some of these membranes are lined with ciliated cells that actively remove foreign materials. Only when these protective barriers are breached can fungi gain access to, colonize, and multiply in host tissues. Fungi gain access to host tissues by traumatic implantation or inhalation. The severity of disease caused by these organisms depends upon the size of the inoculum, magnitude of tissue destruction, the ability of the fungi to multiply in tissues, and the immunologic status of the host.

Fungal Factors

Most of the fungi that infect humans and cause disease are classified by tissue or organ levels that are primary sites of colonization. These are discussed below.

Superficial Fungal Infections

Superficial fungal infections involve only the outermost layers of the stratum corneum of the skin (*Phaeoannellomyces werneckii* [syn. *Exophiala werneckii*] and *M furfur*) or the cuticle of the hair shaft (*Trichosporon beigelii* and *Piedraia hortae*). These infections usually constitute cosmetic problems and rarely elicit an immune response from the host (except occasionally *M furfur* infections). Recently *T beigelii* and *M furfur* were implicated as opportunistic agents of disease, particularly in immunosuppressed or otherwise debilitated patients. Patients are accidentally infected with these common organisms via indwelling catheters or intravenous lines. Virtually nothing is known concerning the pathogenic mechanisms of these fungi.

Dermatophyte Infections

The dermatophytes are fungi that colonize skin, hair, and nails on the living host. These fungi possess greater invasive properties than those causing superficial infections, but they are limited to the keratinized tissues. They cause a wide spectrum of diseases that range from a mild scaling disorder to one that is generalized and highly inflammatory. Studies have shown that the disease-producing potential of these agents depends on various parasite and host factors, such as the species of organism, immunologic status of the host, type of clothing worn, and type of footwear used.

Trauma plays an important role in infection. These organisms gain entry and establish themselves in the cornified layers of traumatized or macerated skin and its integument and multiply by producing keratinase to metabolize the insoluble, tough fibrous protein. The reason why these agents spread no deeper is not known, but it has been speculated that factors such as cell-mediated immunity and the presence of transferrin in serum inhibit fungal propagation to the deeper tissue layers and systemic disease does not occur. Some dermatophytes have evolved a commensal relationship with the host and are isolated from skin in the absence of disease. Little is known about specific pathogenic mechanisms of the dermatophytes, but they do not cause systemic disease.

Subcutaneous Mycoses

The fungi that have been implicated in the subcutaneous mycoses are abundant in the environment and have a low degree of infectivity. These organisms gain access to the subcutaneous tissues through traumatic implantation. Again, little is known about mechanisms of pathogenesis. Histopathologic evidence indicates that these organisms survive in the subcutaneous tissue layers by producing proteolytic enzymes and maintaining a facultative microaerophilic existence because of the lowered redox potential of the damaged tissue. In eumycotic mycetoma there is extensive tissue damage and production of purulent fluid, which exudes through numerous intercommunicating sinus tracts. Microabscesses are common in chromoblastomycosis, but the clinical manifestation of disease indicates a vigorous host response to the organism, as seen by the intense tissue reaction that characterizes the disease (pseudoepitheliomatous hyperplasia).

Although most of the fungi implicated in this category of disease exist in a hyphal morphology, the agents of chromoblastomycosis and sporotrichosis are exceptions. Chromoblastomycosis is caused by a group of fungi that have several features in common. They are all darkly pigmented (dematiaceous) and exhibit a pleomorphism consisting of two distinct morphologies: the organism may exist in a mycelial state or as a thick-walled spherical cell that divides by cleavage. The latter cell morphology, called a muriform cell, sclerotic cell or Medlar body, is the pathologic morphology seen in tissue sections.

However, transition to the sclerotic morphology may not be a crucial requirement for pathogenesis. Several dematiaceous fungi cause a disease called phaeohyphomycosis, which clinically consists of a broad group of diseases characterized by the presence of various darkly pigmented yeastlike to hyphal elements, but not sclerotic cells, in pathologic specimens. Alternatively, the immune reaction of the host may dictate the morphology that the organism assumes. Again, there is no information about mechanisms or the role of morphogenesis in the pathogenesis of this group of fungi.

Sporotrichosis is caused by *Sporothrix schenckii,* which grows as a mold in nature or when cultured at 25°C, but as yeastlike cells when found in tissues. The clinical manifestations of disease caused by *S schenckii* vary, depending on the immune status of the patient. The classic condition, subcutaneous lymphanigitic sporotrichosis, is characterized by numerous nodules, abscesses, and ulcerative lesions that develop along the lymphatics that drain the primary site of inoculation. The disease does not extend beyond the regional lymph nodes that drain the site of the original infection. Alternatively, infection may result in solitary lesions or pulmonary disease. Clinical manifestations of pulmonary infections vary depending on the immune status of the patient. The immunocompetent individual has a high degree of innate resistance to disease, and when infection occurs the organism is often a secondary colonizer of old infarcted or healed cavities of the lungs. If the patient is immunocompromised, dissemination can occur. There is no information about mechanisms of pathogenesis of this dimorphic fungus.

Systemic Mycoses

Of all the fungi that have been implicated in human disease, only the six agents that cause the systemic mycoses have the innate ability to cause infection and disease in humans and other animals. The primary site of infection is the respiratory tract.

Conidia and other infectious particles are inhaled and lodge on the mucous membrane of the respiratory tree or in the alveoli, where they encounter macrophages and are phagocytosed. To successfully colonize the host these organisms must be able to survive at the elevated temperature of the body and either elude phagocytosis, neutralize the hostility they encounter, or adapt in a manner that will allow them to multiply.

Several factors contribute to infection and pathogenesis of these organisms. Of the six systemic agents, five, *Histoplasma capsulatum, Blastomyces dermatitidis, Paracoccidioides brasiliensis, Coccidioides immitis,* and *Penicillium marneffei* are dimorphic, changing from a mycelial to a unicellular morphology when they invade tissues, except *C immitis* that forms spherules. The change from mycelial to yeast morphology in *H. capsulatum* appears critical for pathogenicity.

Several physiologic changes occur in the fungus during the transition, which is induced by the temperature shift to 37°C. The triggering event is a heat-related insult: the temperature rise causes a partial uncoupling of oxidative phosphorylation and a consequent decline in the cellular ATP level, respiration rate, and concentrations of electron transport components.

The cells enter a period of dormancy, during which spontaneous respiration is maintained at a decreased level. Then there is a shift into a recovery phase, during which transformation to yeast morphology is completed. Mycelial cells of *H capsulatum* that are unable to undergo this morphologic transition are avirulent. Similar observations have been made when mycelia of *B dermatitidis* and *P brasiliensis* are shifted from 25°C to 37°C, and it has been implied that transformation to the yeast morphology is critical for infection.

Coccidioides immitis is also dimorphic, but its parasitic phase is a spherule. Little is known about the role of morphologic transformation in infection and disease of this organism. Dimorphism does not appear to play a role in *C neoformans* pathogenesis since the organism is an encapsulated yeast both at 25°C and in host tissues. The sexual phase of *C neoformans, Filobasidiella neoformans,* is known, and the organism assumes a filamentous morphology, producing small basidiospores. It has been suggested that these propagules are relevant in infection.

In addition to adjustment to the elevated temperature of the host, the infectious propagules must deal with the hostile cellular environment of the lungs. Studies with mutants of *C neoformans* have shown that the acidic mucopolysaccharide capsule is important in pathogenesis. Acapsular variants of the yeast are either avirulent or markedly deficient in pathogenicity. Since these mutants were obtained by mutagenesis, it is difficult to rule out the contribution of other genetic defects to their decreased pathogenicity. However, at the cellular level, the capsular polysaccharide inhibits phagocytosis of the yeast. Encapsulated *C neoformans* cells are highly resistant to phagocytosis by human neutrophils, whereas acapsular variants are effectively phagocytosed. The active component of the capsular polysaccharide has been identified as glucoronoxylomannan. In addition, the capsular polysaccharide is poorly immunogenic in humans and laboratory animals, and the glucoronoxylomannan component persists for extended periods in the host.

In addition to the capsular polysaccharide, elaboration of phenyl oxidase (an enzyme that catalyzes the oxidation of various phenols to dopachrome) by *C neoformans* appears to be a determinant of virulence, although the role of this enzyme in virulence is unknown. The infectious propagules of *H capsulatum, B dermatitidis, P brasiliensis,* and *C immitis* are readily phagocytosed by alveolar macrophages. To survive phagocytosis and to multiply, these fungi must neutralize the effects of the phagocytes. The production of reactive oxygen metabolites by phagocytic cells is an important host defense against microorganisms.

Studies have shown that the yeast phase of *H capsulatum* fails to trigger release of reaction oxygen metabolites in unprimed murine macrophages despite extensive phagocytosis. How they avoid destruction by the fungicidal mechanisms within lysosomes is unclear. Arthroconidia of *C immitis* inhibit phagosome-lysosome fusion and survive within normal murine peritoneal macrophages. Phagosome-lysosome fusion takes place after *H capsulatum* infection, but the yeast cells survive in the phagolysosome. It has been speculated that the fungus neutralizes the fungicidal components of the lysosome by a mechanism not yet elucidated.

There is very little information about mechanisms of fungal pathogenicity, in contrast to what is known about molecular mechanisms of bacterial pathogenesis. Fungal pathogenesis is complex and involves the interplay of many factors. Studies to elucidate these mechanisms are needed because of the increasing incidence of opportunistic infections.

SPECTRUM OF MYCOSES

Fungal infections or mycoses cause a wide range of diseases in humans. Mycoses range in extent from superficial infections involving the outer layer of the stratum corneum of the skin to disseminated infection involving the brain, heart, lungs, liver, spleen, and kidneys. The range of patients at risk for invasive fungal infections continues to expand beyond the normal host to encompass patients with the acquired immunodeficiency syndrome; those immunosuppressed due to therapy for cancer and organ transplantation, and those undergoing major surgical procedures. Each of these patient populations has a high risk of developing invasive fungal infections. As the population at risk continues to expand so also does the spectrum of opportunistic fungal pathogens infecting these patients also continue to increase. Many of the deeply invasive mycoses are difficult to diagnose early and often difficult to treat effectively. The development of new approaches to diagnosis and treatment of invasive fungal infections is the subject of intensive research.

Concepts of Classification

Fungal infections may be classified according to the site of infection, route of acquisition, and type of virulence. When classified according to the site of infection, fungal infections are designated as superficial, cutaneous, subcutaneous, and deep. Superficial mycoses are limited to the stratum corneum and essentially elicit no inflammation. Cutaneous infections involve the integument and its appendages, including hair and nails. Infection may involve the stratum corneum or deeper layers of the epidermis. Inflammation of the skin is elicited by the organism or its products. Subcutaneous mycoses include a range of different infections characterized by infection of the subcutaneous tissues usually at the point of traumatic inoculation. An inflammatory response develops in the subcutaneous tissue frequently with extension into the epidermis. Deep mycoses involve the lungs, abdominal viscera, bones and or central nervous system. The most common portals of entry are the respiratory tract, gastrointestinal tract, and blood vessels.

When classified according to the route of acquisition, a fungal infection may be designated as exogenous or endogenous in origin. If classified as exogenous, an infecting organism may be transmitted by airborne, cutaneous, or percutaneous routes. An endogenously-acquired fungal infection may be acquired from colonization or reactivation of a fungus from a latent infection. Fungi may be classified also according to virulence, as primary pathogens or as opportunistic pathogens. A primary pathogen may establish infection in an immunologically normal host; whereas, an opportunistic pathogen requires some compromise of host defenses in order for infection to become established.

Superficial and Cutaneous Mycoses

Superficial Mycoses include the following fungal infections and their etiological agent: black piedra (*Piedraia hortae*), white piedra (*Trichosporon beigelii*), pityriasis versicolor (*Malassezia furfur*), and tinea nigra (*Phaeoannellomyces werneckii*). Pityriasis versicolor is a common superficial mycosis, which is characterized by hypopigmentation or hyperpigmentation of skin of the neck, shoulders, chest, and back. Pityriasis versicolor is due to *Malassezia furfur* which involves only the superficial keratin layer. Black piedra is a superficial mycosis due to *Piedraia hortae* which is manifested by a small firm black nodule involving the hair shaft. By comparison, white piedra due to *T beigelii* is characterized by a soft, friable, beige nodule of the distal ends of hair shafts. Tinea nigra most typically presents as a brown to black silver nitrate-like stain on the palm of the hand or sole of the foot.

Cutaneous Mycoses may be classified as dermatophytoses or dermatomycoses. Dermatophytoses are caused by the agents of the genera *Epidermophyton, Microsporum,* and *Trichophyton*. Dermatomycoses are cutaneous infections due to other fungi, the most common of which are *Candida* spp. The dermatophytoses are characterized by an anatomic site-specificity according to genera. For example, *Epidermophyton floccosum* infects only skin and nails, but does not infect hair shafts and follicles. Whereas, *Microsporum* spp. infect hair and skin, but do not involve nails. *Trichophyton* spp. may infect hair, skin, and nails.

Subcutaneous Mycoses

There are three general types of subcutaneous mycoses: chromoblastomycosis, mycetoma, and sporotrichosis. All appear to be caused by traumatic inoculation of the etiological fungi into the subcutaneous tissue. Chromoblastomycosis is a subcutaneous mycosis characterized by verrucoid lesions of the skin (usually of the lower extremities); histological examination reveals muriform cells (with perpendicular septations) or so-called "copper pennies" that are characteristic of this infection. Chromoblastomycosis is generally limited to the subcutaneous tissue with no involvement of bone, tendon, or muscle. By comparison, mycetoma is a suppurative and granulomatous subcutaneous mycosis, which is destructive of contiguous bone, tendon, and skeletal muscle. Mycetoma is characterized by the presence of draining sinus tracts from which small but grossly visible pigmented grains or granules are extruded. These grains are microcolonies of fungi causing the infection.

Chromoblastomycosis and mycetoma are caused by only certain fungi. The most common causes of chromoblastomycosis are *Fonsecaea pedrosoi*, *Fonsecaea compacta*, *Cladosporium carionii*, and *Phialophora verrucosa*. The causes of mycetoma are more diverse but can be classified as eumycotic and actinomycotic mycetoma. Within the United States, the most common agent of eumycotic mycetoma is *Pseudallescheria boydii* and the most common cause of actinomycotic mycetoma is *Nocardia brasiliensis.*

Many of the fungi causing mycetoma are pigmented brown to black. These organisms are known as dematiaceous (melanized) fungi. The melanin pigment is deposited in the cell walls of these organisms. These fungi may produce a range of infections from superficial to subcutaneous to deep (visceral) infection characterized by the presence of dematiaceous hyphal and/or yeast-like cells in tissue..Such deep infections due to dematiaceous fungi are termed phaeohyphomycosis.

Sporotrichosis is the third general class of subcutaneous mycoses. This infection is due to *Sporothrix schenckii* and involves the subcutaneous tissue at the point of traumatic inoculation. The infection usually spreads along cutaneous lymphatic channels of the extremity involved.

Deep Mycoses

Deep mycoses are caused by primary pathogenic and opportunistic fungal pathogens. The primary pathogenic fungi are able to establish infection in a normal host; whereas, opportunistic pathogens require a compromised host in order to establish infection (e.g., cancer, organ transplantation, surgery, and AIDS). The primary deep pathogens usually gain access to the host via the respiratory tract. Opportunistic fungi causing deep mycosis invade via the respiratory tract, alimentary tract, or intravascular devices.

The primary systemic fungal pathogens include *Coccidioides immitis, Histoplasma capsulatum, Blastomyces dermatitidis,* and *Paracoccidioides brasiliensis.* The opportunistic fungal pathogens include *Cryptococcus neoformans, Candida* spp., *Aspergillus* spp., *Penicillium marneffei,* the Zygomycetes, *Trichosporon beigelii,* and *Fusarium* spp.

Dimorphism in the Pathogenic Fungi

Fungal dimorphism is the morphological and physiological conversion of certain fungi from one phenotype to another when such fungi change from one environment to another. Dimorphic fungi include *C in........, H capsulatum, B dermatitidis, P brasiliensis, P marneffei,* and *S schenckii,* and certain opportunistic fungi such as *Candida albicans* and *Penicillium marneffei.*

Various environmental host factors control fungal dimorphism. These factors include amino acids, temperature, carbohydrates, and trace elements (e.g. zinc). Among the primary pathogens and *S schenckii,* the morphological transformation is from a hyphal form to a yeast-like form (or spherule in the case of *C immitis*) in tissue. However, the dimorphism of *Candida albicans* is somewhat different in that the organism transforms from a budding yeast-like structures (blastoconidia) to filamentous structures known as germ tubes. Other filamentous structures may later develop as pseudohyphae and hyphae. *Penicillium marneffei* is unique in being the only *Penicillium* species pathogenic to humans. It undergoes dimorphic conversion in vivo to transversely dividing sausage-shaped cells.

Primary Mycoses

Most cases of primary deep mycoses are asymptomatic or clinically mild infections occurring in normal patients living or traveling in endemic areas. However, patients exposed to a high inoculum of organisms or those with altered host defenses may suffer life-threatening progression or reactivation of latent foci of infection.

The arthroconidia of *C immitis* are inhaled and convert in the lung to spherules. Most cases of coccidioidomycosis are clinically occult or mild infections in patients who inhale infective arthroconidia. However, some patients have progressive pulmonary infection and also may suffer dissemination to the brain, bone, and other sites. *Coccidioides* meningitis is a life-threatening infection requiring lifelong treatment.

Histoplasmosis is a primary pulmonary infection resulting from inhalation of conidia of *Histoplasma capsulatum* which convert in vivo into the blastoconidial (budding yeast) form.

Dissemination to the hilar and mediastinal lymph nodes, spleen, liver, bone marrow, and brain may be life-threatening in infants and other immunocompromised patients. Histoplasmosis (like tuberculosis) is characterized by intracellular growth of the pathogen in macrophages and a granulomatous reaction in tissue. These granulomatous foci may reactivate and cause dissemination of fungi to other tissues. These patterns of primary infection and reactivation are similar to those of *Mycobacterium tuberculosis*. Histoplasmosis also may be associated with a chronic inflammatory process known as fibrosing mediastinitis, where scar tissue (formed in response to *H capsulatum*) encroaches on vital structures in the mediastinum.

Blastomycosis, similar to histoplasmosis, is a primary pulmonary infection resulting from inhalation of conidia from the mycelial phase of *Blastomyces dermatitidis* which convert in vivo to the parasitic yeast phase. Blastomycosis (due to *B dermatitidis*) in the blastoconidial phase also causes a primary pulmonary infection. The organism elicits a granulomatous reaction often associated with a marked fibrotic reaction. The clinical pattern of pulmonary blastomycosis is one of chronic pneumonia. Dissemination occurs most commonly to the skin, bone, and, in males, prostate.

Opportunistic Mycoses

Candidiasis. Candidiasis (due to *C albicans* and other *Candida* spp.) is the most common opportunistic fungal infection. *Candida albicans* is the most common cause of candidiasis. Candidiasis may be classified as superficial or deep. Superficial candidiasis may involve the epidermal and mucosal surfaces, including those of the oral cavity, pharynx, esophagus, intestines, urinary bladder, and vagina. The alimentary tract and intravascular catheters are the major portals of entry for deep (or visceral) candidiasis. The kidneys, liver, spleen, brain, eyes, heart, and other tissues are the major organ sites involved in deep or visceral candidiasis. The principal risk factors predisposing to deeply invasive candidiasis are protracted courses of broad spectrum antibiotics, cytotoxic chemotherapy, corticosteroids, and vascular catheters.

Aspergillosis. Invasive aspergillosis most frequently involves the lungs and paranasal sinuses. This fungus may disseminate from the lungs to involve the brain, kidneys, liver, heart, and bones. The main portal of entry for aspergillosis is the respiratory tract, however, injuries to the skin may also introduce the organism into susceptible hosts. Quantitative and functional defects in circulating neutrophils are key risk factors for development of invasive aspergillosis. For example, neutropenia due to cytotoxic chemotherapy and systemic corticosteroids are common predisposing factors for invasive aspergillosis.

Zygomycosis. Zygomycosis due to *Rhizopus, Rhizomucor, Absidia, Mucor* species, or other members of the class of Zygomycetes, also causes invasive sinopulmonary infections. An especially life-threatening form of zygomycosis (also known as mucormycosis), is known as the rhinocerebral syndrome, which occurs in diabetics with ketoacidosis. In addition to diabetic ketoacidosis, neutropenia and corticosteroids are other major risk factors for zygomycosis.

Cryptococcosis. Cryptococcosis is most typically an opportunistic fungal infection that most frequently causes pneumonia and/or meningitis. Defective cellular immunity, especially that associated with the acquired immune deficiency syndrome, is the most common risk factor for developing cryptococcosis.

Phaeohyphomycosis. Phaeohyphomycosis is an infection by brown to black pigmented fungi of the cutaneous, superficial, and deep tissues, especially brain. These infections are uncommon, life-threatening, and occur in various immunocompromised states.

Hyalohyphomycosis. Hyalohyphomycosis is an opportunistic fungal infection caused by any of a variety of normally saprophytic fungi with hyaline hyphal elements. For example, *Fusarium* spp. infect neutropenic patients to cause pneumonia, fungemia, and disseminated infection with cutaneous lesions.

Environmental Epidemiology

The epidemiology of dimorphic primary pathogens may be contrasted with that of the opportunistic fungal pathogens. The primary pathogens have a relatively well-defined geographic range of endemic infection in immunocompromised hosts. By comparison, the opportunistic fungi (e.g. *Aspergillus* spp.) are ubiquitously distributed with the frequency of infection being dependent upon a population of immunocompromised hosts. *Penicillium marneffei*, an opportunistic pathogen, appears to be geographically restricted to the East Asia, particularly Thailand and China.

Control and Treatment

Hospital-acquired fungal infections may be reduced by maintaining the lowest possible concentration of fungal spores in the ambient air of the institution. Ideally, a "spore-free" environment should be sought. Antifungal therapy, which is reviewed in depth elsewhere, is an area of intense investigation. New antifungal compounds will hopefully improve the efficacy and reduce toxicity of treatment of invasive fungal infections.

ANTIFUNGAL AGENTS

The development of antifungal agents has lagged behind that of antibacterial agents. This is a predictable consequence of the cellular structure of the organisms involved. Bacteria are prokaryotic and hence offer numerous structural and metabolic targets that differ from those of the human host. Fungi, in contrast, are eukaryotes, and consequently most agents toxic to fungi are also toxic to the host. Furthermore, because fungi generally grow slowly and often in multicellular forms, they are more difficult to quantify than bacteria. This difficulty complicates experiments designed to evaluate the *in vitro* or *in vivo* properties of a potential antifungal agent.

Despite these limitations, numerous advances have been made in developing new antifungal agents and in understanding the existing ones. The polyene compounds are so named because of the alternating conjugated double bonds that constitute a part of their macrolide ring structure. The polyene antibiotics are all products of *Streptomyces* species. These drugs interact with sterols in cell membranes (ergosterol in fungal cells; cholesterol in human cells) to form channels through the membrane, causing the cells to become leaky. The polyene antifungal agents include nystatin, amphotericin B, and pimaricin.

Amphotericin B is the mainstay antifungal agent for treatment of life-threatening mycoses and for most other mycoses, with the possible exception of the dermatophytoses. Discovered by Gold in 1956, it can truly be said to represent a gold standard. Its broad spectrum of activity includes most of the medically important moulds and yeasts, including dimorphic pathogens such as *Coccidioides immitis, Histoplasma capsulatum, Blastomyces dermatitidis, and Paracoccidioides brasiliensis*. It is the drug of choice in treating most opportunistic mycoses caused by fungi such as *Candida* species,

Cryptococcus neoformans, Aspergillus species, and the Zygomycetes. Resistance to this agent is rare, but is noteworthy for *Pseudallescheria boydii, Fusarium* spp., *Trichosporon* spp., certain isolates of *Candida lusitaniae* and *Candida guilliermondii.*

The drug must be administered intravenously and is associated with numerous side effects, ranging from phlebitis at the infusion site and chills to renal toxicity, which may be severe. A major advance in the use of this agent has resulted from an understanding of the mechanism of its renal toxicity, which is presumed to involve tubuloglomerular feedback. The suppression of glomerular filtration can be reduced by administering sodium chloride.

Nystatin was the first successful antifungal antibiotic to be developed, and it is still in general use. It is representative of the polyene antifungal agents developed later. The promise of its broad-spectrum antifungal activity is offset by host toxicity. Therefore, it is limited to topical use, where it has activity against yeasts such as the *Candida* species.

Pimaricin (natamycin), another polyene, is used topically to treat superficial mycotic infections of the eye. It is active against both yeasts and moulds.

Azole Antifungal Drugs

The azole antifungal agents have five-membered organic rings that contain either two or three nitrogen molecules (the imidazoles and the triazoles respectively). The clinically useful imidazoles are clotrimazole, miconazole, and ketoconazole. Two important triazoles are itraconazole and fluconazole. In general, the azole antifungal agents are thought to inhibit cytochrome P450-dependent enzymes involved in the biosynthesis of cell membrane sterols.

Ketoconazole set the stage for the orally administered antifungal azoles. It can be administered both orally and topically and has a range of activity including infections due to *H capsulatum* and *B dermatitidis*, for which it is often used in nonimmunocompromised patients. It is also active against mucosal candidiasis and a variety of cutaneous mycoses, including dermatophyte infections, pityriasis versicolor, and cutaneous candidiasis. It is not indicated for treatment of aspergillosis or of systemic infections caused by yeasts.

The triazoles (fluconazole, itraconazole) have become the standard for the azoles, and have replaced amphotericin B for managing certain forms of the systemic mycoses. Fluconazole is now routinely used to treat candidemia in non-neutropenic hosts, and is gaining acceptance for use in cryptococcosis and selected forms of coccidioidomycosis. Itraconazole has proven to be effective for histoplasmosis, blastomycosis, sporotrichosis, coccidioidomycosis, consolidation treatment for cryptococcosis, and certain forms of aspergillosis. Fluconazole can be administered either orally, or intravenously. The licensed formulation for itraconazole is oral, but an intravenous formulation is under study, and could be a significant addition directed at bioavailability problems relating to absorption of the oral formulation.

Side effects are not as common with the azoles as with amphotericin B, but life-threatening liver toxicity can arise with long-term use. Liver toxicity noted with ketoconazole has been less problematic with the triazoles. Other side effects include nausea and vomiting. Drug interactions are a potential problem between azoles and other drug classes and include cyclosporin, certain antihistamines, anticoagulants, and antiseizure, oral hypoglycemic and other medications that are metabolized via similar pathways in the liver.

5-Fluorocytosine

In contrast to the situation with antibacterial agents, few antimetabolites are available for use against fungi. The best example is 5-fluorocytosine, a fluorinated analog of cytosine. It inhibits both DNA and RNA synthesis via intracytoplasmic conversion to 5-fluorouracil. The latter is converted to two active nucleotides: 5-fluorouridine triphosphate, which inhibits RNA processing, and 5-fluorodeoxyuridine monophosphate, which inhibits thymidylate synthetase and hence the formation of the deoxythymidine triphosphate needed for DNA synthesis. As with other antimetabolites, the emergence of drug resistance is a problem. Therefore, 5-fluorocytosine is seldom used alone. In combination with amphotericin B it remains the treatment of choice for cryptococcal meningitis and is effective against a number of other mycoses, including some caused by the dematiaceous fungi and perhaps even by *C albicans*.

Other Antifungal Agents

Griseofulvin is an antifungal antibiotic produced by *Penicillium griseofulvum*. It is active *in vitro* against most dermatophytes and has been the drug of choice for chronic infections caused by these fungi (e.g., nail infections with *Trichophyton rubrum*) since it is orally administered and presumably incorporated into actively growing tissue. It is still used in such instances but is being challenged by some of the newer azole antifungal agents. The drug inhibits mitosis in fungi.

Potassium iodide given orally as a saturated suspension is uniquely used to treat cutaneous and lymphocutaneous sporotrichosis. This compound, interestingly, is not active against *Sporothrix schenckii in vitro*. It appears to act by enhancing the transepidermal elimination process in the infected host.

Two other classes of antifungal agents represent new additions to topical treatment of the dermatomycoses in Europe. The two allylamines (naftifine and terbinafine) inhibit ergosterol synthesis at the level of squalene epoxidase; one morpholene derivative (amorolfine) inhibits at a subsequent step in the ergosterol pathway.

Selection of Antifungal Agents

In vitro susceptibility testing with the fungi is not yet standardized, and the results of *in vitro* tests do not always compare to the results obtained *in vivo*. Therefore, preliminary selection of an antifungal agent for clinical use is made primarily on the basis of the specific fungal pathogen involved. The spectrum of activity for the licensed antifungal agents is well defined through the results of preclinical and clinical testing with the most common fungal pathogens. This approach is useful in avoiding selection of antifungals for species of fungi that are known to have primary resistance to the agent, but less useful in selecting antifungals for species that are known to develop secondary (drug induced) resistance to a particular agent.

Antifungal drug resistance has become an increasing problem with the development of a larger compendium of antifungal agents. Drug resistance to the polyene antifungals is almost always primary resistance rather than secondary resistance. That is, the susceptibility profiles for the species are characteristic and inherent, and rarely change in response to exposure to the agent. For example, amphotericin B-resistant species such as *Pseudallescheria boydii* and *Candida lusitaniae* are well known, and do not appear to have originated from exposure to the antifungal. Despite decades of widespread clinical use of amphotericin B in *Candida albicans* infections, the development of secondary resistance has been exceedingly rare. In contrast, both primary and secondary resistance to 5-fluorocytosine are known to occur for strains of *Candida* species, serving as the basis for restricting use of this agent to combination therapy with other antifungal drugs.

The question of fungal resistance to the azole drugs is considerably more complex and is currently under evaluation. Examples of both primary and secondary resistance are known for the medically important yeasts and selected azole antifungals. *Candida krusei* as a species is typically resistant to fluconazole. *Candida albicans* strains are typically susceptible to fluconazole and certain other azole antifungals, but there are increasing reports of resistance, especially in HIV-infected hosts having undergone repeated courses of azole antifungal therapy. The question of drug resistance is complicated by the limitations in the available susceptibility testing methodology and the ability to distinguish between microbiological and clinical drug resistance. The latter typically occurs when an inhibitory antifungal agent reaches the limits of its activity in a host with a decreasingly efficient immune system.

With the advent of the polyenes, azoles, and fluorocytosine, previously fatal infections can now be treated. However, as modern medicine continues to extend life through aggressive therapy of other life-threatening diseases such as cancer, there is an increasing population at risk for opportunistic fungal infections. Such patients represent a special challenge because they often are left with little host immune function. Therefore, chemotherapeutic agents should be fungicidal and not just fungistatic. The search continues for fungicidal agents that are nontoxic to the host. Research is also directed toward immunomodulating agents that can reverse the defects of native host immunity.

REFERENCES

Bartnicki-Garcia S. 1968.*Cell wall chemistry, morphogenesis, and taxonomy of fungi*. Ann Rev Microbiol 22:87.

Cole GT, Samson RA. 1979. *Patterns of Development in Conidial Fungi*. Pittman, London.

Fleet GH. 1985. "Composition and structure of yeast cell walls" In McGinnis MR (ed): *Current Topics in Medical Mycology*. Vol. 1. Springer-Verlag, New York.

Gow NAR, Gadd GM (ed). 1994. *The Growing Fungus,* Chapman and Hall, London.

Kwon-chung KJ, Bennett JE. 1992. *Medical Mycology*. Lea and Febiger, Philadelphia.

McGinnis MR. 1980. *Laboratory Handbook of Medical Mycology*. Academic Press, San Diego.

Vanden Bossche H, Odds FC, Kerridge D (eds). 1993. *Dimorphic Fungi in Biology and Medicine*. Plenum Press, New York.

Yoshida Y. 1988. "Cytochrome P-450 of fungi: primary target for azole antifungal agents" In McGinnis MR (ed): *Current Topics in Medical Mycology*. Vol. 2. Springer-Verlag, New York.

15

Microbiology of Protozoans

The Protozoa are considered to be a subkingdom of the kingdom Protista, although in the classical system they were placed in the kingdom Animalia. More than 50,000 species have been described, most of which are free-living organisms; protozoa are found in almost every possible habitat. The fossil record in the form of shells in sedimentary rocks shows that protozoa were present in the Pre-cambrian era. Anton van Leeuwenhoek was the first person to see protozoa, using microscopes he constructed with simple lenses.

Between 1674 and 1716, he described, in addition to free-living protozoa, several parasitic species from animals, and *Giardia lamblia* from his own stools. Virtually all humans have protozoa living in or on their body at some time, and many persons are infected with one or more species throughout their life. Some species are considered commensals, i.e., normally not harmful, whereas others are pathogens and usually produce disease.

Protozoan diseases range from very mild to life-threatening. Individuals whose defenses are able to control but not eliminate a parasitic infection become carriers and constitute a source of infection for others. In geographic areas of high prevalence, well-tolerated infections are often not treated to eradicate the parasite because eradication would lower the individual's immunity to the parasite and result in a high likelihood of reinfection.

Many protozoan infections that are inapparent or mild in normal individuals can be life-threatening in immunosuppressed patients, particularly patients with acquired immune deficiency syndrome (AIDS). Evidence suggests that many healthy persons harbor low numbers of *Pneumocystis carinii* in their lungs. However, this parasite produces a frequently fatal pneumonia in immunosuppressed patients such as those with AIDS. *Toxoplasma gondii,* a very common protozoan parasite, usually causes a rather mild initial illness followed by a long-lasting latent infection.

AIDS patients, however, can develop fatal toxoplasmic encephalitis. *Cryptosporidium* was described in the 19th century, but widespread human infection has only recently been recognized. *Cryptosporidium* is another protozoan that can produce serious complications in patients with AIDS. Microsporidiosis in humans was reported in only a few instances prior to the appearance of AIDS. It has now become a more common infection in AIDS patients. As more thorough studies of patients with AIDS are made, it is likely that other rare or unusual protozoan infections will be diagnosed.

Acanthamoeba species are free-living amebas that inhabit soil and water. Cyst stages can be airborne. Serious eye-threatening corneal ulcers due to *Acanthamoeba* species are being reported in individuals who use contact lenses. The parasites presumably are transmitted in contaminated lens-cleaning solution. Amebas of the genus *Naegleria*, which inhabit bodies of fresh water, are responsible for almost all cases of the usually fatal disease primary amebic meningoencephalitis. The amebas are thought to enter the body from water that is splashed onto the upper nasal tract during swimming or diving. Human infections of this type were predicted before they were recognized and reported, based on laboratory studies of Acanthamoeba infections in cell cultures and in animals.

The lack of effective vaccines, the paucity of reliable drugs, and other problems, including difficulties of vector control, prompted the World Health Organization to target six diseases for increased research and training. Three of these were protozoan infectionsmalaria, trypanosomiasis, and leishmaniasis. Although new information on these diseases has been gained, most of the problems with control persist.

STRUCTURE

Most parasitic protozoa in humans are less than 50 μm in size. The smallest (mainly intracellular forms) are 1 to 10 μm long, but *Balantidium coli* may measure 150 μm. Protozoa are unicellular eukaryotes. As in all eukaryotes, the nucleus is enclosed in a membrane. In protozoa other than ciliates, the nucleus is vesicular, with scattered chromatin giving a diffuse appearance to the nucleus, all nuclei in the individual organism appear alike. One type of vesicular nucleus contains a more or less central body, called an endosome or karyosome. The endosome lacks DNA in the parasitic amebas and trypanosomes. In the phylum Apicomplexa, on the other hand, the vesicular nucleus has one or more nucleoli that contain DNA. The ciliates have both a micronucleus and macronucleus, which appear quite homogeneous in composition.

The organelles of protozoa have functions similar to the organs of higher animals. The plasma membrane enclosing the cytoplasm also covers the projecting locomotory structures such as pseudopodia, cilia, and flagella. The outer surface layer of some protozoa, termed a pellicle, is sufficiently rigid to maintain a distinctive shape, as in the trypanosomes and *Giardia*. However, these organisms can readily twist and bend when moving through their environment. In most protozoa the cytoplasm is differentiated into ectoplasm (the outer, transparent layer) and endoplasm (the inner layer containing organelles); the structure of the cytoplasm is most easily seen in species with projecting pseudopodia, such as the amebas. Some protozoa have a cytosome or cell "mouth" for ingesting fluids or solid particles. Contractile vacuoles for osmoregulation occur in some, such as *Naegleria* and *Balantidium*.

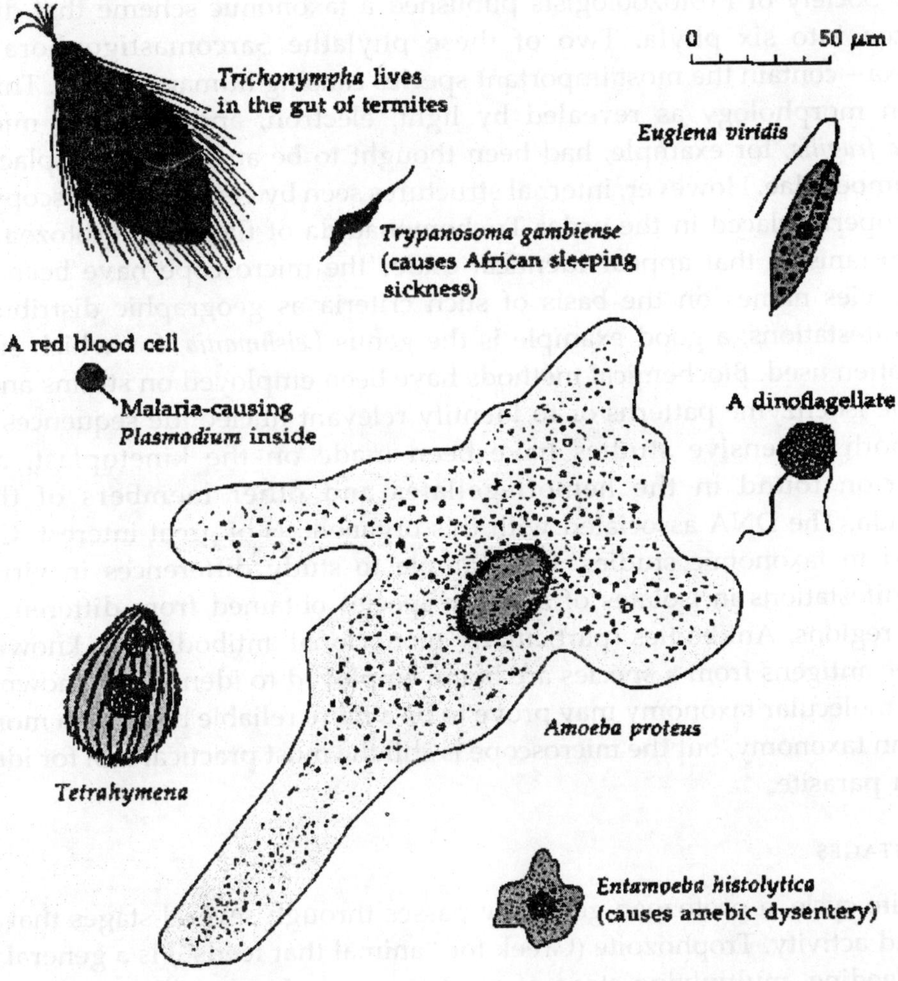

Figure 1. Structure of some well-known protozoams

Many protozoa have subpellicular microtubules; in the Apicomplexa, which have no external organelles for locomotion, these provide a means for slow movement. The trichomonads and trypanosomes have a distinctive undulating membrane between the body wall and a flagellum. Many other structures occur in parasitic protozoa, including the Golgi apparatus, mitochondria, lysosomes, food vacuoles, conoids in the Apicomplexa, and other specialized structures. Electron microscopy is essential to visualize the details of protozoal structure. From the point of view of functional and physiologic complexity, a protozoan is more like an animal than like a single cell.

CLASSIFICATION

In 1985 the Society of Protozoologists published a taxonomic scheme that distributed the Protozoa into six phyla. Two of these phylathe Sarcomastigophora and the Apicomplexa—contain the most important species causing human disease. This scheme is based on morphology as revealed by light, electron, and scanning microscopy. *Dientamoeba fragilis*, for example, had been thought to be an ameba and placed in the family Entamoebidae. However, internal structures seen by electron microscopy showed that it is properly placed in the order Trichomonadida of flagellate protozoa. In some instances, organisms that appear identical under the microscope have been assigned different species names on the basis of such criteria as geographic distribution and clinical manifestations; a good example is the genus *Leishmania*, for which subspecies names are often used. Biochemical methods have been employed on strains and species to determine isoenzyme patterns or to identify relevant nucleotide sequences in RNA, DNA, or both. Extensive studies have been made on the kinetoplast, a unique mitochondrion found in the hemoflagellates and other members of the order Kinetoplastida. The DNA associated with this organelle is of great interest. Cloning is widely used in taxonomic studies, for example to study differences in virulence or disease manifestations in isolates of a single species obtained from different hosts or geographic regions. Antibodies (particularly monoclonal antibodies) to known species or to specific antigens from a species are being employed to identify unknown isolates. Eventually, molecular taxonomy may prove to be a more reliable basis than morphology for protozoan taxonomy, but the microscope is still the most practical tool for identifying a protozoan parasite.

LIFE CYCLE STAGES

During its life cycle, a protozoan generally passes through several stages that differ in structure and activity. Trophozoite (Greek for "animal that feeds") is a general term for the active, feeding, multiplying stage of most protozoa. In parasitic species this is the stage usually associated with pathogenesis. In the hemoflagellates the terms amastigote, promastigote, epimastigote, and trypomastigote designate trophozoite stages that differ in the absence or presence of a flagellum and in the position of the kinetoplast associated with the flagellum. A variety of terms are employed for stages in the Apicomplexa, such as tachyzoite and bradyzoite for *Toxoplasma gondii*. Other stages in the complex asexual and sexual life cycles seen in this phylum are the merozoite (the form resulting from fission of a multinucleate schizont) and sexual stages such as gametocytes and gametes. Some protozoa form cysts that contain one or more infective forms. Multiplication occurs in the cysts of some species so that excystation releases more than one organism. For example, when the trophozoite of *Entamoeba histolytica* first forms a cyst, it has a single nucleus. As the cyst matures nuclear division produces four nuclei and during excystation four uninucleate metacystic amebas appear. Similarly, a freshly encysted *Giardia lamblia* has the same number of internal structures (organelles) as the trophozoite. However, as the cyst matures the organelles double and two trophozoites are formed. Cysts passed in stools have a protective wall, enabling the parasite to survive in the outside environment for a period ranging from days to a year, depending on the species and environmental conditions. Cysts formed in tissues do not usually have a heavy protective wall and rely upon carnivorism for transmission. Oocysts are stages resulting from sexual reproduction in the Apicomplexa. Some apicomplexan oocysts are passed in the feces of the host, but the oocysts of *Plasmodium*, the agent of malaria, develop in the body cavity of the mosquito vector.

REPRODUCTION

Reproduction in the Protozoa may be asexual, as in the amebas and flagellates that infect humans, or both asexual and sexual, as in the Apicomplexa of medical importance. The most common type of asexual multiplication is binary fission, in which the organelles are duplicated and the protozoan then divides into two complete organisms. Division is longitudinal in the flagellates and transverse in the ciliates; amebas have no apparent anterior-posterior axis. Endodyogeny is a form of asexual division seen in *Toxoplasma* and some related organisms. Two daughter cells form within the parent cell, which then ruptures, releasing the smaller progeny which grow to full size before repeating the process. In schizogony, a common form of asexual division in the Apicomplexa, the nucleus divides a number of times, and then the cytoplasm divides into smaller uninucleate merozoites. In *Plasmodium*, *Toxoplasma*, and other apicomplexans, the sexual cycle involves the production of gametes (gamogony), fertilization to form the zygote, encystation of the zygote to form an oocyst, and the formation of infective sporozoites (sporogony) within the oocyst.

Some protozoa have complex life cycles requiring two different host species; others require only a single host to complete the life cycle. A single infective protozoan entering a susceptible host has the potential to produce an immense population. However, reproduction is limited by events such as death of the host or by the host's defense mechanisms, which may either eliminate the parasite or balance parasite reproduction to yield a chronic infection. For example, malaria can result when only a few sporozoites of *Plasmodium falciparum*perhaps ten or fewer in rare instancesare introduced by a feeding *Anopheles* mosquito into a person with no immunity. Repeated cycles of schizogony in the bloodstream can result in the infection of 10 percent or more of the erythrocytesabout 400 million parasites per milliliter of blood.

NUTRITION

The nutrition of all protozoa is holozoic; that is, they require organic materials, which may be particulate or in solution. Amebas engulf particulate food or droplets through a sort of temporary mouth, perform digestion and absorption in a food vacuole, and eject the waste substances. Many protozoa have a permanent mouth, the cytosome or micropore, through which ingested food passes to become enclosed in food vacuoles. Pinocytosis is a method of ingesting nutrient materials whereby fluid is drawn through small, temporary openings in the body wall. The ingested material becomes enclosed within a membrane to form a food vacuole.

Protozoa have metabolic pathways similar to those of higher animals and require the same types of organic and inorganic compounds. In recent years, significant advances have been made in devising chemically defined media for the in vitro cultivation of parasitic protozoa. The resulting organisms are free of various substances that are present in organisms grown in complex media or isolated from a host and which can interfere with immunologic or biochemical studies. Research on the metabolism of parasites is of immediate interest because pathways that are essential for the parasite but not the host are potential targets for antiprotozoal compounds that would block that pathway but be safe for humans. Many antiprotozoal drugs were used empirically long before their mechanism of action was known. The sulfa drugs, which block folate synthesis in malaria parasites, are one example.

The rapid multiplication rate of many parasites increases the chances for mutation; hence, changes in virulence, drug susceptibility, and other characteristics may take place. Chloroquine resistance in *Plasmodium falciparum* and arsenic resistance in *Trypanosoma rhodesiense* are two examples.

Competition for nutrients is not usually an important factor in pathogenesis because the amounts utilized by parasitic protozoa are relatively small. Some parasites that inhabit the small intestine can significantly interfere with digestion and absorption and affect the nutritional status of the host; *Giardia* and *Cryptosporidium* are examples. The destruction of the host's cells and tissues as a result of the parasites' metabolic activities increases the host's nutritional needs. This may be a major factor in the outcome of an infection in a malnourished individual. Finally, extracellular or intracellular parasites that destroy cells while feeding can lead to organ dysfunction and serious or life-threatening consequences.

PROTOZOA: PATHOGENESIS AND DEFENSES

Resistance to parasitic protozoa appears to be similar to resistance against other infectious agents, although the mechanisms of resistance in protozoan infections are not yet as well understood. Resistance can be divided into two main groups of mechanisms: (1) nonspecific mechanism(s) or factor(s) such as the presence of a nonspecific serum component that is lethal to the parasite; and (2) specific mechanism(s) involving the immune system. Probably the best studied nonspecific mechanisms involved in parasite resistance are the ones that control the susceptibility of red blood cells to invasion or growth of plasmodia, the agents of malaria.

Individuals who are heterozygous or homozygous for the sickle cell hemoglobin trait are considerably more resistant to *Plasmodium falciparum* than are individuals with normal hemoglobin. Similarly, individuals who lack the Duffy factor on their red blood cells are not susceptible to *P vivax*. Possibly both the sickle cell trait and absence of the Duffy factor have become established in malaria-endemic populations as a result of selective pressure exerted by malaria.

Epidemiologic evidence suggests that other inherited red blood cell abnormalities, such as thalassanemia and glucose-6-phosphate dehydrogenase deficiency, may contribute to survival of individuals in various malaria-endemic geographical regions. A second well-documented example of a nonspecific factor involved in resistance is the presence in the serum of humans of a trypanolytic factor that confers resistance against *Trypanosoma brucei brucei*, an agent of trypanosomiasis (sleeping sickness) in animals. There is evidence that other nonspecific factors, such as fever and the sex of the host, may also contribute to the host's resistance to various protozoan parasites. Although nonspecific factors can play a key role in resistance, usually they work in conjunction with the host's immune system.

Different parasites elicit different humoral and/or cellular immune responses. In malaria and trypanosome infections, antibody appears to play a major role in immunity. In both *T cruzi* and *T brucei gambiense* infections, antibody-dependent cytotoxic reactions against the parasite have been reported. Although antibody has been shown to be responsible for clearing the Afric ypanosomes from the blood of infected animals, recent evidence suggests that the survival time of infected mice does not necessarily correlate with the ability of the animal to produce trypanosome-specific antibody. In other words, resistance as measured by survival time may not solely involve the specific humoral immune system. Recent data suggest that cellular immunity is required for resistance to malaria. for example, vaccine trials with a sporozoite antigen indicated that both an active cellular response and sporozoite-specific antibody may be needed for successful immunization.

Cellular immunity is believed to be the single most important defense mechanism in leishmaniasis and toxoplasmosis. In animals infected with *Toxoplasma*, the activated macrophage has been shown to play an important role in resistance. Accordingly, resistance to the protozoan parasites most likely involves nonspecific factors as well as specific humoral and/or cellular mechanisms. Cytokines are involved in the control of both the immune response and pathology. It has become apparent that there are subsets of both helper (h) and cytotoxic (c) T-cells that produce different profiles of cytokines. For example, the Th-1 subset produces gamma interferon (IFN-a), and interleukin-2 (IL-2) and is involved in cell-mediated immunity. In contrast the Th-2 subset produces IL-4 and IL-6, and is responsible for antibody-mediated immunity. The induction of a particular T-cell subset is key to recovery and resistance. The Th-1 subset and increased IFN-g are important in resistance to *Leishmania, T cruzi* and *Toxoplasma* infections, whereas the Th-2 response is more important in parasitic infections in which antibody is a key factor. It is important to recognize that the cytokines produced by one T-cell subset can up or downregulate the response of other T-cell subsets. IL-4 will downregulate Th-1 cells and exacerbate infection and/or susceptibility of mice to *Leishmania*. The cytokines produced by T and other cell types do not act directly on the parasites but influence other host cell types. The response of cells to cytokines includes a variety of physiological changes, such as changes in glucose, fatty acid and protein metabolism.

For example, IL-1 and tumor necrosis factor will increase gluconeogenesis, and glucose oxidation. It should be noted that cytokines influence the metabolism not only of T-cells, but also a variety of other cell types and organ systems. Cytokines can also stimulate cell division and, therefore, clonal expansion of T and B-cell subsets. This can lead to increased antibody production and/or cytotoxic T-cell numbers. The list of cytokines and their functions is growing rapidly, and it would appear that these chemical messages influence all phases of the immune response. they are also clearly involved in the multitude of physiological responses (fever, decreased food intake, etc.) observed in an animal's response to a pathogen, and in the pathology that results.

Unlike most viral and bacterial infections, protozoan diseases are often chronic, lasting months or years. When associated with a strong host immune response, this type of chronic infection is apt to result in a high incidence of immunopathology. The question also arises of how these parasites survive in an immunocompetent animal.

PATHOLOGY

The protozoa can elicit humoral responses in which antigen-antibody complexes in the region of antibody excess activate Hageman blood coagulation factor (Factor XII), which in turn activates the coagulation, fibrinolytic, kinin and complement systems. It has been suggested that this type of immediate hypersensitivity is responsible for various clinical syndromes in African trypanosomiasis, including blood hyperviscosity, edema, and hypotension. Similar disease mechanisms would be expected in other infections by protozoa involving a strong humoral immune response.

Immune complexes have been found circulating in serum and deposited in the kidneys and other tissues of humans and animals infected with protozoans. These parasite antigen-antibody complexes, plus complement, have been eluted from kidney tissue in cases of malaria and African trypanosomiasis. Antigen and antibody have been directly visualized in the glomeruli of infected animals by light and electron microscopy. Inflammatory cell infiltrates accompany these deposits, and signs of glomerulonephritis are usually seen. African trypanosomes and presumably their antigens are also found in a variety of extravascular locations. Immune complexes, cellular infiltrates, and tissue damage have been detected in these tissues.

Another important form of antibody-mediated pathology is autoimmunity. Autoantibodies to a number of different host antigens (for example, red blood cells, laminin, collagen, and DNA) have been demonstrated. These autoantibodies may play a role in the pathology of parasitic diseases in two ways. First the antibodies may exert a direct cytotoxic effect on the host cells; for example, autoantibodies that coat red blood cells produce hemolytic anemia. Alternatively, autoantibodies may be pathogenic through a buildup of antigen-antibody complexes in the kidneys or other tissues, leading to glomerulonephritis or other forms of immediate hypersensitivity. A particularly good example of a protozoan infection in which autoimmunity appears to be an important contributor to pathogenesis is *T cruzi* infection. In this case, there is substantial evidence that host and parasite share cross-reacting antigens. Antibodies and cytotoxic lymphocytes to these antigens appear to be harmful to host tissue. This type of experimental data, combined with the fact that the parasite itself seems not to cause the tissue pathology, lead one to conclude that autoimmunity may play a key role in pathogenesis.

Cellular hypersensitivity is also observed in protozoan diseases. For example, in leishmaniasis (caused by *Leishmania tropica*), the lesions appear to be caused by a cell-mediated immune response and hav[illegible] any, if not all, of the characteristics of granulomas observed in tuberculosis or schistosomiasis. In these lesions, a continuing immune response to pathogens that are able to escape the host's defense mechanisms causes further influx of inflammatory cells, which leads to sustained reactions and continued pathology at the sites of antigen deposition. During a parasitic infection, various host cell products (cytokines, lymphokines, etc.) are released from activated cells of the immune system. These mediators influence the action of other cells and may be directly involved in pathogenesis. An example is tumor necrosis factor (TNF), which is released by lymphocytes. TNF may be involved in the muscle wasting observed in the chronic stages of African trypanosomiasis. TNF has also been implicated in the cachexia and wasting in *Leishmania donovani* infection, cerebral malaria in *P falciparum* in children and decreased survival in *T cruzi*-infected mice. It is apparent that mediators involved in resistance to protozoan parasites may also lead to pathology during a chronic infection. There appears to be a delicate balance between the factors involved in resistance to infectious agents and those which ultimately produce pathology and clinical disease.

Numerous authors have suggested that toxic products produced by parasitic protozoa are responsible for at least some aspects of pathology. For example, the glycoproteins on the surface of trypanosomes have been found to fix complement. This activation of complement presumably results in the production of biologically active and toxic complement fragments. In addition, trypanosomes are known to release proteases and phospholipases when they lyse. These enzymes can produce host cell destruction, inflammatory responses, and gross tissue pathology. Furthermore, it has been hypothesized that the trypanosomes contain a B-cell mitogen that may alter the immune response of the host by eliciting a polyclonal B-cell response that leads to immunosuppression. Finally it has recently been shown that the African trypanosomes also contain an endotoxin which is presumably released during antibody- mediated lysis. Parasitic protozoa have also been reported to synthesize (or contain) low-molecular-weight toxins. For example, the trypanosomes produce several indole catabolites; at pharmacologic doses, some of these catabolites can produce pathologic effects, such as fever, lethargy, and even immunosuppression. Similarly, enzymes, B-cell mitogen, etc., are presumably released by many if not all of the other parasitic protozoa. There has been limited work on the role of these protozoal products in pathogenesis. However, parasitic protozoa are generally not known to produce toxins with potencies comparable to those of the classic bacterial toxins (such as the toxins responsible for anthrax and botulism). One possible exception is the African trypanosomes which are suggested to contain an endotoxin.

IMMUNE ESCAPE

Parasite escape mechanisms may include a number of different phenomena. In antigenic masking, the parasite becomes coated with host components and so fails to be recognized as foreign. In blocking, noncytotoxic antibody combines with parasite antigens and inhibits the binding of cytotoxic antibodies or cells. The parasite may pass part of its life cycle in an intracellular location, for example, in erythrocytes or macrophages, in which it is sheltered from intracellular digestion and from the cytotoxic action of antibody and/ or lymphocytes. Some parasites practice antigenic variation, altering their surface antigens during the course of an infection and thus evading the host's immune responses. Finally, the parasite may cause immunosuppression, reducing the host's immune response either to the parasite specifically or to foreign antigens in general. These strategies are discussed in more detail below.

Masking and Mimicry

Various species of trypanosomes have host immunoglobulins associated with their cell surfaces. There are several reports that these antibodies are not bound to the trypanosomes through their variable regions, but presumably through the Fc portion of their molecule. These antibodies may mask the parasite-that is, prevent immune recognition by the host. However, no evidence other than the presence of immunoglobulins on the surface of the trypanosomes supports this hypothesis. Mimicry, in which the parasite has the genetic information to synthesize antigens identical to those of its host, has not been demonstrated in parasitic protozoa.

Blocking

It has been hypothesized that in some cases antigen-antibody complexes in serum of infected animals bind to the parasite's surface, mechanically blocking the actions of cytotoxic antibodies or lymphocytes and directly inhibiting the actions of lymphocytes. This type of immune escape mechanism has been proposed for tumor cells and for the parasitic helminths. Because the trypanosomes carry immunoglobulins on their cell surfaces, they may use a similar mechanism; however, no direct evidence has yet been reported.

Intracellular Location

Many protozoan parasites grow and divide within host cells. For example, *Plasmodium* parasites grow first in hepatocytes and then in red blood cells. *Leishmania* and *Toxoplasma* organisms are capable of growing in macrophages; one genus of parasitic protozoa, *Theilera,* not only multiplies in lymphocytes but appears even to stimulate the multiplication of the infected lymphocytes. Although some parasites, such as Plasmodium, are restricted to a limited number of host cell types, others, such as *T cruzi* and *Toxoplasma,* appear to be able to grow and divide in a variety of different host cells.

An intracellular refuge may protect a parasite from the harmful or lethal effects of antibody or cellular defense mechanisms. For example, *Plasmodium* may be susceptible to the actions of antibody only during the brief extracellular phases of its life cycle (the sporozoite and merozoite stages). It should be remembered that *Plasmodium* actually resides in a membrane-bound vacuole in the host cell. Thus, plasmodia are shielded from the external environment by at least two host membranes (the outer cell membrane and an inner vacuole membrane). Although intracellular plasmodia are very well protected from the host's immune response early in their growth, this strategy does create physiologic problems for the parasite. For example, the parasite must obtain its nutrients for growth through three membranes (two host and one parasite), and must eliminate its waste products through the same three membranes. Plasmodia solve this problem by appropriately modifying the host cell membranes. Parasitic proteins are incorporated into the red blood cell outer membrane. The host eventually responds to these antigens, and this response ultimately leads to the increased removal of infected host cells.

The existence of extracellular phases in the malaria life cycle is important, since immunization against these stages is the rationale for the development of our current vaccine candidates. The protective antigens on these extracellular stages have been purified as potential antigens for a vaccine. However, this approach has problems. For example, the sporozoite stage is exposed to protective antibody for only a brief period, and even a single sporozoite that escapes immune elimination will lead to an infection. Second, the antigenic variability of different isolates and the ability of different strains to undergo antigenic variation are not fully known. Therefore, the effectiveness of the vaccine candidates must still be demonstrated. However a large synthetic peptide containing antigenic sequences from 3 different proteins of *P falciparum* has been shown to reduce the clinical incidence of malaria by 31% in field trials. There is therefore optimism that a vaccine against *P falciparum* may be available in the near future.

A number of parasitic protozoa reside in macrophages. Although these organisms are protected from external immune threats, they must still evade digestion by the macrophage. Three strategies have been suggested. First, the parasite may prevent the fusion of lysosomes with the phagocytic vacuole. The actual mechanism responsible for this inhibition is not yet understood, but it has been shown to occur in cells infected with *Toxoplasma*. A second mechanism is represented by the ability of *T cruzi* to escape from the phagocytic vacuole into the cytoplasm of the macrophage. Finally, it is possible that some parasites can survive in the presence of lysosomal enzymes, as can the leprosy bacillus. One of the best-studied examples of a protozoan parasite able to survive in the phagolysosome is *Leishmania*. It has been suggested that the resistance of this parasite to the host's hydrolytic enzymes is due to surface components that inhibit the host's enzymes and/or to the presence of parasitic enzymes that hydrolyze the host's enzymes. As previously noted, at least one protozoan parasite, *Theilera*, is capable of growing directly in lymphocytes. Therefore, this parasite may escape the host's immune response by growing inside the very cells required for the response.

Antigenic Variation

Three major groups of parasitic protozoa are known to be able to change the antigenic properties of their surface coat. The African trypanosomes can completely replace the antigens in their glycocalyx each time the host exhibits a new humoral response. These alterations in serotype are one important way in which the African trypanosomes escape their host's defense mechanism. Although less well-characterized, similar changes are reported to occur in *Plasmodium, Babesia, and Giardia.*

It has been estimated that African trypanosomes have approximately 1,000 different genes coding for surface antigens. These genes are located on various chromosomes; however, to be expressed, the gene must be located at the end of a chromosome (telomeric site). The rate at which variation occurs in a tsetse-fly-transmitted population appears quite high. It has been shown that 1 in 10 cells appears to be capable of switching its surface antigen. The order in which the surface coat genes are expressed is not predictable. Much information is available on the nucleotide sequence of the genes coding the coat proteins; however, neither the factor(s) that induces a cell to switch its surface antigens nor the specific genetic mechanisms) involved in the switch are fully understood. The antibody response does not induce the genetic switch, but merely selects variants with new surface antigens out of the original population. Considerably less information is available on the phenomenon of antigenic variation in malaria or babesiosis. However, antigen variation could be a major problem in reference to the development of a blood stage (merozoite) vaccine for malaria. Finally, antigenic variation has been observed in *Giardia* lamblia. A number of different gene families coding for surface proteins in Giardia have been identified. Antigenic variation has been suggested to assist Giardia in escaping the host's immune response.

Immunosuppression

Immunosuppression of the host has been observed with almost every parasitic organism carefully examined to date. In some cases the suppression is specific, involving only the host's response to the parasite. In other cases the suppression is much more general, involving the response to various heterologous and nonparasite antigens. It has not yet been proven that this immunosuppression allows the parasites to survive in a normally immunocompetent host. However, one can postulate that immunosuppression could permit a small number of parasites to escape immune surveillance, thus favoring establishment of a chronic infection. This mechanism might be particularly effective in parasites thai undergo antigenic variation, since it could allow the small number of parasites with new surface antigens to go undetected initially. Immunosuppression experimentally induced by various extraneous agents has certainly been shown to produce higher parasitemias, higher infection rates, or both. Therefore, the hypothesis that parasite-induced immmosuppression increases the chance for a parasite to complete its life cycle makes sense.

It should be noted that immunosuppression can be pathogenic itself. A reduced response to heterologous antigens could favor secondary infections. Humans suffering from malaria or trypanosomiasis have been shown to be immunosuppressed to a variety of heterologous antigens. Secondary infections may often be involved in death from African trypanosomiasis.

A variety of mechanisms have been suggested to explain the immunosuppression observed in protozoan infections. The most common mechanisms proposed are (1) the presence in the infected host of parasite or host substances that nonspecifically stimulate the growth of antibody-producing B cells, rather than stimulating the proliferation of specific antiparasite B-cells; (2) proliferation of suppressor T-celis and/or macrophages that inhibit the immune system by excretion of regulatory cytokines; and (3) production by the parasite of specific immune suppressor substances.

INTESTINAL PROTOZOA: AMEBAS

Amebas are unicellular organisms common in the environment: many are parasites of vertebrates and invertebrates. Relatively few species inhabit the human intestine and only *Entamoeba histolytica* is identified as a human intestinal pathogen. A second pathogen of the human colon is *Dientamoeba fragilis*, which looks like an ameba under the light microscope, and was classified as a true ameba for many years, but which is now identified as a flagellate.

Pathogenic and non-pathogenic strains of *E histolytica* inhabit the human digestive tract. Even pathogenic strains may live in the lumen as benign commensals. If mucosal invasion occurs, it may be limited to a few simple superficial erosions or it may progress to total involvement of the colonic mucosa with ulceration. The clinical manifestations vary with the extent of involvement. Mucosal erosion causes diarrhea, which increases in severity with increasing area and depth of involvement. Symptoms are also affected by the site of the infection. The more distal the lesion in the colon, the greater the likelihood and severity of symptoms; thus small rectal lesions are more likely to be symptomatic than larger cecal lesions. Rectal bleeding is only slightly less common than diarrhea and is usually, but not invariably, associated with diarrhea. Such bleeding may be grossly apparent or may be occult and demonstrable only by chemical testing for blood. Urgency, tenesmus, cramping abdominal pain and tenderness may be present.

The intestinal syndromes caused by *E histolytica* form a continuum ranging in severity from mild diarrhea to hemorrhagic dysentery. The span from mild to severe diarrhea is classified as non-dysentery colitis. Amebic dysentery has a dramatically different clinical presentation. The diarrhea is replaced by dysenteric stools consisting largely of pus and blood without feces. There is evidence of systemic toxicity with fever, dehydration, and electrolyte abnormalities. Tenesmus and abdominal tenderness are regular features. This fulminant presentation may occur suddenly or evolve from less severe, pre-existing disease.

Occasionally, and for no apparent reason, colonic infection with *E histolytica* will evoke a proliferative granulomatous response at an ulcer site. This infectious pseudotumor, called an *ameboma,* may become the leading point of an intussusception or may cause intestinal obstruction. This complication is uncommon.

Peritonitis as a result of perforation has been reported in connection with severe amebic colitis and, much less often, in patients with few or no symptoms. Other complications of intestinal amebiasis include colocutaneous fistula, perianal ulceration, urogenital infection, colonic stricture, intussusception, and hemorrhage. Most of these complications are uncommon and therefore may prove difficult to diagnose. The term post-amebic colitis is used for nonspecific colitis following a bout of severe acute amebic colitis. In such cases, the colon is free of parasites and the clinical findings resemble those of chronic ulcerative colitis.

Extraintestinal amebiasis begins with hepatic involvement. Many patients with acute intestinal infection also have hepatomegaly, but in these cases amebas are not demonstrable in the liver and the pathogenesis of this hepatomegaly is not clear. A focal amebic abscess in the liver represents metastasis from intestinal infection. Symptomatic intestinal infection need not be present. The abscess appears as a slowly enlarging liver mass. Often the patient will have right upper quadrant pain, which may be referred to the right shoulder. If the abscess is located in a palpable portion of the liver, the area will be tender. Occasionally the enlarging abscess presses on the common bile duct and causes jaundice. If located under the dome of the diaphragm, the abscess may cause elevation of the dome of the diaphragm which presses on the right lung base, causing atelectasis and physical findings of consolidation. As the abscess nears the diaphragm the inflammation may stimulate pleural effusion.

Pleural, pulmonary, and pericardial infection occurs as a result of direct extension from the liver. Lung involvement is far more common than pericardial infection. Infection metastatic from the liver can involve other viscera or can give rise to a brain abscess. However, these complications are uncommon.

Structure

E histolytica has a relatively simple life cycle that alternates between trophozoite and cyst stages. The trophozoite is the actively metabolizing, mobile stage, and the cyst is dormant and environmentally resistant. Diagnostic concern centers on both stages. Trophozoites vary remarkably in size-from 10 to 60 μm or more in diameter, and when they are alive they may be actively motile. Amebas are anaerobic organisms and do not have mitochondria. The finely granular endoplasm contains the nucleus and food vacuoles, which in turn may contain bacteria or red blood cells. The parasite is sheathed by a clear outer ectoplasm. Nuclear morphology is best seen in permanent stained preparations. The nucleus has a distinctive central karyosome and a rim of finely beaded chromatin lining the nuclear membrane.

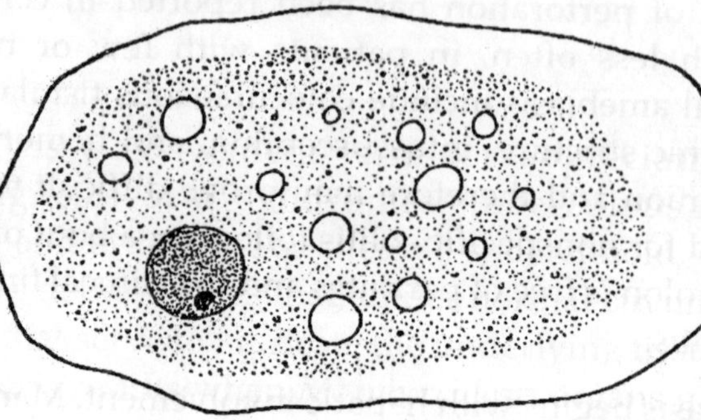

Figure 2. Structure of Entamoeba histolytica

The cyst is a spherical structure, 10-20 μm in diameter, with a thin transparent wall. Fully mature cysts contain four nuclei with the characteristic amebic morphology. Rod-like structures (chromatoidal bars) are present variably, but are more common in immature cysts. Inclusions in the form of glycogen masses also may be present. A number of non-pathogenic amebae can parasitize the human gastrointestinal tract and may cause diagnostic confusion. These include *Entamoeba hartmanni, Entamoeba gingivalis, Entamoeba coli, Endolimax nana,* and *Iodamoeba butschlii.*

Classification

Many infections with *E histolytica* occur without evidence of invasion of the intestinal lining. Virulence in the amebathe ability to produce intestinal invasion or extraintestinal diseaseis a heritable characteristic. Morphologically identical amebas may be identified as pathogenic or non-pathogenic on the asis of size, cultural characteristics, virulence in a rat model or in tissue culture, selective agglutination by lectins, reaction with monoclonal antibodies, or isoenzyme pa terns. A pathogen-specific galactose adhesion epitope is described. Ribosomal RNA sequence analysis and restriction fragment length polymorphism analysis also can separate pathogenic from non-pathogenic strains.

A number of non-pathogenic but apparently genuine *E histolytic* strains have been isolated from human carriers. These amebas can be cultured at room temperature as well at 37° C and will grow in hypotonic media, whereas pathogenic amebas require isotonic media and 37° C for growth. These low-temperature strains have isoenzyme patterns identical with the sewage- associated, non-pathogenic *Entamoeba moshkovskii*. Two classic tests to identify pathogenic strains are the ability to cause cecal ulceration in weanling rats and agglutination by the lectin concanavalin A. These tests of virulence have been supplanted by isoenzyme analysis and the use of monoclonal antibodies to identify pathogenic strains of *E histolytica*, but the clinical applicability of this technique is pending.

Isoenzyme patterns are known for four amebic enzymes: glucose phosphate isomerase (GPI), hexokinase (HK), malate:NADP+ oxidoreductase (ME), and phosphoglucomutase (PGM). The isoenzyme patterns of three of these, GPI, HK, and PGM, can be used to define 20 zymodemes of *E histolytica*. The enzyme markers associated with pathogenicity are the presence of a b band and the absence of an a band for PGM.

Zymodemes II, VI, VII, XI, XII, XIII, XIV, XIX, and XX are pathogenic. Zymodemes II and XI are responsible for liver abscesses. There have been several reports of cultured amebas undergoing a change in zymodeme pattern after manipulation of associated bacterial flora. Attempts to reproduce these observations have not been successful. Zymodeme patterns are of epidemiologic and research interest but their limited availability makes them less useful clinically. A number of other factors, primarily environmental, that affect virulence are discussed below.

It is possible to distinguish with monoclonal antibodies the galactose-specific adhesions from pathogenic and non-pathogenic ameba. This offers the possibility of simplified laboratory determination of pathogenicity.

Multiplication and Life Cycle

Amebas multiply in the host by simple binary fission. Most multiplication occurs in the host, and survival outside the host depends on the desiccation-resistant cyst form. Encystment occurs apparently in response to desiccation as the ameba is carried through the colon. After encystment, the nucleus divides twice to produce a quadrinucleate mature cyst. Excystment occurs after ingestion and is followed by rapid cell division to produce four amebas which undergo a second division. Each cyst thus yields eight tiny amebas.

Pathogenesis

The fecal-oral transmission of the ameba usually involves contaminated food or water. The parasite can also be transmitted directly by ano-genital or oro-anal sexual contact. Latent infections can become invasive in a setting of impaired host immunity. Ingested cysts of *E histolytica* excyst in the small intestine. Trophozoites are carried to the colon, where they mature and reproduce. The parasite may lead a commensal existence on the mucosal surface and in the crypts of the colon. Successful colonization depends on factors such as inoculum size, intestinal motility, transit time, the presence or absence of specific intestinal flora, the host's diet and the ability of the ameba to adhere to the colonic mucosal cells. The ameba adherence molecule has been identified as a lectin which can bind to either of two common carbohydrate components of cell membrane, galactose and *N*-acetyl glactoseamine. Binding to colonic mucins blocks adherence to mucosal cells. Depletion of mucus results in binding to the mucosa, an essential step in the development of the disease. If amebas pass down the colon they encyst under the stimulus of desiccation, and then are evacuated with the stool.

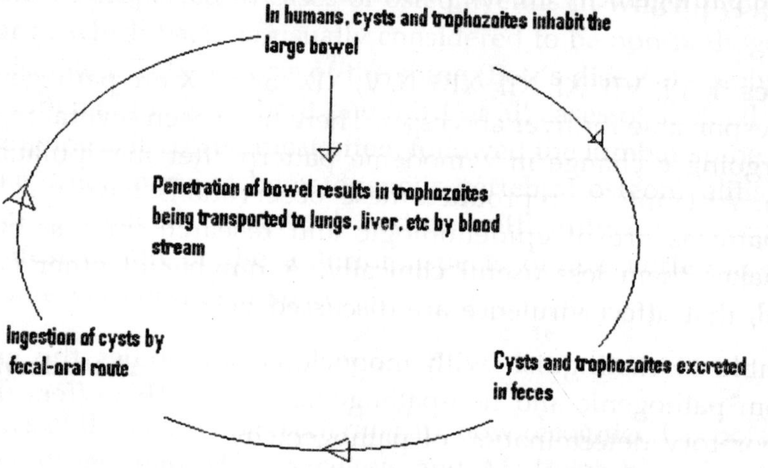

Figure 3. Life cyle of E histolytica

The factors that lead to tissue invasion by *E histolytica* are poorly understood. The genetic virulence factors mentioned above play a major role but several environmental factors are also important. Although the mechanisms of action are not clear, both changes in the intestinal flora and the nature of the host's diet have been implicated. All virulence factors come to a final common path where the ameba attacks and kills the host cell. Binding involving the galactose adherence lectin is essential for the cytolytic effect. Blocking adherence_blocks the killing. This cytolytic event is a result of incorporation in the host cell membrane of an ameba-produced, pore-forming protein, amoebapore. This protein forms ion channels in lipid cell membranes and results in cell death within minutes of cell contact with the ameba. Amoebapore has been isolated, synthesized and well characterized. Non-pathogenic strains of *E. histolytica* can also produce amoebapore but are much less efficient at its production and the molecule is not exactly similar to that produced by virulent strains.

The initial lesion is in colonic mucosa, most often in the cecum or sigmoid colon. The slow transit of the intestinal contents in these two locations seems an important factor in invasion of the mucosa, both because it affords the ameba greater mucosal contact time and because it permits changes in the intestinal milieu that may facilitate invasion. The initial superficial ulcer may deepen into the submucosa and muscularis to become the characteristic flask-shaped, chronic amebic ulcer. Spread may occur by direct extension, by undermining of the surrounding mucosa until it sloughs, or by penetration that can lead to perforation or fistulous communication to other organs or the skin. If the amebas gain access to the vascular or lymphatic circulation, metastases may occur first to the liver and then by direct extension or further metastasis to other organs, including the brain.

Virulent *E histolytica* strains are capable of penetrating intact intestinal mucosa. The infection is not opportunistic and does not require pre-existing mucosal damage. Numerous proteases have been isolated in *E histolytica*; however, the mechanism of penetration remains unclear. Metastatic foci present as abscesses with a central zone of lytic necrosis surrounded by a zone of inflammatory cell infiltration. Metastatic abscesses behave as space-occupying lesions unless they become secondarily infected or rupture.

The clinical presentation of intestinal infections depends on the extent and anatomic location of the ulceration and mucosal damage. Small, sparse ulcerations may be asymptomatic. As the involved area of the mucosa increase in size and/or in depth, motility disturbances occur, primarily diarrhea with cramping pain. Exudation from the denuded mucosa adds to intestinal content. When the mucosal involvement becomes extensive, diarrhea is replaced by dysentery, with the passage of exudate, blood and mucus. Toxic megacolon and perforation are rare complications of extensive involvement. Systemic signs of infection include fever, rigor, and polymorphonuclear leukocytosis.

Host Defenses

The gastric acid "barrier" and the steady movement of food through the intestine are nonspecific defense mechanisms invoked to explain both the experimental observation that large inocula are required to produce consistent infection in animals and the pathologic observation that few lesions are found in the small intestine, a zone of rapid transit. A role for colonic mucins in protection and depletion of these mucins in infection has been suggested.

Usually amebas alone stimulate little or no direct cellular response. Primary intestinal lesions elicit little reaction until secondary bacterial infection occurs. Amebic abscesses similarly elicit only a mild leukocytic response, which may be largely a response to the host cellular debris in the abscess. Amebas are antigenic and stimulate an antibody response and cellular sensitivity. In vivo studies have yielded contradictory results regarding the response of amebas to exposure to humoral antibodies. The occurrence of progressive and/or recurrent infection in the face of established immune sensitivity suggests that the host immune response is relatively ineffective against established infections.

Epidemiology

Fecal-oral transmission occurs when food preparation is not sanitary or when drinking water is contaminated. Contamination may come directly from infected food handlers or indirectly from faulty sewage disposal. Endemic or epidemic disease may result. The prevalence of amebiasis in underdeveloped countries reflects the lack of adequate sanitary systems.

Amebas are found in all climates, arctic to tropical. Symptomatic infections (amebic disease) are far more prevalent in certain geographic foci, and this uneven prevalence of disease, as opposed to infection, is now explained by the variable geographic predominance of pathogenic zymodemes. Similar environments thus are likely to have a comparable infection rate but may have a widely different disease prevalence.

Diagnosis

Amebic infections are diagnosed definitively by identifying the ameba in stool or exudate. Under some circumstances, however, the physician must settle for a presumptive diagnosis based on serologic or clinical evidence alone. Diagnosis may be difficult if few organisms are shed in the stool. Effective methods exist for concentrating cysts but not trophozoites in stool specimens. Fortunately, a direct relationship is usually seen (although there are exceptions) between the severity of disease and the number of amebas shed in the stool; hence, the more severe the infection the easier the diagnosis. Unfortunately, a number of substances that may be administered to the patient in the course of diagnosis or therapy can impair the ability to make a direct diagnosis. These compounds can suppress the shedding of amebas into the stool but may not interfere with the course of invasion infection. Such compounds include barium, bismuth, kaolin, soapsuds (as enemas), and antimicrobials that can reach the intestinal lumen. The suppression of shedding may be short-lived (soapsuds enema), or may last weeks or months (broad-spectrum antibiotics). These compounds render timely direct diagnosis unreliable and often impossible.

Amebas may be identified in direct smears, but specific diagnosis usually depends upon obtaining a fixed stained preparation. Trophozoites deteriorate rapidly in stool specimens, and therefore preservatives, either polyvinyl alcohol or the merthiolate-iodine-formaldehyde (MIF) combination, are important diagnostic aids. Finally, it is unrewarding to search for trophozoites in formed stool because most trophozoites encyst as the stool desiccates.

Trophozoites can be found in diarrhea. Most infections in formed stool specimens will be detected by examining three specimens passed over a 7- to 10-day period. A negative examination of single stool specimen does not rule out infection. Trophozoites may be obtained by administering a purgative agent or by scraping suspicious lesions at the time of sigmoidoscopy.

Amebas are difficult to demonstrate in aspirates from extraintestinal abscesses unless special precautions are taken. The contents of most amebic abscesses are relatively free of the organism. Instead, the organisms concentrate adjacent to the wall of the abscess cavity. If care is taken during aspiration to separate serial aliquots of aspirate, amebas may be found in the last syringe that empties the cavity. Cysts or trophozoites are only found in approximately one-half of all patients with amebic liver abscess.

Serologic studies may be useful, particularly when direct diagnosis is not possible. Such methods include gel diffusion, immunoelectrophoresis, countercurrent electrophoresis, indirect hemagglutination, indirect fluorescent antibody, skin tests, enzyme-linked immunosorbent assay (ELISA) and latex agglutination. Many of these techniques are best suited for immunoepidimiology, but gel diffusion, countercurrent electrophoresis, and latex agglutination are available for clinical studies because they are readily run on a single serum sample. A positive result on these tests indicates only prior experiences with invasive amebiasis. In environments where the incidence of amebiasis is low, such as in the United States, a positive antibody test often indicates active disease, an impression strengthened if the clinical findings agree. In areas of high prevalence a single positive antibody test is less significant. The physician rarely observes the patient long enough to measure a rising titer as evidence of active ongoing invasive infection.

Amebas may be cultured from the stool. However, because the techniques involved are somewhat more cumbersome than those routinely used for bacterial organisms, culturing is not widely used as a diagnostic tool. It is essential for virulence testing.

Testing with monoclonal antibodies demonstrates ameba in the stool, and, if the galactose adhesion epitopes are tested for, pathogenicity may be determined as well. Broad scale application is proposed.

A number of nonpathogenic amebas that can inhabit the human intestinal tract may confuse direct diagnosis. These include *Entamoeba hartmanni, Entamoeba gingivalis, Entamoeba coli, Endolimax nana,* and *Iodamoeba butschlii.* Although these parasites do not cause illness, they indicate that the patient has ingested feces-contaminated food or water, so their presence may prompt careful study of additional specimens.

Control

Preventive measures are limited to environmental and personal hygiene. Treatment depends on drug therapy, which in the case of some abscesses must be supplemented with drainage, either open or by aspiration. Effective drugs are available for liver abscess but intestinal infection is less successfully treated. No single drug is completely effective in eradicating amebas from the gut, so reliance is often placed on combination therapy.

Acute intestinal disease is best treated with metronidazole at a dose of 750 mg three times a day orally for 10 day. In children the dose is 40 mg/kg/day divided into three doses and given orally for 10 days. While this treatment is effective against invasive intestinal disease, it is less effective in clearing amebas from the intestine. Patients unable to take metronidazole may be given a broad spectrum antibiotic for two weeks. It too is relatively ineffective at clearing the amebas from the gut. There are two choices for a drug to clear amebas from the lumen of the gut: iodoquinol at an adult dose of 650 mg orally three times daily for 20 days or diloxanide furoate at an adult dose of 500 mg orally three times daily for 10 days.

Amebic liver abscess is best treated with metronidazole at several possible dose regimens, but cases of drug failure have been reported. Chloroquine or dehydroemetine are less desirable alternatives. Aspiration of the abscess is not helpful except for diagnostic purposes unless rupture is imminent. Amebic abscesses heal at the same rate with or without aspiration. Abscesses with secondary bacterial infection must be drained surgically. Abscesses involving other organs respond less well to drugs and require drainage.

REFERENCES

Englund PT, Sher A (eds). 1988. *The Biology of Parasitism. A Molecular and Immunological Approach.* Alan R. Liss, New York,

Frenkel JK. 1988. "Pathophysiology of toxoplasmosis". *Parasitol Today* 4:273.

Goldsmith R, Heyneman D (eds). 1989. *Tropical Medicine and Parasitology. Appleton and Lange,* East Norwalk, CT.

Lee JJ, Hutner SH, Bovee EC (eds). 1985. *An Illustrated Guide to the Protozoa. Society of Protozoologists,* Lawrence, KS.

Kotler DP, Orenstein JM. 1994 . "Prevalence of Intestinal Microsporidiosis in HIV-infected individuals referred for gastrointestinal evaluation". *J Gastroentero* l 89.

Neva FA, Brown H. 1994. *Basic Clinical Parasitology,* 6th edition, Appleton & Lange, Norwalk, CT.

Mock BA, Nacy CA. 1988. "Hormonal modulation of sex differences in resistance to Leishmania major systemic infections". *Infect Immun* 56:3316.

16

Infectious Diseases

Infections still cause about one-third of all deaths worldwide and are the leading cause of death, mainly because of disease in developing countries. In developed countries including the United States, improvements in sanitation and hygiene during the 19th century lowered the death rates from infectious diseases even before the dramatic impact of antimicrobial agents and new vaccines. However, recent data suggest that mortality due to infectious diseases in the United States is actually increasing. Between 1980 and 1992, mortality from infections increased by 58% and age-adjusted death rates increased by 39%, so that taken as a group infectious diseases became the third leading cause of death (up from fifth in 1980). The epidemiology and pathogenesis of infection can be discussed in several complementary ways.

First, consider the formula for infection:

$$\text{Likelihood of infection} \approx \frac{\text{virulence of microorganism X number of microorganisms}}{\text{host resistance}}$$

Microorganisms are virulent to the extent that they cause disease in previously healthy individuals; that is, when host resistance is high. Virulence is sometimes defined in terms of the percentage of infected persons who develop serious disease or, in some instances, as the case-fatality rate. Host resistance can be reduced because of localised or systemic disease or injury. Damaged or abnormal tissue that is easily infected is sometimes called a *locus minoris resistentiae* (place of least resistance).

A second framework for understanding infections is the epidemiologic triad or chain of infection:

Reservoir → Means of transmission → Susceptible host

A third framework for understanding infections concerns exogenous versus endogenous microorganisms. Exogenous infections arise from the animate or inanimate environment, whereas endogenous infections arise from the patient's flora.

Infectious disease is usually an accidental event in a world in which each of us lives intimately with billions of microorganisms. In many cases we depend on them, as they on us, for survival. Death is undesirable from the microbe's perspective as well as ours

NORMAL AND COLONISING BACTERIAL FLORA

Colonisation begins at birth and is of two types:

— Permanent colonisation by bacteria that are more-or-less expected to be part of the normal flora at all times

— Transient colonisation by potential pathogens.

Examples of the former include "diphtheroid" bacteria (*Propionibacterium* and *Corynebacterium* species) on the skin, viridans streptococci in the oral cavity, and *E. coli*, enterococci, and *Bacteroides* species in the colon. Examples of the latter include *Streptococcus_pneumoniae, Haemophilus influenzae,* and *Neisseria meningitidis* in the upper respiratory tract. Colonisation is by no means a haphazard event. Microorganisms have on their surfaces specialised molecules called adhesins that bind with specific receptors on host epithelial cells or with extracellular matrix materials. An appreciation of the major components of the body?s normal and colonising bacterial flora promotes understanding of many of the common infections encountered in clinical medicine.

In respiratory infections, for example, viral infection facilitates invasion by colonising bacteria such as *S. pneumoniae* and *H. influenzae,* which in turn facilitates superinfection by "normal flora" anaerobes such as *Prevotella, Fusobacteria,* and *Peptostreptococcus* species.

Staphylococcus Aureus

Potential sites of colonisation by *S. aureus* include the skin, the perineal area, and the gastrointestinal tract, but by far the most important site is the nasal mucosa (anterior nares). *S. aureus* nasal carriage is extremely common. At a given time, 20% to 40% of adults are likely to have positive nasal swab cultures for *S. aureus*. Many persons (up to 25% of the population) are permanent nasal carriers of *S. aureus*. Most people (about 60% of the population) exhibit intermittent colonisation with *S. aureus,* while some persons never show colonisation. Because most of us (whether we admit it or not) put our fingers in our noses on a regular basis, person-to-person transmission of *S. aureus* by hand contact is a universal phenomenon. Viral upper respiratory infection in patients with *S. aureus* nasal colonisation sometimes results in wide dispersal of the organism in the immediate environment; this phenomenon has long been recognised in pediatrics as "cloud babies," but recently "cloud adults" have also been reported. Some persons with staphylococcal nasal colonisation are prone to develop styes, folliculitis, or furunculosis (boils). Most, however, remain asymptomatic until the organism is given the opportunity to invade the skin on account of a wound, abrasion, vascular access line, or surgical procedure. Occasional persons develop *S. aureus* pneumonia, especially during influenza epidemics since influenza A increases the density of staphylococcal colonisation. Staphylococcal bacteremia can arise from colonisation, local infection, trauma, foreign bodies, or pneumonia. Complications of staphylococcal bacteremia, which often presents as a nonspecific flu-like illness, include septic shock, endocarditis, and metastatic infection.

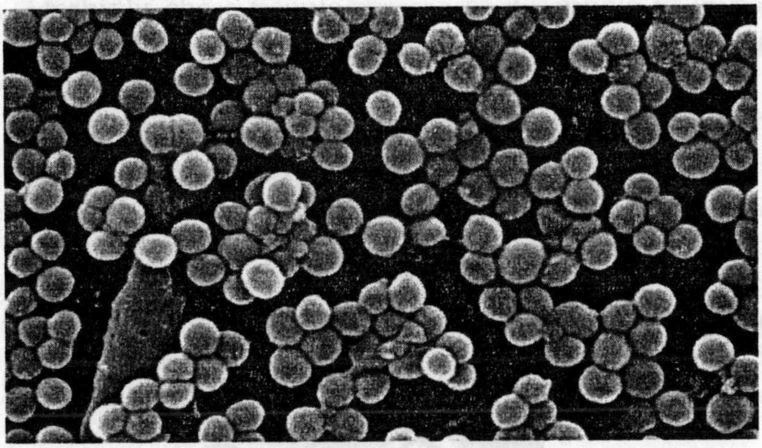

Figure 1. Colony of Staphylococcus Aureus

Of great current concern is the spread in communities throughout the United States of *S. aureus* strains that are resistant to antibiotics (these are commonly called "methicillin-resistant" or MRSA strains) and that are associated especially with skin abscesses and furuncles and with severe necrotising pneumonia. These strains typically express the Panton-Valentine leukocidin, an exotoxin that induces pore formation in polymorphonuclear neutrophils and monocytes, leading to activation, degranulation, and release of inflammatory mediators.

Viridans Streptococci

Most of the α-hemolytic streptococci present in the normal flora are loosely known as "viridans" (Latin *viridis*, "green") because they cause green hemolysis on blood agar plates, but numerous individual species are now recognised on the basis of physiological, biochemical, and molecular typing methods. Under normal circumstances these bacteria comprise the major aerobic component of the flora of the human mouth, making up nearly 50% of all bacteria that can be cultured from saliva. Individual species of viridans streptococci occupy distinct ecologic niches. *Streptococcus sanguis* and *S. mitis* adhere preferentially to the buccal mucosa; *S. salivarius* and *S. mitis* to the dorsal surface of the tongue; while *S. sanguis, S. mitis, S. oralis, S. gordonii,* and *S. anginosus* are frequently found in dental plaques. *S. mutans,* which adheres to teeth in large numbers and ferments dietary sugars into acids, is strongly associated with dental caries¯no doubt the world's most prevalent bacterial infection. One group of viridans streptococci, variably known as the "*S. milleri* group" or "*S. anginosus,*" is associated with purulent infections including brain abscess. Nearly all of the viridans streptococci occasionally cause endocarditis, usually in persons with diseased heart valves. With these 3 exceptions, colonisation by viridans streptococci is nearly always harmless; indeed, it is highly beneficial since it provides resistance to colonisation by more virulent microorganisms.

Streptococcus Pyogenes

S. pyogenes is a major pathogen not only because of the frequency of streptococcal pharyngitis and impetigo but also because of the potential for life-threatening complications. Complacency engendered by the declining incidence of acute rheumatic fever has given way to renewed concern because of the streptococcal toxic shock syndrome and necrotising fasciitis.

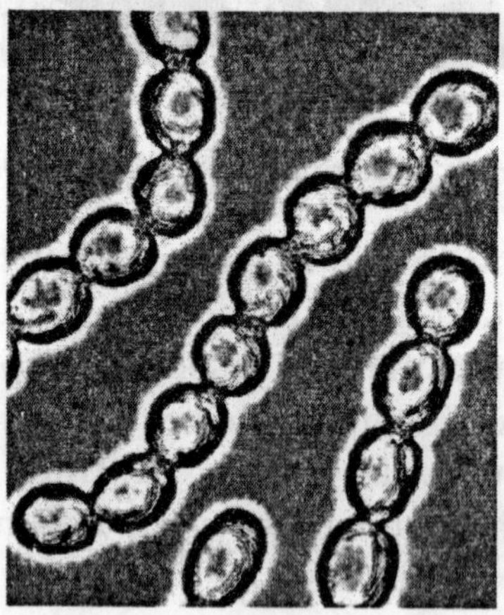

Figure 2. Streptococcus Pyogene

S. pyogenes is a frequent coloniser of the human pharynx, especially in children where carriage rates of up to 20% have been reported. Asymptomatic colonisation is less common in adults. Streptococcal M protein enables the organism to resist phagocytosis and multiply in blood. The development of antibodies against M protein confers lasting immunity, but unfortunately the immunity is type-specific and more than 90 types of M protein have been identified.

The streptococcal toxic shock syndrome is, simply put, any streptococcal infection associated with the sudden onset of shock and organ failure. Portals of entry are apparent in most cases. Some cases are associated with necrotising fasciitis. Both pyrogenic exotoxins and certain M proteins seem to be capable of acting as "superantigens," causing massive release of monocyte cytokines and lymphokines.

Streptococcus Pneumoniae

Increasing resistance to β-lactam antibiotics and other drugs makes the pneumococcus a common cause of otitis media, sinusitis, pneumonia, meningitis, and other serious infections more dangerous today than at any time since the pre-antibiotic era. Most humans are intermittently colonised in the nasopharynx by this organism, especially during midwinter. Prevalence studies indicate that 20% to 40% of children and 5% to 10% of adults are colonised at a given time. From the nasopharynx, pneumococci have access to the eustachian tubes, the ostia to the paranasal sinuses, and the tracheobronchial tree.

Enterococci, Streptococcus bovis, and Group B Streptococci

Group D streptococci formerly included the enterococci, *S. bovis*, and *S. equinus*. Newer classifications give enterococci their own genus with at least 12 species, of which *Enterococcus faecalis* (80% to 90%) and *E. faecium* are the major isolates from humans. Enterococci form a major part of the normal flora of the lower gastrointestinal tract, being the predominant aerobic gram-positive bacteria in stools. Enterococci occasionally cause community-acquired urinary tract infection and endocarditis. However, the ability of enterococci to cause disease by themselves (that is, as sole pathogens) is limited. Enterococci are commonly involved in hospital-acquired infections are frequently isolated from urine of patients with obstructive uropathy and from wounds including decubitus ulcers. Resistance to numerous antibiotics complicates treatment of enterococcal infection.

Streptococcus bovis is occasionally found in the human gastrointestinal tract, especially in patients with cancer or precancerous lesions of the bowel; documented *S. bovis* bacteremia is an indication for colonoscopy. Group B streptococci (*S. agalactiae*), found in genital tract or colon in 5% to 40% of women, are of concern primarily because of neonatal and puerperal sepsis but also cause disease in adults with impaired host defenses

Neisseria Meningitidis

The meningococcus is one of the few microorganisms capable of killing a previously-healthy person within a few hours. However, asymptomatic colonisation is relatively common, being found in 18% of "normal" family members over a 32-month period. Asymptomatic carriage of meningococci leads to the development of protective antibodies directed against the organism's polysaccharide capsule. Most cases of invasive meningococcal disease occur among the newly colonised. Studies suggest that adult males often bring the organism into a household; respiratory transmission leads to colonisation of other family members, with children being the most likely victims of invasive disease.

Haemophilus influenzae and Moraxella catarrhalis

Haemophilus influenzae is a small, pleomorphic, aerobic, gram-negative coccobacillus found mainly in the upper respiratory tract. Some strains contain a polysaccharide capsule, the major virulence factor, and are typed (a through f) according to the nature of the capsule. Most instances of life-threatening disease such as meningitis are caused by type b strains. Wide deployment of the conjugate vaccine against *Haemophilus influenzae* type b has greatly diminished the importance of this scourge of early childhood. Non-encapsulated strains are frequently associated with sinusitis, otitis media, exacerbations of chronic bronchitis in patients with COPD, and conjunctivitis. About 30% to 80% of healthy persons have nasopharyngeal colonisation by non-encapsulated strains of *H. influenzae*. About 2% to 4% of children were colonised by type B strains prior to the vaccine. *H. influenzae*, like *N. meningitidis*, preferentially colonises non-ciliated epithelial cells in the nasopharynx.

Moraxella catarrhalis, previously known as *Neisseria catarrhalis* and then *Branhamella catarrhalis* is a gram-negative diplococcus associated with upper and lower respiratory infections in both children and adults. Up to two-thirds of infants, but only 1% to 5% of healthy adults, are colonised by this microorganism, which causes a spectrum of disease similar to that caused by *H. influenzae*.

E. coli and Other Aerobic Gram-Negative Rods

Escherichia coli, the major aerobic gram-negative rod (bacillus) found in the lower gastrointestinal tract, is of enormous importance in primary care because of its role in (1) the great majority of cases of community-acquired urinary tract infection (UTI); (2) occasional deep tissue infectious such as vertebral osteomyelitis in patients with underlying medical problems; (3) rare cases of colitis caused by enteropathogenic or enterohemorrhagic strains; and (4) the well-publicised problem of hemorrhagic colitis and hemolytic syndrome due to strains belonging to the 0157:H7 serotype. All humans (excepting those who have received broad-spectrum antimicrobial therapy) are probably colonised with *E. coli,* but asymptomatic colonisation with enteropathogenic or enterohemorrhagic strains is rare if it occurs at all.

Proteus mirabilis causes up to 10% of community-acquired UTIs and presumably colonises the normal human gastrointestinal tract. Other aerobic gram-negative rods cause infections in patients with underlying diseases who have received broad-spectrum antimicrobial therapy. *Klebsiella, Enterobacter,* and *Serratia* species are often found in the stool flora of patients who have received broad-spectrum antibiotics. *Pseudomonas aeruginosa* can be part of the normal fecal flora but, unlike *E. coli* or *Proteus mirabilis,* is rarely associated with community-acquired UTI in the absence of a predisposing factor such as urologic instrumentation. *Acinetobacter* species, which often resist the action of soap, can be found in the skin flora in up to 25% of persons but rarely cause community-acquired disease. *Salmonella* and *Shigella* species are not considered part of the normal intestinal flora.

ANAEROBIC BACTERIA

Anaerobic bacteria are operationally defined by their failure to grow on solid media in the presence of 10% carbon dioxide (or 18% oxygen). Most of the common aerobic bacteria encountered in medicine can grow under anaerobic conditions as well, and are therefore sometimes called "facultative" (that is, they can grow either aerobically or anaerobically).

The term "anaerobic" is usually reserved for *strict* anaerobes. Quantitatively, these bacteria are the most important component of the normal human flora. Thus, saliva contains 10^7 to 10^8 anaerobic bacteria/mL; the terminal ileum 10^4 to 10^6/mL; and the colon, where anaerobes outnumber aerobes by a ratio of about 1000:1, 10^{11} or more per gram of stool (dry weight). Anaerobes are also highly prevalent in the normal flora of the skin, vagina, and periurethral tissues.

Anaerobic bacteria are commonly found in odontogenic infections including infected root canals, chronic sinusitis, chronic otitis media, and pelvic inflammatory disease. Otherwise, anaerobic bacteria rarely assume importance in primary care unless (1) the patient has serious underlying disease or (2) the infection is of such severity that hospitalisation is clearly indicated. This is the case because anaerobic bacteria cause serious infection only when there has been a major disruption of tissue (for example, a wound or perforated bowel) or when the oxidation-reduction potential has been lowered (for example, by ischemia, necrotic tumors, or foreign bodies). To the contrary, anaerobic bacteria are of major importance to human well being since they protect against colonisation by more pathogenic organisms.

When anaerobic bacteria cause disease, they generally arise from the indigenous body flora. The major exception is the clostridial syndromes such as tetanus (*Clostridium tetani*) and botulism (*Cl. botulinum*). The species most commonly isolated from deep tissue infections include peptostreptococci ("anaerobic streptococci"), which are normally present in all of the sites mentioned above; *Prevotella, Porphyromonas,* and *Fusobacterium* species, which are normally present in the oral cavity; and the *Bacteroides fragilis* group of bacteria, which make up the bulk of the normal fecal flora. The most important clue to an anaerobic infection is its foul odor. This finding is diagnostic though present in only about one-half of cases. Other clues include tissue gas (observed as bullae or as crepitation on physical examination, or found on x-ray); tissue necrosis, the presence of multiple bacterial morphologies on Gram's stain of a specimen, and the failure of bacteria to grow on a routine aerobic culture ("sterile pus"). Settings in which anaerobic bacteria should always be suspected include bite wounds, aspiration pneumonia, lung abscess, pleural empyema, brain abscess, necrotising fasciitis, myonecrosis, diabetic foot ulcers, decubitus ulcers, and septic thrombophlebitis.

DIAGNOSIS AND CLINICAL REASONING

Diagnosis is of two types: presumptive and definitive. Presumptive diagnosis is usually based on the history and physical examination, sometimes supported by laboratory and radiographic findings. Definitive etiologic diagnosis usually requires cultures and serologic methods. In primary care, most diagnoses of infectious diseases are presumptive. This is understandable, since the conditions most commonly encountered tend to be self-limited and often involve the upper respiratory tract. For seriously ill patients including hospitalised patients, definitive diagnosis is usually desirable although sometimes difficult to achieve. Some principles of diagnosis include the following:

- Assume the worst-case scenario
- Search for a syndrome
- Look for atypical features
- Pay attention to the peripheral blood smear
- Perform diagnostic testing only when the results will alter patient management, but arrange for close follow-up
- Arrange for follow-up ("tincture of time")
- Document the level of diagnostic certainty
- "Think" tuberculosis and endocarditis

The major categories of clinical reasoning are, in ascending order of importance: (1)

pattern recognition; (2) probabilistic thinking; and (3) pathophysiology. Consider, for example, an 18-year-old woman with chief complaint of burning on urination. Pattern recognition and probabilistic thinking suggests uncomplicated UTI and hence a quick prescription for a 3-day course of antimicrobial therapy. The pathophysiologic approach, however, would be to ask further questions directed at determining whether the dysuria is external rather than internal and whether risk factors for sexually transmitted disease (STD) are present. One then determines whether to perform a pelvic examination and obtain studies for STD as well as a urine culture before beginning therapy.

What diagnostic tests should be obtained, and when? Generations of medical students have learned such gems as Sutton's law, Occam's razor, and Anselm's ass. A better and more sophisticated approach is to consider the properties of tests within the context of probability theory.

All clinicians must be familiar with "sensitivity," "specificity," and related terms. Sensitivity basically means "positive in disease"¯we say a test is 99% sensitive if positive test results are obtained in 99% of persons to have the disease by one or another gold standard, such as biopsy or autopsy. Specificity basically means "negative in health"¯we say that a test is 99% specific if positive test results are obtained in only 1% of persons who clearly do not have the disease. However, it is extremely important to consider "sensitivity," "specificity," and related concepts in the context of pre-test probability¯the likelihood that the patient has the disease.

This concept is best captured by Bayes's theorem, which holds that the likelihood that a positive test result is actually false-positive rather than true-positive varies inversely with the prevalence of the disease in the population. The likelihood that a positive test result is false-positive rather than true-positive is 100% if nobody in the population represented by the patient has the disease, but 0% if everybody has the disease. Between these extreme cases, the relationship is described not by a straight line but rather than by a hyperbolic curve. The upshot is that if the pre-test probability is extremely low, then the chances are overwhelming that a positive screening test result is actually a false-positive result even if the sensitivity and specificity of the test are superb. Increasingly, the concept of pre-test probability is being expressed as the likelihood ratio, and nomograms are available for evaluating of the usefulness of a test.

INFECTION CONTROL

Rigorous infection control is a moral imperative and legal requirement. All medical personnel should know the basic principles of disease transmission and control. Let us briefly review disease transmission as it applies to preventing infection in the office setting:

- Contact transmission involves person-to-person or object-to-person touching of mucous membranes or open skin. This is an important means of transmission of staphylococci, *Clostridium difficile,* and some respiratory viruses including respiratory syncytial virus. Frequent handwashing is the major defense against contact transmission, but attention should also be paid to routine disinfection of stethoscopes, toys, bathroom fixtures, and other objects with patient care areas.
- Droplet transmission involves coughing, sneezing, or suctioning procedures (as in bronchoscopy), resulting in a spray of secretions capable of contacting conjunctiva, nasal mucosa, and lips within a 3-foot radius. This is an important means of transmission of meningococci, influenza viruses, and pertussis. The use of eye protection including goggles and shields during certain procedures is a defense against droplet transmission.
- Airborne transmission involves inhalation of particles that are much smaller than droplets, often referred to as "droplet nuclei." This is an important means of transmission when organisms remain suspended in the air after coughing in the form of "droplet nuclei," as in tuberculosis (pulmonary and laryngeal), chickenpox, and measles (rubeola). Masks, ultraviolet lights, and immunisation constitute some of the defenses.
- Vehicle transmission by contaminated items, although now uncommon in health care settings as a result of tight regulations, still occurs and can cause outbreaks of even epidemics. Causes include use of expired medications or antiseptics, irrigation fluids that have been left in open containers, and use of diluted bleach solution that is over 24 hours old. Disease frequently involved organisms that survive well in water such as *Pseudomonas* species. Defenses include monitoring refrigerator temperatures, checking for expired medications, discarding irrigation solutions without preservatives at the end of the day of opening, and selecting disinfectants that do not require dilution.
- Vector transmission by insects or animals is extremely rare in today's health care facilities

UPPER RESPIRATORY TRACT INFECTIONS

Upper respiratory tract infection (URI) causes at least of one-half of all symptomatic illness in the community, exacting huge tolls that can be measured as morbidity, absenteeism from school and work, direct health care costs, and overuse of antibiotics leading to the emergence of drug-resistant bacteria. This disease burden is largely explained by anatomy. The nose, mouth, and pharynx are exposed to circulating viruses and are normally colonised by large numbers of bacteria including potential pathogens such as *S. aureus, S. pneumoniae, H. influenzae,* and group A streptococci. Mucosal injury caused by viral infection, allergy, or other factors compromises the mucociliary barriers designed to maintain sterility of the middle ears, paranasal sinuses, and lungs. Most URIs are self-limited but progression to life-threatening acute illness occurs and progression to chronic disease is common.

The Common Cold

The common cold is as an acute, self-limited catarrhal (Latin *catarrhus,* to flow down) syndrome limited to the mucosal membranes of the upper respiratory tract. Recently, "viral rhinosinusitis" has been suggested as the technically correct term, since CT scanning shows abnormalities of the paranasal sinuses in most cases. However, "cold" is preferred for clinical use since "rhinosinusitis" suggests to patients the need for antibiotics. The common cold accounts for up to three-quarters of all illnesses in young infants and up to one-half of illnesses in adults. Elimination of the common cold would also eliminate in the United States each year an estimated 27 million office visits, 23 million days of work absenteeism, 26 million days of school absenteeism, and nearly $2 billion worth of over-the-counter remedies.

Rhinoviruses, of which there are more than 100 serotypes, cause an estimated 30% to 50% of colds. Coronaviruses account for perhaps 10% of cases. Respiratory syncytial virus (RSV) is an important cause of cold symptoms in young children and among the elderly. Influenza virus can cause colds, but more often cause lower respiratory tract infection or systemic symptoms. Adenoviruses, of which there are at least 47 antigenic types, cause 5% to 10% of colds. Although echoviruses and coxsackieviruses have been associated with colds, they more commonly cause an undifferentiated flu-like illness or distinctive syndromes such as aseptic meningitis or pharyngitis. Rhinoviruses and parainfluenza viruses cause outbreaks of cold symptoms during the fall and late spring; RSV, adenoviruses, and coronaviruses cause outbreaks during the winter and early spring. Viruses that remain to be discovered probably cause a substantial fraction of colds.

Rhinoviruses are transmitted most efficiently by direct contact. Hand contact with the eyes and nose is very common in everyday life. Rhinoviruses remain viable on skin and also on objects (fomites) for at least 2 hours. Rhinoviruses can be recovered from hands in 40% to 90% of persons with colds and from up to 15% of objects near persons with colds. However, brief exposure such as a handshake or even being around an infected person for 36 hours causes transmission in less than 10% of subjects and in one study transmission was only 38% between spouses. Rhinoviruses can also be transmitted by aerosolisation, for example, by being in a crowded room where people are sneezing. Kissing does not seem to be a common mode of transmission, probably because only about 10% of persons with colds have demonstrable virus in their saliva. Studies carried out in Antarctica dispel the popular idea that cold weather increases susceptibility to rhinovirus infection.

The nasal epithelium of persons with colds is remarkably intact even when studied with the electron microscope. Symptoms are best explained by physiologic responses:

— release of chemical mediators of inflammation;

— sensitisation and irritation of airway receptors, with stimulation of the parasympathetic nervous system.

After an incubation period of 24 to 72 hours, most patients develop a sore or scratchy throat which is followed by nasal obstruction, rhinorrhea, and sneezing. A green or yellow nasal discharge should not be construed as evidence of secondary bacterial infection. By the second and third day of the illness, rhinitis with nasal congestion replaces sore throat as the major complaint. By the fourth and fifth day, nasal symptoms have usually decreased but in about 30% of cases are replaced by cough or "chest cold".

Acute Bacterial Sinusitis

Few common problems in primary care are as confusing as sinusitis. Acute bacterial sinusitis is vastly over diagnosed, but chronic sinusitis can be frustrating and disabling. The paranasal sinuses are accessible to direct examination only by sophisticated instruments. Adequate specimens for cultures can be obtained only by invasive procedures. Low-grade sinusitis is an intrinsic feature of the common cold. Clinicians must distinguish between self-limited viral rhinosinusitis and acute bacterial sinusitis, which usually calls for antibiotic therapy.

Sinusitis is usually caused by obstruction of the ostia, as from edema, damage to ciliated epithelial cells, and/or increased volume or viscosity of the mucous secretions. The pathogenesis of acute sinusitis can be discussed in 3 complementary ways: anatomy, physiology, and microbiology.

Anatomically, the outflow tract of the maxillary sinus sits in an awkward position high on the medial wall of the sinus cavity and is connected to the nasal cavity by a narrow tubular passage known as the infundibulum. Gravitational drainage of the maxillary sinus is therefore tenuous and easily disrupted. A small area between the middle and inferior nasal turbinates where drainage from the maxillary, ethmoid, and frontal sinuses converges is known as the ostiomeatal complex. Anatomic or physiologic compromise of the ostiomeatal complex predisposes not only to maxillary sinusitis but also to infection of multiple sinuses (pansinusitis). Causes of mechanical obstruction of sinus drainage include deviated nasal septum, polyps, foreign bodies, tumors, concha bullosa (enlarged middle turbinates from pneumatisation, present in 10% of the population), ethmoid bullae, choanal atresia, and most commonly mucosal swelling.

Mucosal swelling is usually due to viral infection or allergic inflammation, but can also be caused by systemic disorders (such as cystic fibrosis and the dyskinetic cilia syndromes) or injury brought about by trauma, swimming or diving, or overuse of topical medication (*rhinitis medicamentosa*) or cocaine.

Physiologically, the sinuses are kept sterile mainly by a mucociliary blanket that changes 2 to 3 times per hour, a rate sufficient to prevent mucus from accumulating in the sinuses. Sinusitis results when mucociliary drainage of the paranasal sinuses fails because of mucosal edema, dysfunction of the ciliated epithelial cells, or both. Viral rhinitis and allergic rhinitis disrupt the epithelium as do cigarette smoking and intranasal cocaine use. Swimming predisposes to sinusitis for reasons that are somewhat unclear, although chlorine is known to irritate the mucosa. Other factors that may help to preserve the normal sterility of the paranasal sinuses include cellular and humoral immunity and nitric oxide.

Microbiologically, acute sinusitis typically starts with viral infection that paves the path for pathogenic bacteria. The major pathogens are *Streptococcus pneumoniae* and non-typeable strains of *Haemophilus influenzae*. Other bacteria associated with community-acquired sinusitis include group A streptococci, *S. aureus*, *Moraxella catarrhalis*, and viridans streptococci. Whether *Chlamydia pneumoniae* commonly causes sinusitis is currently being studied. Patients who have had nasogastric tubes in place are vulnerable to sinusitis caused by aerobic gram-negative rods such as *Pseudomonas aeruginosa* and *Klebsiella pneumoniae*. Anaerobic bacteria are associated with acute sinusitis mainly in the setting of dental disease. Anaerobic bacteria possibly play a role in chronic sinusitis.

Sinusitis usually begins with symptoms of the common cold such as runny nose, nasal obstruction, sore throat, cough, and the sensation of "pressure" or "tightness" in the face. The symptoms differ somewhat between children and adults. Children with sinusitis can have either of two presentations. The more common presentation consists of *persistent* cold symptoms that is, symptoms lasting more than 10 days. Children with persistent sinusitis seldom complain of headache or facial pain. Parents of young children often report malodorous breath. The less common presentation consists of *severe* cold symptoms-that is, cold symptoms that are accompanied by high fever (> 39° C) and purulent nasal discharge. Some of these children experience headaches usually located behind or around the eye, occasionally with periorbital edema. Adults with sinusitis, compared with children, tend to have more prominent facial pain, sometimes with local tenderness, swelling, and erythema. Pain patterns and other findings vary according to which sinuses are involved:

— *Maxillary sinusitis* (the most common location): Pain is over the cheekbones or above the maxillary 2nd molar teeth.

— *Ethmoid sinusitis* (common): Pain is between the orbits and the nasal bridge.

— *Frontal sinusitis* (uncommon): Pain is over the frontal bones.

— *Sphenoid sinusitis* (least common, but dangerous): Pain can be frontal, temporal, orbital, or occipital; facial pain due to Vth nerve involvement is characteristic; pain is occasionally perceived over the vertex of the skull.

Low-grade fever is often present. Physical examination may reveal tenderness on percussion of the maxillary or frontal sinuses, and pinching the bridge of the nose may bring out tenderness if the ethmoid sinuses are involved. However, the sensitivity and specificity of these findings are unknown, and the clinical findings are often subtle.

Acute bacterial sinusitis is often self-limited, but the frequency with which this condition resolves spontaneously is unknown. Direct sinus puncture for accurate diagnosis has not been carried out in placebo-controlled clinical trials. Serious complications occur often enough to justify close follow-up, especially in the clinically more severe cases. The most common complication is progression to chronic sinusitis. Cavernous sinus thrombosis can result from ethmoid, frontal, or sphenoid sinusitis. Ethmoid sinusitis can also cause orbital cellulitis. Frontal sinusitis can also cause osteomyelitis of the frontal bone with swelling and edema of the forehead and subdural empyema. Sphenoid sinusitis can be a medical emergency.

Chronic Sinusitis

When acute sinusitis fails to resolve and becomes chronic, cultures may reveal a variety of opportunistic pathogens including anaerobic bacteria. Some authorities feel that the problem is no longer mainly "infectious" but rather reflects permanent mucosal injury. It has been estimated that chronic sinusitis causes morbidity, measured as absenteeism from school, work, or social activities, of the same magnitude as heart disease and arthritis.

The prevailing view holds chronic sinusitis to be a disorder of abnormal anatomy and physiology of the paranasal sinuses with one or more causes:

— previous acute sinusitis

— nasal polyposis (as in the triad of asthma, allergies, and aspirin sensitivity)

— previous sinus surgery

— cystic fibrosis

These processes lead to anatomic changes including obstruction of the infundibula and ostia, mucosal edema and scarring, bone hypertrophy, polypoid degeneration, and/or mucosal fibrosis, rendering the mucociliary clearance mechanism defunct.

Numerous microorganisms can be isolated from patients with chronic sinusitis, but correlation between culture results and the disease process is often poor. Mixtures of aerobic and anaerobic bacteria are common. The general conclusion at this time is that in most patients, no single microorganism can be assigned a pathogenic role. In some patients, however, *Pseudomonas aeruginosa* or *Staphylococcus aureus* seems to be clearly pathogenic, and there are data suggesting roles for *Haemophilus influenzae* and *Moraxella catarrhalis* (in children). Patients with chronic sinusitis often have exacerbations analogous to the acute exacerbations of chronic obstructive lung disease. In these instances, especially in children, *Streptococcus pneumoniae* and *Haemophilus influenzae* may be important. Numerous bacteria including gram-negative rods have been isolated with patients with post-operative sinusitis. Some investigators believe that many patients with chronic sinusitis have *allergic fungal sinusitis,* the disease manifestations being caused by an immune response to extramucosal fungi.

The typical history consists of nasal drainage, obstruction, and postnasal drip lasting for at least several months, often against a background of chronic "sinus trouble." Patients often complain of headache or "sinus pain," nocturnal cough, and bad breath. Loss of smell may also be present. Fever is unusual. A history of inhalant allergy is 4.5 times more common in patients with chronic sinusitis than in persons without chronic sinusitis.

Although physical examination can reveal findings such as nasal septal deviation or mucosal changes, imaging studies are necessary to make a correct diagnosis. The coronal CT scan represents the current gold standard, but axial CT scans are often useful especially in children. Sinus endoscopy an also provide invaluable information.

Chronic sinusitis may have a relapsing or remitting course, but, untreated, patients seldom become entirely free of symptoms related to the paranasal sinuses. Complications include remodeling of the facial bones, osteomyelitis, and occasionally invasive disease of the CNS caused by bacteria or fungi.

Fungal Sinusitis

Fungal sinusitis is relatively uncommon, but should be considered in patients with chronic sinusitis because of its potentially serious complications. *Aspergillus* species are the most common causes of fungal sinusitis, and fungi of the order Mucorales are the most dangerous (rhinocerebral mucormycosis). Various widely distributed pigmented fungi that are collectively known as dematiaceous molds can cause a variety of syndromes that include life-threatening disease; examples of these organisms include *Alternaria, Bipolaris, Cladosporium, Curvularia,* and *Exserohilum.* Numerous other fungi sometimes cause sinusitis.Five syndromes are currently recognised.

— Simple colonisation of the paranasal sinuses by fungi may be relatively common, although the incidence is unknown.

— Sinus mycetoma (fungus ball) presents most often as a mass in the maxillary sinus. Underlying disease is usually not present although some patients have nasal polyps and chronic bacterial sinusitis. Patients usually seek medical attention for nasal obstruction, facial pain, symptoms of chronic sinusitis, or fetid breath (cacosmia). Seizures have been reported as a presenting manifestation.

— Allergic fungal sinusitis usually presents as intractable sinusitis with nasal polyposis in patients with atopy. Some patients also have allergic bronchopulmonary aspergillosis. Children with allergic sinusitis may develop hypertelorism or proptosis when the frontal or ethmoid sinuses are involved. As mentioned above, some investigators believe that allergic fungal sinusitis is present in a majority of patients who carry the diagnosis of "chronic rhinosinusitis".

— Acute (fulminant) invasive fungal sinusitis is essentially synonymous with rhinocerebral mucormycosis, which usually occurs in patients with diabetes mellitus or severe immunosuppression from other causes. The sinuses have been called "way stations to the brain" in this medical emergency, which classically presents as a painless black eschar on the palate or a nasal turbinate followed by epistaxis, headache, changes in mental status, and focal neurologic symptoms and signs (such as diplopia). Other fungi including *Aspergillus* species, *Fusarium* species, and *Pseudallescheria boydii* can cause an identical syndrome, typically in patients who are severely immunosuppressed from disease (including AIDS) or chemotherapy for cancer or organ transplantation.

— *Chronic invasive fungal sinusitis* can occur not only in immunocompromised patients but also in patients who are immunologically normal. Most of these latter patients have chronic sinusitis and nasal polyposis. Causative organisms include the dematiaceous molds noted above; the condition is known as phaeohyphomycosis. Dense masses of fungal elements resembling mycetoma are found, but there is also invasion into the mucosa and then into bone. Patients present with headache and localising symptoms such as decreased vision and loss of eye movement (orbital apex syndrome) or behavioral changes (mycetoma of the frontal lobe).

Otitis Externa

Otitis externa, a spectrum of conditions caused by infection, allergy, or primary skin disease, affects up to 10% of all people during their lifetimes. Necrotising (malignant) otitis externa is a medical emergency seen usually in patients with diabetes mellitus or a compromised immune system or in patients who have had prior irradiation of the head.

Otitis externa begins with breakdown in the cerumen barrier. Cerumen, although commonly considered a nuisance, protects against infection by (1) creating an acidic environment hostile to bacterial and fungal growth; (2) promoting a dry environment through its hydrophobic properties; and (3) trapping debris by its sticky nature. Excessive cleaning or scratching of the ear canal promotes breakdown of the cerumen barrier. Swimming notoriously predisposes to otitis externa, the main effect being promotion of a more alkaline pH, which in turn promotes bacterial growth. However, increased moisture of any origin leads to maceration of the skin and breakdown. Mechanical trauma from devices such as ear plugs (used for hearing conservation in many industries), headphones, hearing aids, and diving caps also predisposes to otitis externa.

Pseudomonas aeruginosa and *S. aureus* are common causes of otitis externa. Group A streptococci and various gram-negative rods sometimes cause this condition, and anaerobic bacteria are involved in up to 25% of cases. About 10% of cases are caused by various fungi, with *Aspergillus* species being the most common followed by *Candida* species.

- Acute localised otitis externa presents as a single or several pustules or furuncles located in the ear canal, usually due to *S. aureus*. Initial symptoms include itching, pain, swelling, redness, and sometime decreased hearing. A small furuncle that would be inconspicuous on most parts of the skin becomes intensely painful in the confines of the narrow ear canal.

- Acute diffuse otitis externa (swimmer's ear) typically presents with pain (otalgia), itching, discharge, and hearing loss. The pain can become intolerable. There is often a history of recent exposure to water. Mobile redwood hot tub systems were associated with a severe hemorrhagic form of the disease. *Pseudomonas aeruginosa* is the most common pathogen. Other patients give a history of ear instrumentation, excessive cleaning, previous infection, otitis media, tinnitus, or vertigo. Examination reveals the ear canal to be diffusely red and edematous.

- Erysipelas involving the concha and canal is caused by group A streptococci, the pathogenesis being similar to that of erysipelas elsewhere. Patients present with a diffusely red and painful ear. Examination may reveal hemorrhagic bullae on the walls of the canal and on the tympanic membrane, and there may be tender regional lymphadenopathy.

- Chronic otitis externa presents with mild discomfort and flaking of the skin of long duration, often with a history Otitis media can be graded in severity from mild (minor pain and pruritus, minimal edema of the ear canal) to severe (severe pain and pruritus, complete occlusion of the ear canal, auricular and periauricular erythema, and, frequently, fever and lymphadenopathy).

— Malignant otitis externa is a medical emergency usually encountered in patients with diabetes mellitus. A history of chronic otitis media with drainage is often obtained. *Pseudomonas aeruginosa* is the usual causative organism. Complications include osteomyelitis of the temporal bone (a defining feature of the disease) and extension to the base of the skull, with cranial nerve palsies, meningitis, and other serious consequences.

Diagnosis of otitis externa is nearly always made by the history and physical examination. Pressure on the tragus or pulling the auricle superiorly causes pain; the latter maneuver is a valuable diagnostic aid. Examination of the ear canal shows erythema and edema and, in severe cases, partial or complete occlusion of the canal. To exclude otitis media, one should demonstrate with pneumatic insufflation that the tympanic membrane is mobile. However, the tympanic membrane is often partially or totally obscured by edema in the ear canal.

Although otitis externa is usually considered a self-limited condition, serious complications can occur. Perforation of the tympanic membrane can be caused by extension of the disease process or by misguided attempts by patients or health care providers to relieve the condition through mechanical manipulation. Other complications include stenosis of the ear canal, auricular cellulitis, or chondritis.

Acute Otitis Media

The importance of otitis media in primary care cannot be overemphasized, especially in pediatrics practice, where it accounts for about 25% of all office visits, 50% of office visits for illness, and 40% of antibiotic prescriptions. By age 5, between 75% and 95% of children have had at least one episode of otitis media. This disease is responsible for some 25 million office visits each year, with annual health care costs estimated to be between 3 and 5 billion dollars. Although 80% of patients with otitis media are less than 15 years of age, more than one-fourth of all oral antibiotic prescriptions in the United States are written for this condition. Heavy prescribing of â-lactam antibiotics for otitis media is thought to be responsible in large measure for the decreasing drug susceptibility of *S. pneumoniae* strains.

Acute otitis media is often preceded by viral URI that causes edema and obstruction of the eustachian tube, causing an *ex vacuo* serous transudate into the middle ear and paving the way for pathogenic bacteria. Children are predisposed because their eustachian tubes are shorter, wider, and straighter compared with those of adults. Appreciation of otitis media requires knowledge of the relationship of the eustachian tube to the nasopharynx, the middle ear, and the mastoid air cells. The eustachian tube normally serves to regulate pressure in the middle ear, to protect against nasopharyngeal sound pressure and secretions, and to afford a pathway for the drainage of secretions produced within the middle ear into the nasopharynx. The latter function requires an intact mucociliary system. Factors predisposing to eustachian tube dysfunction and otitis media include allergy, cleft palate, ciliary dysmotility, immunodeficiency, exposure to tobacco smoke, exposure to frequent upper respiratory tract infections (notoriously, in day care centers), early age of first infection, and race. Native Americans are markedly predisposed to otitis media for reasons that are unclear. Adults can be predisposed to otitis media on account of diabetes mellitus, cancer, immune deficiencies, and injection drug use. The adenoids have long been implicated in the pathogenesis of otitis media, for better or worse, since

— inflammation of the adenoids causes inflammatory obstruction of the adjacent eustachian tube orifices;
— colonisation of the adenoids by pathogenic bacteria promotes invasion of the middle ear by these same bacteria;
— yet the adenoids provide local immunity in the form of secretory IgA secretion.

The microbiology in otitis media is similar in adults and children when tympanocentesis is carried out. In about 40% of cases, culture of middle ear fluid fails to show a bacterial pathogen. This might reflect a viral etiology or sterile inflammation. Respiratory syncytial virus has been found relatively commonly, with a special tendency to cause the disease in children. Influenza viruses and parainfluenza viruses also predispose to otitis media. *Streptococcus pneumoniae* causes about 30% to 40% of cases, and the prevalence of strains with reduced susceptibility to penicillin is increasing. *H. influenzae* causes between 20% and 30% of cases, and *M. catarrhalis* between 10% and 15% of cases, especially in children. Group A streptococci cause less than 5% of cases but can cause up to 10% during the winter months.

Acute otitis media in children usually presents with rapid onset of otalgia, fever, and/or irritability. Otalgia in young infants is manifest by pulling on the ear. Young children can also have anorexia, loose stools, and vomiting. Otalgia tends to be the major symptom in adults. A minority of patients experience spontaneous perforation of the tympanic membrane.

The tympanic membrane is abnormal, often bulging, with loss of the usual landmarks. Erythema of the tympanic membrane alone is not diagnostic; it can be caused, for example, by crying. Purulent fluid is sometimes seen behind the tympanic membrane. The key procedures are pneumatic otoscopy and, when indicated, tympanocentesis. Pneumatic otoscopy, which is done by gently squeezing and then releasing a rubber bulb attached to the otoscope, provides information about the mobility of the tympanic membrane. Tympanocentesis provides fluid for culture, which is becoming more important due to the emergence of drug-resistant bacteria. Cultures of the nasopharynx correlate poorly with cultures of fluid obtained by tympanocentesis, and are therefore of limited usefulness.

Acute otitis media must be distinguished from otitis media with effusion (serous otitis media). The latter consists of an asymptomatic or hyposymptomatic middle ear effusion, which can be acute (less than 3 weeks), subacute (3 weeks to 3 months), or chronic. Although hearing loss is frequently present in both acute otitis media and otitis media with effusion, patients with otitis media and effusion lack systemic signs and symptoms such as otalgia and fever.

The natural history of untreated otitis media continues to prompt debate whether most cases should be treated with antibiotics. A meta-analysis of the literature based on data obtained from 5400 children in 33 studies indicates that 81% percent of children have spontaneous resolution. However, it is difficult to predict on clinical grounds whether an individual patient?s disease will resolve spontaneously, and the current consensus opinion in the United States is that all patients should be treated.

Chronic Suppurative Otitis Media and Mastoiditis

Chronic suppurative otitis media is a complication of acute otitis media, usually occurring when there is a defect in the tympanic membrane, such as a "central" perforation or a tympanostomy tube. It is accompanied by purulent discharge (otorrhea). Mastoiditis is invariably present. The associated bacteria seem to vary depending on whether an infected cholesteatoma is present. Cholesteatoma is often associated with anaerobic bacteria and "skin flora" microorganisms, and the otorrhea often has a foul odor. When cholesteatoma is not present, gram-negative rods including *Pseudomonas aeruginosa* and *E. coli* are often found.

In the pre-antibiotic era, mastoiditis was often a dramatic and severe illness with retroauricular inflammation and serious intracranial complications. Today, mastoiditis is more typically an indolent, low-grade, often painless infection of the temporal bone that tends to be clinically silent ("masked mastoiditis") unless a complication such as brain abscess develops. Patients at high risk of complications include newborn infants, persons with diabetes mellitus, the elderly, and the immunocompromised.

Spontaneous resolution is rare, if it occurs at all. Local complications of chronic suppurative otitis media and mastoiditis include bone destruction, subperiosteal abscess, facial paralysis, labyrinthitis, and petrositis. Intracranial complications include brain abscess, subdural abscess, epidural abscess, septic thrombosis of the lateral sinus, meningitis, and hydrocephalus. Patients with chronic suppurative otitis media and/or mastoiditis should be referred to an otolaryngologist, as effective treatment usually requires surgical intervention.

Acute Suppurative Parotitis

Acute suppurative parotitis, an ûncommon condition, usually results from decreased salivary flow, allowing retrograde ascent of bacteria from Stenson's duct. It can also result from ductal obstruction from mucus or fibrinous plugs, tumors, or foreign bodies). The clinical signs and symptoms include an acutely swollen, indurated check with fever and pain. Pus can be expressed from Stenson's duct. More than 80% of cases are caused by *Staphylococcus aureus*. Most patients are elderly, debilitated, and dehydrated. Mortality is high.

Acute Pharyngitis

Acute pharyngitis (sore throat) is one of the most common problems encountered in clinical practice. Viruses cause most cases as part of the common cold. However, about 15% of cases, and up to 50% of cases in children during some periods, are caused by group A â-hemolytic streptococci (*S. pyogenes*). Although usually self-limited, streptococcal pharyngitis demands respect as a cause of acute rheumatic fever and¯less commonly¯major suppurative complications, acute glomerulonephritis, and even the streptococcal toxic-shock syndrome. The clinician's task is to determine in a cost-effective manner which patients need treatment and which do not.

Acute pharyngitis has many known etiologies, and pathogens remain to be discovered for an estimated 30% of cases. Viral infections cause sore throat, it is thought, by generating bradykinin and lysyl bradykinin, which stimulate nerve endings. Group A streptococci and certain other pathogens including some of the respiratory viruses cause pain by invading the mucosa.

Group A streptococci are carried in the human nasopharynx and transmitted from person to person usually by direct contact with saliva or nasal secretions. Acquisition is greatest in school-aged children, suggesting the gradual development of immunity over time. Children also serve as a reservoir for spread among family members. Asymptomatic pharyngeal carriage of group A streptococci is relatively common, and the factors that cause some persons to develop acute pharyngitis and other complications are poorly understood. Group C and group G streptococci cause a pharyngitis syndrome clinically indistinguishable from that caused by group A streptococci, sometimes recognised as outbreaks related to a common food source. Group C streptococci (*S. dysgalactiae* subspecies *equisimilis*) appear to be a frequent cause of pharyngitis in college-aged students.

Pharyngitis due to group A streptococci occurs most frequently in children between 5 and 15 years of age, usually during the winter and early spring. In its severe form the disease starts abruptly with fever, sore throat, and odynophagia. Chills, headache, and abdominal pain are sometimes present. Examination reveals diffuse erythema of the pharynx and tonsils accompanied by a patchy, purulent tonsillar and pharyngeal exudate, hypertrophy of the lymphoid nodules in the posterior pharyngeal mucosa, and tender cervical lymphadenopathy. Occasional strains of *S. pyogenes* elaborate the erythrogenic toxin of scarlet fever, resulting in a striking rash and "red strawberry tongue" with enlargement of the papillae. Rhinorrhea and cough are usually not present, but may occur. However, these dramatic manifestations are absent in many, perhaps most cases of streptococcal pharyngitis. Because the features of group A streptococcal pharyngitis blend imperceptibly with those of other causes of sore throat, numerous students of the disease have concluded that the diagnosis *must* be secured by laboratory methods prior to definitive treatment. Other syndromes include the following:

— The common cold is often accompanied by sore throat, which is frequently the first symptom but is usually not the main complaint when patients seek medical care. Rhinorrhea, postnasal drainage, and cough are usually more prominent symptoms than sore throat. Fever is seldom prominent and severe sore throat with odynophagia are uncommon.

— Influenza sometimes presents with sore throat as the chief complaint, but is usually accompanied by other symptoms suggesting influenza such as myalgia, headache, and cough. In pharyngitis due to the common cold or to influenza, purulent pharyngeal or tonsillar exudates and tender cervical lymphadenopathy are not present. Tracheal tenderness may be present in influenza, indicating diffuse viral infection of the respiratory mucosa.

— Adenoviral pharyngitis, on the other hand, often presents with sore throat as the chief complaint. Fever, chills, headache, malaise, and myalgias can be prominent. About one-third to one-half of patients with adenoviral pharyngitis also have follicular conjunctivitis; this syndrome is known as pharyngoconjunctival fever. Patients with adenoviral pharyngitis often have pharyngeal exudate, so the disease can mimic streptococcal pharyngitis. Epidemics of pharyngoconjunctival fever occur during the summer months in civilian populations and during the winter months in military recruits.

— Infectious mononucleosis due to the Epstein-Barr virus causes exudative tonsillitis or pharyngitis in about one-half of cases. Tonsillar and pharyngeal exudates can be prominent. If examined with Wright's stain, the exudates of mononucleosis consist mainly of mononuclear cells in contrast to the exudates of streptococcal pharyngitis, which consist mainly of polymorphonuclear neutrophils. Tender cervical lymphadenopathy is often prominent in the posterior triangles of the neck (spinal accessory chain of nodes), contrasting with the prominent anterior nodal enlargements (jugulodigastric chain) typical of bacterial pharyngitis. Patients with mononucleosis usually have headache, fatigue, and other features of the disease such as palpable splenomegaly (about one-half of cases). Peripheral blood smear typically shows lymphocytosis with atypical lymphocytes. Cytomegalovirus mononucleosis can also cause sore throat, but pharyngeal exudate is rare.

— Primary infection with HIV, known as the acute retroviral syndrome, sometimes presents with fever and pharyngitis. Pharyngeal erythema can be marked but exudates do not seem to occur. This illness usually occurs within 3 to 6 weeks of the initial infection, during a phase of initial viral multiplication and prior to the appearance of HIV antibodies. Fever, lethargy, arthralgia, and myalgia are usually prominent and many patients have a nonpruritic maculopapular rash. Lymphadenopathy, which can be in the anterior and/or posterior triangles, appears about one week after the onset of pharyngitis.

— Herpes simplex virus infection can cause pharyngitis, which can sometimes resemble viral or streptococcal pharyngitis. Vesicles and shallow ulcers on the palate suggest the herpetic etiology. These can be extensive and confluent, causing severe oral pain. Herpangina is an uncommon syndrome caused by coxsackieviruses and is seen mainly in children. It is characterised by small (1 to 2 mm) vesicles on the soft palate, uvula, and anterior tonsillar pillars, which rupture to form small white ulcers. Fever, sore throat, and dysphagia can be severe and occasional patients experience anorexia and abdominal pain suggesting acute appendicitis.

— *Chlamydia pneumoniae* can cause pharyngitis with or without infection of the lower respiratory tract. No distinctive features have been described. There is some epidemiologic evidence that *Mycoplasma pneumoniae* causes pharyngitis, usually mild and again without distinctive features.

— Diphtheria should be mentioned, as it still causes occasional cases of pharyngitis in the United States in patients who have not been vaccinated. Classical diphtheria (*Corynebacterium diphtheriae*) has a slow onset followed by marked systemic toxicity. Sore throat is usually not severe despite the finding of a gray "pseudomembrane" adherent to the tonsillar and pharyngeal mucosa. *Arcanobacterium hemolyticum* (formerly known as *Corynebacterium hemolyticum*) has been increasingly recognised as a cause of exudative pharyngitis in adolescents and young adults, associated with a diffuse and sometimes pruritic maculopapular rash on the trunk and extremities. Rarely, this disease can mimic diphtheria. *Corynebacterium ulcerans* is a rare cause of pharyngitis associated with the ingestion of raw milk.

— Anaerobic pharyngitis (Vincent's angina) presents with a purulent exudate and often a foul odor to the breath. This uncommon infection is caused by a mixture of anaerobic bacteria and spirochetes, with group A streptococci and *S. aureus* sometimes playing a role. This infection sometimes progresses to peritonsillar abscess (quinsy) or to septic thrombophlebitis of the internal jugular vein (Lemierre syndrome).

- *Yersinia enterocolitica* can cause exudative pharyngitis with a fulminant course associated with high mortality. Fever, tender cervical lymphadenopathy, and abdominal pain with or without diarrhea are prominent features. This illness can occur in outbreak form due to ingestion of contaminated food or beverages.
- Tularemia can present as pharyngitis in the typhoidal form of the disease due to inhalation of the organisms, which can occur while skinning an infected rabbit
- Kawasaki syndrome, a systemic vasculitis affecting mainly infants and young children, can present with fever and sore throat. Diffuse oropharyngeal erythema without exudate is found during the acute febrile phase of the illness. Other features include bilateral, nonpurulent conjunctivitis, erythema with fissuring, cracking, and bleeding of the lips, strawberry tongue, edema of the hands and feet with erythema of the palms and soles, and an erythematous rash.

Miscellaneous causes of sore throat include juvenile rheumatoid arthritis, systemic lupus erythematosus, bullous pemphigoid, Behçet's disease, paraquat ingestion, and drug reactions.

Acute Laryngitis

Acute laryngitis is extremely common, usually occurring as part of upper respiratory tract infection. Treatment is symptomatic, but prolonged hoarseness mandates the search for other etiologies. Acute laryngitis is most often caused by respiratory viruses, but vocal abuse or gastroesophageal reflux must also be considered. Parainfluenza viruses are the usual cause in patients between ages 5 and 15. Hoarseness complicates up to 29% of rhinovirus infections, 35% of influenza virus and adenovirus infections, and 63% of coronavirus infections according to various studies. Hoarseness can also complicate acute streptococcal pharyngitis. *H. influenzae* and *Moraxella catarrhalis* are often isolated, but their pathogenic roles are unclear. Rarely, fungi such as *Candida* species, *Cryptococcus neoformans* and *Coccidioides immitis* can cause laryngitis. Uncommonly, laryngitis can also be caused by tuberculosis and blastomycosis.

Acute laryngitis presents as hoarseness. Speech and/or swallowing may be painful. The voice is hoarse, harsh, broken, or nearly absent. There are often concomitant symptoms of common cold including sore throat. Diagnosis of acute laryngitis is made on clinical grounds, typically by the history. Direct or indirect laryngoscopy shows erythema and edema of the vocal folds, sometimes with submucosal bruising or microhemorrhages (if the patient continues major voice use or has a bad cough). Viral cultures and special studies are seldom indicated.

Odontogenic Infections

Odontogenic infections, the most common infections of the oral cavity, begin in and around the teeth and can spread to cause life-threatening local or systemic complications.

The normal oral cavity contains dense masses of bacteria, of which more than 80% or more are anaerobic. Viridans streptococci (such as *S. mitis*, *S. sanguis*, *S. salivarius*, and *S. mutans*) preferentially colonize one or another anatomic site. *Streptococcus mutans* is the major cause of dental caries (cavities). As is well known, poor oral hygiene facilitates development of dental plaque, composed mainly of anaerobic bacteria. Diet high in simple sugars and carbohydrates predisposes to plaque. Periodontal disease, on the other hand, is unrelated to diet but is associated with poor oral hygiene, increasing age, and with various congenital immunodeficiency diseases and juvenile diabetes mellitus. Gingivitis and periodontitis are caused mainly by anaerobic bacteria. Once established, suppuration arising in and around the teeth can spread along fascial planes, invade bone, or enter the bloodstream. Deeper infections requiring surgical drainage, such as periapical abscesses and deep fascial space infections, are usually associated with numerous anaerobic bacterial species.

Odontogenic infections present as one of several syndromes:

— Periapical abscess and acute alveolar abscess represent infection of the tooth pulp. In the early stages, the tooth is sensitive to touch and also to both hot and cold. In the later stages, the tooth is extremely painful to heat, the pain being relieved by cold.

— Gingivitis is often first manifest by friability and bleeding of the gums. A more severe form, known as acute necrotising gingivitis (Vincent's angina or trench mouth), presents with sudden onset of severe pain and necrosis of the gingiva, typically with a grayish pseudomembrane, halitosis, and systemic symptoms including fever, malaise, and generalised lymphadenopathy.

— Periodontitis, broadly defined as infections of the periodontium (the supporting structures of the teeth, which include alveolar bone, cementum, and the periodontal ligament as well as the gingiva) begins gradually, usually in early adulthood, and is associated with plaque below the gingival margin. Patients sometimes complain of itchy sensations of the gums in between the teeth, bad taste in the mouth, vague jaw pains, and sensitivity to both hot and cold. Pus can sometimes be expressed by pressure on the gingival margin. Periodontal abscess can present as a red, fluctuant swelling of the gingival margin. Pus can readily be expressed after probing.

— Pericoronitis consists of acute localised pain over a partially erupted or impacted wisdom tooth. The gums are red and swollen, and a small amount of pus can usually be expressed by pressing on the gum flap overlying the partially erupted tooth.

Trismus caused by irritation of the masseter or medial pterygoid muscle is often marked.

Diagnosis of dentoalveolar infections, gingivitis, and periodontal infections is based upon symptoms, examination, and, when indicated, dental x-rays. Dental x-rays are especially valuable for visualising periapical abscess and acute alveolar abscess.

Mouth Ulcers

Careful examination of the oral cavity often provides evidence of important systemic and local disease. More commonly, primary care clinicians evaluate patients with self-limited diseases such as stomatitis due to the herpes simplex viruses and aphthous stomatitis (canker sores). Here we will briefly review these latter conditions and their differential diagnosis.

Herpes simplex viruses 1 and 2 cause fever blisters and, less commonly, lesions elsewhere in the mouth including the palate. The cause of aphthous stomatitis is unknown. Current opinion favors an immunopathogenesis involving T-cell immunity, possibly a delayed-type hypersensitivity reaction to an antigen residing within the epithelium. Some cases of aphthous stomatitis have been attributed to drugs. Stress, smoking, hormonal factors, and food allergy have also been invoked, but the evidence is unconvincing.

Herpes simplex stomatitis begins with small, 1 to 2 mm vesicles that rupture, leading to ulceration. Confluent lesions can give rise to large areas of ulceration. Lesions typically occur on keratinised or attached oral mucosal surfaces such as the lips, the gingiva, the lateral surfaces of the tongue, or the palate. Aphthous stomatitis begins with ulcers of various sizes, usually from several mm to 1 cm or more in diameter. Lesions typically occur on non-keratinised, unattached mucosal surfaces. Both types of lesions are painful.

LOWER RESPIRATORY TRACT INFECTIONS

Acute Bronchitis

Tracheobronchial infections without pneumonia comprise a spectrum of disorders with different clinical implications: acute bronchitis, chronic bronchitis, and bronchiectasis. Acute bronchitis or "acute simple bronchitis" in otherwise-healthy persons is extremely common, usually of viral etiology, and a common reason for overuse of antibiotics. The term "acute infectious bronchitis" is sometimes used to distinguish this entity from other causes of cough, and the term "tracheobronchitis" is sometimes used for accuracy since the trachea is also inflamed. However, "chest cold" is probably the best term for daily practice since it implies that antibiotics are seldom necessary. *Bordetella pertussis*, the agent of whooping cough, is now recognized as a cause of acute bronchitis in adults.

Infection by either *Mycoplasma pneumoniae* or *Chlamydia pneumoniae* accounts for many of the stubborn cases in which symptoms fail to resolve or recur soon after treatment has been discontinued.

Infection of the tracheobronchial mucosa causes local inflammation, increased secretion of mucus, and damage to ciliated cells. Symptoms result both from the inflammatory response and also from the interruption of the mucociliary blanket that normally cleanses the lower respiratory tract. Most cases of acute bronchitis (95% by some estimates) are caused by viruses. All of the common viruses affecting the upper respiratory tract have been implicated: rhinoviruses, coronavirus, respiratory syncytial virus, adenoviruses, coxsackieviruses, influenza viruses A and B, and parainfluenza virus. In 2 studies in which attempts were made to establish a precise diagnosis, the etiology was established in only 16% and 29% of cases with viruses being the most common causes.

Mycoplasma pneumoniae and *Chlamydia pneumoniae* probably play minor roles in this illness, at least in most populations. However, *M. pneumoniae* probably causes more cases of bronchitis than pneumonia, and *C. pneumoniae* may be an important cause of acute bronchitis in college-aged students. Whether *S. pneumoniae, H. influenzae,* and *M. catarrahalis* cause chest cold in otherwise-healthy persons is unclear, but there is little support for the concept of "acute bacterial bronchitis" as a community-acquired disease. Recently it has been emphasized that *Bordetella pertussis* can infect adults, even when vaccinated, providing a reservoir for causing whooping cough among infants. *Bordetella parapertussis* causes a protracted illness similar to whooping cough but without systemic toxicity. Whether these observations can be generalized to other populations is undetermined.

The onset is typically preceded by a prodrome of at least 24 hours with symptoms of coryza and pharyngitis. A dry cough, signifying early inflammation of the upper airway, often evolves into a cough productive of moderate amounts of mucopurulent sputum. Fever, headache, myalgias, and retrosternal chest pain or discomfort may be present. Fever is most common when an influenza virus or *Mycoplasma pneumoniae* is the causative agent. The patient rarely looks toxic. Tracheal tenderness is often present. Auscultation may reveal a few coarse crackles with occasional wheezes in the chest, but there are no signs of consolidation.

Acute Infectious Exacerbations of Chronic Bronchitis

Chronic bronchitis is defined by the American Thoracic Society (ATS) as excessive sputum production with cough, present on most days for at least 3 months a year and not less than 2 successive years, without an underlying etiology such as tuberculosis or bronchiectasis. This common disorder, affecting up to 25% of the adult population, can lead to full-blown chronic obstructive pulmonary disease (COPD), the fourth-leading cause of death in the United States. The extent to which acute exacerbations are due to treatable infections remains controversial.

Chronic bronchitis is caused mainly by cigarette smoking. Air pollution, cold and damp climates, heredity, frequent lower respiratory tract infections, and immunodeficiency disorders (such as common variable hypogammaglobulinemia or isolated IgA deficiency) play a role in some patients. The essential feature is anatomic change in the larger airways, including an increased number of mucus-producing goblet cells and mucosal gland hypertrophy in the bronchial walls. Increased bronchial secretions and impaired ability to handle them lead to chronic cough and disabling complications.

Current opinion holds that most acute exacerbations of chronic bronchitis are caused by viruses or by non-infectious agents. Viruses have been found in as few as 7% to as many as 64% of cases in which they were sought. By conservative estimate viruses cause about one-third of cases, the more common ones being influenza viruses A and B, parainfluenza virus, coronaviruses, and rhinoviruses. Cultures of sputum often show non-typable strains of *Haemophilus influenzae, Streptococcus pneumoniae,* and/or *Moraxella catarrhalis*. However, the extent to which these bacteria explain exacerbations in a given patient is hard to determine since they often colonize the damaged lower respiratory tract on a more-or-less permanent basis. Evidence suggests that repeated episodes of bacterial infections—especially when caused by *H. influenzae*—contribute to deterioration of pulmonary function. *S. aureus* and aerobic gram-negative rods occasionally cause exacerbations of chronic bronchitis. The pathogens associated with "atypical pneumonia" such as *M. pneumoniae, C. pneumoniae,* and *L. pneumophila* probably cause fewer than 10% of exacerbations. Evidence to date suggests that *Chlamydia pneumoniae* is more strongly associated with the underlying chronic bronchitis than with its acute exacerbations. However, *Chlamydia pneumoniae* can cause a stubborn respiratory illness lasting several weeks or longer and tending to relapse after each course of antibiotics.

Bronchiectasis

Bronchiectasis is an acquired disorder characterized anatomically by abnormal dilatation of bronchi and bronchioles and clinically by chronic productive cough and frequent lower respiratory tract infections. Its prevalence fell dramatically after the introduction of broad-spectrum antibiotics and widespread immunization against measles and pertussis. Although bronchiectasis is now uncommon, it often goes undiagnosed until far-advanced. Newer imaging studies now enable earlier diagnosis, and our understanding of its causes continues to improve.

Cigarette smoking, the major cause of chronic bronchitis, plays little role in bronchiectasis except for predisposing to recurrent infections. The basic problem in bronchiectasis is permanent structural damage to the walls of bronchi and bronchioles brought about by the concerted action of (1) infection and (2) impairment of the pulmonary toilet, airway obstruction, and/or a defect in host defenses.

In the past, bronchiectasis was associated especially with frequent or severe lower respiratory infections during childhood. Bronchiectasis continues to be associated with such infections - especially necrotizing pneumonias in which treatment is delayed - but the list of known causes has expanded. Bronchiectasis can be the earliest clue to cystic fibrosis presenting during adolescence or early adulthood. *Staphylococcus aureus, Pseudomonas aeruginosa,* and *Pseudomonas cepacia* are often isolated from these patients. *Mycobacterium avium-intracelluare* complex (MAC) infection is not infrequently associated with bronchiectasis, especially in older women and/or thin women. Allergic bronchopulmonary aspergillosis often leads to bronchiectasis, which might be prevented by early recognition of this syndrome. Immunodeficiency disorders, both congenital (hypogammaglobulinemia) and acquired (AIDS) predispose to bronchiectasis. The dyskinetic cilia syndromes are sufficiently common (about 1 in every 20,000 to 60,000 persons) that a case is likely to occur in every medium-sized city. Patients with advanced bronchiectasis experience daily cough productive of large amounts of mucopurulent, thick, tenacious sputum. However, most patients produce lesser amounts of sputum, at least during the early stages, and cough may be nonproductive ("dry bronchiectasis") or even absent. Dyspnea and hemoptysis are common. Patients often give a history of repeated respiratory infections and sometimes give a history of recurrent pleuritic chest pain. Hard crackles are heard locally over the lung fields in about 70% of patients. Rhonchi and widespread expiratory wheezes are also common. Clubbing is present in only about 3% of patients. Plain chest x-rays (PA and lateral views) are usually abnormal.

Acute Community-acquired Pneumonia

Pneumonia accounts for an estimated 45,000 deaths in the United States each year. It is the 6th most common cause of death and the most common infectious cause of death. Since it is not a reportable disease, the precise incidence is unknown. Estimates suggest that 4 million cases occur each year, prompting 10 million physician visits and 600,000 to 1.2 million hospitalizations and adding $23 billion to health care costs. Data suggest a 28-fold increased cost for managing the disease on an inpatient basis ($7,517 versus $264 for outpatient therapy). However, the mortality rate is 1% or less for patients managed as outpatients versus 14% to 25% for those admitted to the hospital. Physicians often overestimate the short-term mortality risk, but erring toward hospitalization is understandable given the potentially fatal nature of the disease.

Microorganisms can enter the lungs by aspiration, inhalation, or by way of the bloodstream (hematogenous pneumonia). Aspiration of bacteria that have colonized the oropharynx is by far the most common mechanism. Most humans aspirate small amounts of oropharyngeal secretions on a nightly basis. Microorganisms that are not removed by the mucociliary blanket are taken up and killed by pulmonary alveolar macrophages, the last line of defense. This process, called pulmonary clearance, is impaired by viral respiratory infections, tobacco smoke, chronic lung disease, alcohol, and many other factors associated with debilitating diseases. One or more chronic diseases are present in the majority of adult patients with pneumonia (58% to 89% of patients in various studies). Alcoholism predisposes to aspiration, but cigarette smoking is the main avoidable risk factor for community-acquired pneumonia in adults.

Inhalation of aerosolized particles is an important route of entry for many viruses including the influenza viruses and, most recently, the Hantaviruses. Bacteria that cause pneumonia by airborne transmission include *M. tuberculosis*, *Yersinia pestis* (plague), *Bacillus anthracis* (anthrax), and probably *Legionella pneumophila* (Legionnaire's disease) and *Francisella tularensis* (tularemia). Spore-producing fungi such as *Histoplasma capsulatum*, *Blastomyces dermatitidis*, and *Coccidioides immitis* also cause inhalation pneumonia.

Hematogenous pneumonia classically develops from septic pulmonary emboli, frequently resulting in patchy or nodular bilateral pulmonary infiltrates sometimes accompanied by pleural effusions. In inner-city populations, a familiar scenario consists of bilateral pneumonia associated with *S. aureus* endocarditis on the tricuspid or pulmonic valves of injecting drug users. Another scenario consists of emboli from septic thrombophlebitis: for example, of the pelvic veins (pelvic inflammatory disease, septic abortion), internal jugular vein (the Lemierre syndrome) or any large vein where a catheter has been inserted. Hematogenous seeding of the lungs possibly explains some pneumonias caused by gram-negative bacteria and by unusually virulent organisms such as *F. tularensis* (tularemia).

The microbial cause of community-acquired pneumonia is usually difficult to determine. In prospective studies of patients requiring hospitalization, a cause is found in only 40% to 70% of cases. In primary care practice, a far greater fraction of cases are never diagnosed. Most of these cases respond to empiric therapy. Published data concerning the causes of pneumonia vary from one region to another, but some generalizations are possible. *Mycoplasma pneumoniae* has been determined to be the most common cause in some communities, when presumptive diagnoses were taken into account, followed by *Streptococcus pneumoniae* and *Chlamydia pneumoniae*. Adults with compromised host defenses are likely to have pneumococcal pneumonia, but can also have pneumonia due to *H. influenzae*, *Moraxella catarrhalis*, *S. aureus*, or aerobic gram-negative rods. There is wide agreement that *S. pneumoniae* is the most common cause of community-acquired pneumonia requiring hospitalization. An emerging and controversial area concerns the frequency of pneumonia caused by more than one microorganism. In one study, a second pathogen was found in about 10% of patients with pneumonia due to a conventional bacterial pathogen, but in 55% of patients in whom an "atypical" pathogen was found.

In 1938, the term "atypical pneumonia" was introduced to describe "an unusual form of tracheobronchopneumonia with severe constitutional symptoms". It later became customary to distinguish between "classical bacterial pneumonia" and "atypical pneumonia". This distinction was challenged during the 1990s when researchers found it difficult if not impossible to differentiate these illnesses on clinical grounds. Some authorities now suggest abandoning the term "atypical pneumonia". Others keep the term since it enriches our appreciation of the disease, forces us to consider unusual etiologies, and reminds us that the "atypical pneumonias" do not respond to â-lactam antibiotics. For these latter reasons and for the sake of clarity, we keep the terms here even while agreeing that in some cases it may be impossible to distinguish between "classic bacterial pneumonia" and "atypical pneumonia" in actual clinical practice.

"Typical pneumonia" is an alveolar disease whereas "atypical pneumonias" affects mainly the tracheobronchial mucosa and interstitium of the lung; hence, the different clinical manifestations.

Classical bacterial pneumonia begins with sudden onset of fever, chills, pleuritic chest pain, and productive cough. In the absence of impaired consciousness or inebriation, patients usually seek medical care within 6 hours of the onset of symptoms. Chills occur in about 50% and chest pain in about 30% of patients. Most patients are febrile although some, especially the elderly, may have normal or subnormal temperatures. The respiratory rate is usually increased. Physical examination often reveals signs of consolidation such as dullness to percussion, pectoriloquy, and egophony (e to a change). Lobar consolidation is present on chest x-ray in about one-third of patients. The white blood count is usually elevated with a shift-to-the-left. However, leukopenia rather than leukocytosis may be present and portends a poor prognosis.

Atypical pneumonia, on the other hand, usually begins gradually. The insidious onset is often brought out by asking, "When was the last time you were in your usual good health?" Constitutional symptoms are usually more prominent than the pulmonary symptoms. Chest pain is experienced as substernal discomfort. Cough is non-productive or productive of only scanty amounts of sputum. Relative bradycardia is frequently present. The trachea may be tender but the lung fields are essentially clear to auscultation, prompting one to be surprised by the extent of infiltrate present on chest x-ray. The white blood count is often normal or near normal. Modest elevation of liver enzymes (specifically, the aminotransferases - AST (SGOT) and ALT (SGPT) - is often present. Atypical pneumonia, in summary, seldom presents as an acute, life-threatening medical problem but forces the physician to expand the differential diagnosis.

Pneumococcal Pneumonia

Streptococcus pneumoniae remains the major cause of severe community-acquired pneumonia and, worldwide, a leading cause of death. It accounts for about two-thirds of cases of bacteremic pneumonia, and is the most common cause of pneumonia leading to hospitalization in all age groups. Some authorities believe that *S. pneumoniae* may cause up to one-half of all community-acquired pneumonias. There is concern that the incidence of pneumococcal disease may be increasing at the same time that drug resistance is becoming much more common. Primary care clinicians should strive to make pneumococcal vaccination an imperative for patients at increased risk.

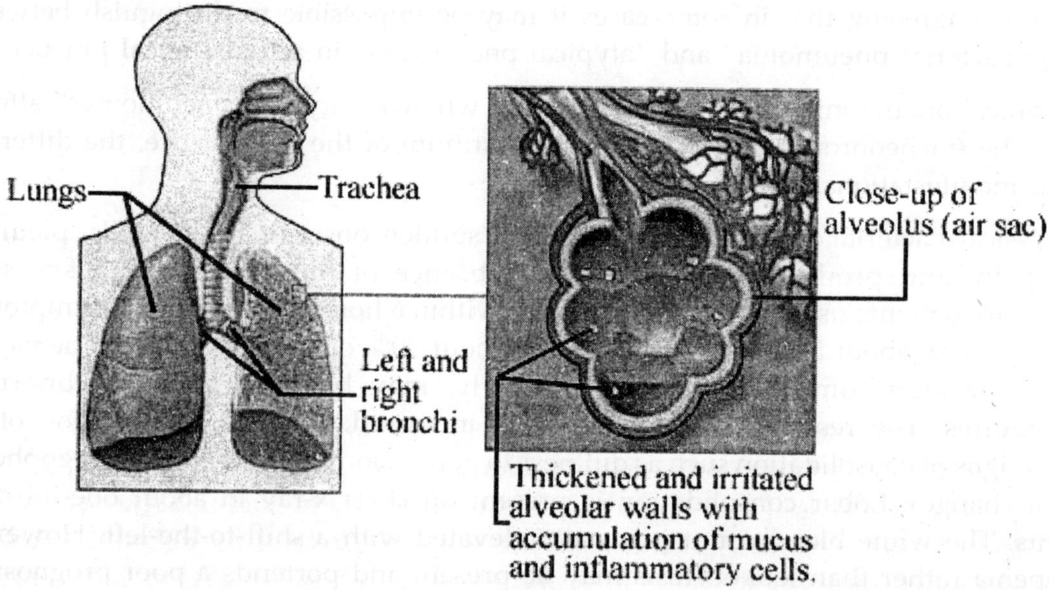

Figure 3. Pneumonia infection in human

S. pneumoniae is a common colonizer of the nasopharynx. Invasive pneumococcal disease occurs most often after a new serotype has been acquired, typically after an incubation of one to three days. Viral illness increases the incidence of disease presumably by interfering with normal host defenses. Risk factors for invasive pneumococcal disease include extremes of age, alcoholism, HIV disease, end-stage renal disease, sickle cell disease, diabetes mellitus, dementia, malnutrition, malignancies, diseases affecting B lymphocyte function (notably, multiple myeloma and hypogammaglobulinemia), and immunosuppressive disorders. Patients with asplenia are susceptible to fulminant pneumococcal disease. The pneumococcus does not invade cells as readily as do some of the other streptococci. However once in the lungs, it passes easily from one alveolus to another through the pores of hence the basis?Cohn until stopped by the dense connective tissue of a fissure for lobar consolidation.

As classically described by previous generations of clinicians, *S. pneumoniae* causes a lobar pneumonia with by the sudden onset of fever with a single, hard-shaking chill, cough productive of rusty-colored mucopurulent sputum, and pleuritic chest pain. The patient presents with systemic toxicity including tachypnea. Physical examination reveals crepitant râles, tubular breath sounds, and signs of lobar consolidation (dullness to percussion, egophony with e to a change, and pectoriloquy). Today, however, pneumococcal disease is often a more subtle illness. Patchy infiltrates and bronchopneumonia are relatively common. It is often difficult to say precisely what represents pneumococcal pneumonia and what does not unless blood cultures are positive.

Bacterial Pneumonia due to Agents other than *S. pneumoniae*

Among the numerous bacteria other than *S. pneumoniae* that sometimes cause acute community-acquired pneumonia, the most common are *H. influenzae, S. aureus, Streptococcus pyogenes,* miscellaneous aerobic gram-negative rods, and anaerobic "mouth flora" bacteria. Patients with these pneumonias often have significant underlying disease, severe pneumonia, or both. Therefore, hospitalization is usually indicated.

H. influenzae is a frequent cause of pneumonia in elderly patients and in patients with serious underlying diseases including chronic obstructive lung disease. The pneumonia usually has a patchy or segmental distribution characteristic of bronchopneumonia as opposed to lobar pneumonia. A sputum Gram's stain showing small, pleomorphic gram-negative coccobacilli can be virtually diagnostic.

Staphylococcus aureus pneumonia, when community-acquired, tends to be an acute, fulminant about 1% of cases, except during influenza epidemics. Influenza virus infection markedly predisposes to staphylococcal colonization of the respiratory mucosa. Staphylococcal pneumonia tends to be a necrotizing process with abscess formation. The chest x-ray sometimes shows air pockets known as pneumatoceles, especially in children.

Streptococcus pyogenes (group A streptococcal) pneumonia is also uncommon except during influenza epidemics. This pneumonia is often accompanied by the rapid development of large empyemas. Chest tube drainage is often necessary, resulting in prolonged hospitalization.

Klebsiella pneumoniae is a relatively common cause of pneumonia in patients suffering from alcoholism. The pneumonia often assumes a lobar distribution. Classically, this pneumonia affects the upper lobes and causes a "bulging fissure" on chest x-ray. E. coli and other aerobic gram-negative rods are relatively common causes of pneumonia in the frail elderly. Pseudomonas aeruginosa, although a common cause of nosocomial pneumonia, is rarely associated with community-acquired pneumonia in patients without underlying lung disease or severe debility.

Pneumonia due to "mouth flora" bacteria - by which is meant a combination of anaerobic and aerobic bacteria with the anaerobes usually predominating - occurs most frequently in patients suffering from alcoholism and poor oral hygiene and results from aspiration. The sputum is usually copious and often foul smelling. "Mouth flora" pneumonia in an edentulous patient should prompt suspicion of underlying lung cancer. A foul odor to the breath is present in many but not all of these patients. This form of pneumonia is often associated with lung abscess and with empyema due to bronchopleural fistula.

Mycoplasma pneumoniae pneumonia

Formerly known as the "Eaton agent", *Mycoplasma pneumoniae* is the most commonly identified cause of atypical pneumonia although its precise incidence is unknown. Various investigators have determined this microorganism to be the cause of 13% to 27% of community-acquired pneumonias. It can also cause hospital-acquired pneumonias, and it has caused as many as 50% of pneumonias during epidemics in closed populations. *Mycoplasma pneumoniae* pneumonia becomes less common after age 40, but older persons may experience more severe manifestations.

M. pneumoniae is a cell-wall-deficient organism with particular affinity for the respiratory tract epithelium. Many of the disease manifestations are now thought to be immune-mediated. Close, prolonged contact promotes transmission by respiratory secretions. There is currently interest in the extent to which *M. pneumoniae* accompanies other agents as a co-pathogen. In one study, an additional pathogen was found in about two-thirds of patients with *M. pneumoniae* pneumonia who required hospitalization; *S. pneumoniae* was most commonly found, but *Legionella* species and *Chlamydia pneumoniae* were also identified.

It is estimated that of persons infected with *M. pneumoniae*, about 20% are symptomatic, about 70% develop a mild respiratory illness (pharyngitis and/or tracheobronchitis), and fewer than 10% develop pneumonia. The disease occurs in all age groups including toddlers and the elderly but peaks between ages 5 to 15 years.

After an incubation period of about 3 weeks, symptoms begin gradually with fever, headache, malaise, chills, sore throat, substernal productive, and cough. The cough is initially non-productive, paroxysmal, and worse at night. It commonly becomes productive later in the illness. Physical examination is usually unimpressive. Bullous myringitis (inflammation of the tympanic membrane with bullae) is uncommon, occurring at most in about 5% of patients, but has a high positive predictive value for *M. pneumoniae* infection. More commonly there is mild tenderness over the paranasal sinuses, mild erythema of the posterior pharyngeal mucosa, soft cervical lymphadenopathy, and tracheal tenderness. Scattered râles and wheezes may be present but are usually unimpressive.

The white blood count is normal in 75% or more of cases. Thrombocytosis can occur as an acute-phase response. Liver enzymes, notably the aminotransferases (AST and ALT), are often mildly elevated. The chest x-ray commonly shows infiltrates that are much more extensive than one would have suspected from physical examination. The most common pattern is a peribronchial pneumonia in which thickened bronchial shadows are surrounded by streaky interstitial infiltrates and patchy atelectasis. Other patterns include nodular infiltrates and hilar lymphadenopathy. The lower lobes are most commonly involved, and pleural effusions - which can be especially severe in patients with sickle cell disease - occur in up to 20% of patients when carefully sought.

Extrapulmonary manifestations of *M. pneumoniae* pneumonia sometimes dominate the clinical picture and include hemolytic anemia, rashes including the life-threatening Stevens-Johnson syndrome, central nervous system complications (about 0.1% of patients, especially children), cardiac complications, and polyarthritis.

Chlamydia Pneumoniae Pneumonia

Chlamydia pneumoniae, described in 1986 as the TWAR agent, has been determined by some researchers to be the third or fourth most common cause of community-acquired pneumonia, explaining perhaps 10% to 14% of cases (up to 28% in some series). Pneumonia is recognized most frequently among the elderly, in whom it can be severe.

Chlamydia pneumoniae is classified as a bacterium on the basis of its cell wall and growth properties. Unlike most bacteria ' vever, it grows only as an intracellular parasite. Serologic studies suggest that most humans gain experience with *C. pneumoniae* at some point in their lives, although immunity is short-lived. About 50% of all persons have antibodies by age 20, and up to 75% of elderly persons are seropositive. It is also thought that most infections (up to 90%) are asymptomatic. Transmission is probably person-to-person by respiratory secretions.

After an incubation period of several weeks, most patients experience gradual onset of non-specific upper and lower respiratory symptoms including those of sinusitis, otitis, and pharyngitis. Sore throat with hoarseness is often prominent among the initial symptoms and tends to be the dominant symptom in college-aged persons. Symptoms of pneumonia tend to develop slowly. Often, patients have experienced symptoms for several weeks before seeking medical care. The history sometimes suggests a biphasic illness, as follows: (1) upper respiratory infection with sore throat that resolved, then (2) lower respiratory infection with cough.

The severity is age-dependent. Children under age 5 seldom have evidence of significant disease. University students often present with a 10-day history of sore throat or hoarseness with minimal fever. The mean age of patients with pneumonia is about 56 years. Ronchi and râles are present on physical examination more frequently than in *M. pneumoniae* pneumonia, even among patients who do not complain of cough. The white blood count is usually normal. Chest x-ray may show one or more infiltrates, the most common finding being a single, patchy, subsegmental infiltrate.

Wheezing is sometimes present. Accumulating evidence suggests that *C. pneumoniae* sometimes precipitates adult-onset asthma. Reported extrapulmonary manifestations of *C. pneumoniae* infection include meningoencephalitis, cerebellar dysfunction, Guillain-Barré syndrome, reactive arthritis, and myocarditis. The possibility that *C. pneumoniae* might cause coronary artery disease has received much attention. High antibodies to *C. pneumoniae* have been observed in patients with chronic obstructive lung disease, sarcoidosis, and lung cancer but an etiologic link is unclear.

Chlamydia psittaci (Psittacosis; Ornithosis)

About 100 to 200 cases of psittacosis are reported in the U.S. each year, but the true incidence is thought to be much higher. Mortality can be high if the diagnosis is not suspected.

Chlamydia psittaci infects many and perhaps all species of birds, which may remain asymptomatic or show symptoms and signs of illness such as anorexia, dyspnea, and ruffled feathers. Strains of *C. psittaci* that are most virulent for humans tend to be those associated with psittacine birds (from the Latin psittacus, or parrot), such as parrots, parakeets, macaws, cockatoos, and budgerigars, and also with turkeys. Humans who develop psittacosis are commonly bird fanciers or work in poultry farms (notably, turkey farms), abattoirs, processing plans, pet shops, or veterinarians' offices. The organism is usually acquired by inhalation, but human-to-human transmission occurs on rare occasions.

After an incubation period of 5 to 15 days, patients develop symptoms and signs of iilness ranging in severity from mild to life threatening. Atypical pneumonia, the most characteristic form of the disease, is manifest by headache, fever, and non-productive cough. Chest x-ray is usually abnormal (75%) of cases, most commonly showing consolidation of one lower lobe. The radiographic findings are usually much more striking than the findings on auscultation of the chest. Psittacosis can also present as a typhoidal illness (fever, malaise, relative bradycardia, hepatosplenomegaly), a non-specific flu-like "viral syndrome", a mononucleosis-like syndrome, or as fever of unknown origin.

TUBERCULOSIS

Tuberculosis holds a special place in medical history, is responsible worldwide for a staggering toll of disease and death, remains relatively common in the United States, can humble master clinicians, and poses formidable challenges to public health authorities, yet paradoxically is both treatable and preventable. The World Health Organization, estimating that over 8 million cases and 2 million deaths from tuberculosis occur worldwide each year, has declared tuberculosis to be a global Public health emergency. Further, persons with tuberculosis *disease* represent only the tip of the iceberg. The estimated 25% to 33% of the world's population (> 1.5 billion persons) with silent *latent infection* with *M. tuberculosis* comprise a formidable reservoir for future cases.

Tuberculosis in the United States is now largely a disease of the disadvantaged: the frail elderly, immigrants from developing nations, the inner-city poor, migrant far workers, injecting drug users, and persons infected with HIV. In 2006, there were 13,767 reported cases of TB in the United States; cases were reported in every state; drug-resistant cases, which present a formidable challenge, were identified in almost every state; co-infection with *M. tuberculosis* and HIV has continued to emerge as a complex problem presenting challenges both for diagnosis and for therapy; and the estimated 10 to 15 million persons who remain latently infected with *Mycobacterium tuberculosis* constitute a reservoir from which, without intervention, up to 10% of persons will eventually develop active TB disease.

Mycobacterium tuberculosis, commonly called *M. TB* or simply the tubercle bacillus, is a slightly curved or straight rod-shaped bacillus that requires special (acid-fast) stains to be visualized by routine microscopy. It is closely related to *M. bovis,* which, as the name implies, is primarily a pathogen of cattle and related animals. *M. tuberculosis* is also related to *M. leprae,* the causative agent of leprosy, as well as to numerous other mycobacterial species, which are referred to collectively as non-tuberculous mycobacteria (NTM, see below). Infections due to NTM are not spread from person to person and thus do not have the same community health importance as cases of tuberculosis.

Tuberculosis is spread from person to person through the air by droplet nuclei 1 to 5 m in diameter that have been expulsed into the air by a person with pulmonary tuberculosis, usually unrecognized and untreated. Cough is the primary means by which tubercle bacilli are aerosolized, but singing, sneezing, or speaking may contribute to a lesser extent. Droplet nuclei, unlike larger respiratory droplets that rapidly fall to surfaces or to the ground, are small enough to remain suspended in the air for relatively long periods of time. Persons who share airspace with an individual with infectious tuberculosis are at risk for infection. The probability of transmission depends on numerous factors relating to the source case, the exposed contact(s), and to the air space they share.

TB pathogenesis begins when a droplet containing viable tubercle bacilli is inhaled, transits the upper and middle airways without impacting on ciliated respiratory epithelium, and reaches the alveolar surface, typically in a peripheral lower lobe location. The alveolar macrophage response often fails to halt bacillary multiplication, resulting in a local focus of infection. Bacilli then spread through the pulmonary lymphatics and reach hilar or mediastinal lymph nodes, which may become enlarged. Efferent lymphatics then carry bacilli into the systemic circulation permitting seeding of any organ in the body. Areas most commonly seeded include the apices of the lungs, the brain, kidneys, and bones. Tubercle bacilli replicate relatively slowly, having a dividing time of the order of 18 to 24 hours (compared to about 20 minutes for most common pathogens). Thus, the process of local, lymphatic, and eventual systemic spread described above typically requires several weeks. By that time the bacillary load has become sufficient to stimulate cell-mediated host defenses, which usually halts bacillary multiplication. Only about 5% of apparently immunocompetent persons will develop disease in the first year or two following infection. The remaining 95% remain infected and carry a lifelong risk of reactivation of latent infection, which occurs in another 5% of infected persons. Thus, on average in most populations, perhaps only 10% of persons infected with *M. tuberculosis* develop clinical disease over their lifetimes. However the risk of progression from infection to disease is considerably higher in certain sub-populations. For example, persons with untreated HIV co-infection may progress from TB infection to TB disease at the rate of 10% per year. Infants, and persons with other immunologic, metabolic, or systemic co-pathologies are also more likely to develop disease once infected. This complex scenario explains why tuberculosis, though acquired by the airborne route, may affect any organ in the body, and why certain vulnerable sub-populations are thus in need of targeted testing and treatment of latent TB infection.

Latent Tuberculosis Infection

Recognition and treatment of selected subgroups from among the 10 to 15 million persons in the U.S. with latent tuberculosis infection may be the primary care clinician's most important role in the national strategy for eventual elimination of TB as a public health problem. Unfortunately, and despite intensive efforts, no convenient serologic test is available for diagnosis of latent tuberculosis. Diagnosis therefore depends upon tuberculin skin testing. Targeted skin testing is now recommended for identifying persons at high risk of developing active tuberculosis. It is now recognized that the interpretation of the test (that is, the extent of induration required to call the test "positive") should vary according to the patient population. The Mantoux method, involving the intradermal administration of Purified Protein Derivative (PPD), is preferred. The reaction to the Mantoux test should be read 48 to 72 hours after injection. The reaction is recorded as millimeters of induration measured across the forearm (that is, perpendicular to the long axis of the limb).

The previous criterion of 10 mm of induration for a positive tuberculin test was based mainly on the use of PPD for population surveys. The new criteria reject this "one size fits all" approach. Different "cut points" for different patient populations are now used as a guide to recommending treatment for latent tuberculosis infection. The mean reaction size of HIV-negative patients with active tuberculosis is about 15 mm, with about 85% of such patients having between 10 mm and 20 mm of induration. In nearly all published series of culture-proven tuberculosis, however, about 20% to 30% of patients failed to react at all to tuberculin. Thus, in a person being tested for latent tuberculosis infection, a large reaction (e.g., 15 mm of induration) is extremely likely to indicate past infection with *M. tuberculosis*; a somewhat smaller reaction (e.g., 10 to 14 mm or even 5 to 10 mm in certain subgroups) is consistent with past infection; while an absent or small (0 to 4 mm) reaction is unhelpful.

Pulmonary Tuberculosis

Worldwide, tuberculosis remains the most common cause of death due to an infectious agent. Pulmonary tuberculosis is the most common manifestation and the form of the disease usually responsible for its transmission. The usual patient with pulmonary tuberculosis presents with a history of several weeks of a progressive illness. The most important pulmonary symptom is cough, which initially may not be productive but which becomes productive of sputum as inflammation and tissue necrosis develop. Hemoptysis is variable and is more suggestive of advanced disease. Chest pain on deep inspiration or coughing suggests pleural involvement. Dyspnea is uncommon unless there is extensive disease or underlying pulmonary pathology. Constitutional complaints coexist and may predominate. These include fever, chills, night sweats, weight loss, appetite loss, and easy fatigability. Not all patients present with all of these manifestations. Because neither the pulmonary nor the constitutional symptoms are specific for tuberculosis, the clinician often entertains more common diagnoses at first such as bacterial pneumonia, carcinoma of the lung, or, especially if constitutional symptoms predominate, occult cancers or other systemic diseases. Often it is only once a patient has failed to improve after receiving one or more courses of oral antibiotics that tuberculosis is considered in the differential diagnosis. HIV has a profound effect on the natural history of tuberculosis infection, greatly increasing the risk that clinically silent latent TB infection will progress to overt TB disease. However, the diagnosis of

TB in patients with HIV, especially if profoundly immunosuppressed, can be difficult since coinfected patients are more likely to have small or even absent tuberculin skin test results, and are less likely to show findings typical of tuberculosis on a chest radiograph. Thus documentation of a positive serological test for HIV can help caution and guide the diagnostic evaluation. Suspicion of TB is an indication for HIV antibody testing.

Findings on physical examination can neither confirm nor exclude pulmonary tuberculosis but may provide information about the patient's overall condition. Tuberculin testing must be carefully performed and interpreted. It should again be noted that about 20% to 30% of persons with active tuberculosis have non-reactive skin tests. Abnormalities on chest x-ray are most commonly seen in the apical and posterior segments of the upper lobe, or in the superior segments of the lower lobe¯hence the rule of thumb learned by students that *TB is a disease of the apices of the lungs*. However, lesions of pulmonary TB can be present in any lung zone and may differ greatly in size, shape, density and appearance. Patients who are coinfected with HIV often have atypical radiographic presentations including isolated mediastinal or hilar lymphadenopathy (more commonly seen in HIV-negative children with early or *primary* TB infection), or disease that involves the mid- or lower-lung zones. Some TB patients with advanced HIV disease have entire normal looking chest radiographs.

All patients suspected of having pulmonary tuberculosis should have 3 or more sputum specimens examined for mycobacteria by smear and culture. Detection of acid-fact bacilli (AFB) in stained smears often provides the first bacteriologic clue of TB. However sputum smears for AFB have two shortcomings. First, they are less sensitive than are sputum cultures; indeed it is not uncommon for TB patients to have one or more positive sputum cultures despite having had negative AFB smears. Second, *M. tuberculosis* cannot be distinguished from non-tuberculous mycobacteria on the basis of the smear alone. Thus, positive AFB smears suggest but do not prove TB. The predictive value of a positive sputum smear depends on the relative prevalence of *M. tuberculosis* and non-tuberculous mycobacteria in the patient population. For example, a positive AFB smear in a young adult household contact of a diagnosed case of pulmonary TB in a major US metropolitan area almost certainly represents *M. tuberculosis.* However a positive smear in a middle-aged smoker from a rural area of the Midwest and who has not been exposed to TB is more likely to indicate a non-tuberculous mycobacterium. However, from a practical point of view, both of these patients would be managed initially as though they had tuberculosis if other evidence pointed toward tuberculosis. Cultures may take up to 8 weeks for isolation of *M. tuberculosis.* Use of molecular methods, specifically PCR, for early and specific diagnosis of TB is at present an area of active investigation.

Extrapulmonary Tuberculosis

Extrapulmonary tuberculosis is less common than pulmonary tuberculosis but is usually more difficult to diagnose. Diagnosis usually requires invasive procedures or biopsies. Molecular methods of diagnosis such as PCR offer the promise of increased sensitivity over traditional culture methods. On the other hand, extrapulmonary tuberculosis usually responds to treatment more readily than pulmonary tuberculosis because the density of *M. tuberculosis* organisms is usually much lower than in the usual pulmonary cavitary lesion. The syndromes of extrapulmonary tuberculosis can be divided into 3 categories: military (disseminated) tuberculosis, serosal tuberculosis (that is, affecting the linings of various spaces), and tuberculosis of solid organs.

Miliary (Disseminated) Tuberculosis

Miliary tuberculosis, so-named because the individual lesions resemble millet seeds, represents lymphohematogenous dissemination of *M. tuberculosis* throughout the body. The clinical presentation can be dramatic or subtle, and autopsy studies indicate that about 20% of cases are never correctly diagnosed. Miliary tuberculosis is especially common in patients with HIV disease. Classically, miliary tuberculosis results from the passage of *M. tuberculosis* from the lungs to the thoracic duct and then into the systemic arterial circulation. This event occurs routinely shortly after primary *M. tuberculosis* infection and, although usually contained by host defenses, sometimes results in clinical disease. Miliary tuberculosis can also result from previous, untreated tuberculosis involving a solid organ. In immunocompetent persons, the typical lesion is a small, well formed, often caseating granuloma containing relatively few microorganisms. In severely immunocompromised patients including those with advanced HIV disease, granulomas may fail to develop and tissues as well as blood may contain numerous bacilli. Patients in whom miliary tuberculosis is more likely to occur include young children recently exposed to the disease, pregnant women, the elderly, persons suffering from alcoholism or liver disease, and persons who are immunosuppressed for any reason.

Acute miliary tuberculosis presents with high fevers, night sweats, and other symptoms and signs of dramatic infectious illness. Occasional patients present with the acute respiratory distress syndrome, septic shock, and multiorgan failure. Localizing symptoms may point to organ involvement; examples include headache (meningitis), chest pain (pleurisy or pericarditis) and abdominal pain (peritonitis). Children tend to present with an acute illness. Young adults often present with a subacute illness. Miliary lesions are often seen on chest x-ray, although they may not be present when the patient is first seen.

. Chronic miliary tuberculosis presents with various combinations of fever, anorexia, weight loss, or, especially in the elderly, simply failure to thrive. The term "cryptic miliary tuberculosis" applies to older patients with extremely subtle symptoms and signs that are usually attributed to another underlying disease process. Occasionally, the disease presents as a hematologic disorder such as a leukemoid reaction, thrombocytopenia, myelofibrosis, or polycythemia. Chronic miliary tuberculosis is an important cause of fever of unknown origin.

Serosal Tuberculosis

"Serosal tuberculosis" is a convenient term for appreciating the pathogenesis, presentation, and diagnosis of 5 syndromes: tuberculous pleurisy, meningitis, pericarditis, peritonitis, and arthritis. The pleural, subarchnoid, pericardial, peritoneal, and synovial membranes define spaces that, under normal circumstances, contain small amounts of sterile fluid that serve to lubricate the underlying tissues or to allow freedom of movement. Tuberculosis results when *M. tuberculosis* organs gain access to the space. This can occur during hematogenous dissemination of *M. tuberculosis* or by extension into the membrane of a localized lesion such as a granuloma or tuberculous lymph node. Thus, careful examination of the brain in fatal cases of tuberculous meningitis invariably reveals a subependymal tubercle (Rich focus) that has ruptured into the subarachnoid space. The clinical manifestations of serosal tuberculosis can reflect either or both of 2 processes: (1) multiplication of *M. tuberculosis* within the space, and (2) host inflammatory response to *M. tuberculosis* antigens (delayed hypersensitivity reaction). It is the latter of these processes that often predominates. Experimentally, injection of *M. tuberculosis* into the pleural spaces of guinea pigs causes large pleural effusions only if the animals have been previously rendered tuberculin-positive; otherwise, the animals succumb to miliary tuberculosis with minimal pleural effusion. Similarly, the symptoms and signs of tuberculous meningitis can be reproduced in tuberculin-positive human volunteers by instillation of tuberculin antigen into the CSF (intrathecal tuberculin reaction). These considerations help to explain the varied presentations of these syndromes and also the relatively low yield, in some instances, of AFB smears and cultures for establishing the diagnosis. Molecular methods, notably PCR, are becoming increasingly useful for diagnosis of these syndromes, most of w⁻ich will warrant referral or hospitalization.

Tuberculous pleurisy occurs in about 5% of all cases of tuberculosis and in 10% to 30% of cases of miliary tuberculosis. This diagnosis must be considered in all cases of unexplained pleural fluid exudates. Tuberculous pleurisy results from rupture of a subpleural focus of tuberculosis into the pleural space. In adolescents and younger adults, the disease usually presents with fever, cough, pleuritic chest pain, and an obvious pleural effusion that is usually unilateral. This form of the disease often reflects primarily a hypersensitivity response to M. tuberculosis antigen. Therefore, it is commonly self-limited. In older adults with underlying diseases such as heart failure or chronic liver disease, the presentation can be subtle and insidious. Tuberculous empyema (frank pus in the pleural space) can also occur, but is rare.

Tuberculous meningitis is a life-threatening disease with serious sequelae that can occur as an isolated syndrome or as part of disseminated tuberculosis (miliary-meningeal tuberculosis). It usually presents as a subacute or chronic meningitis, but can be an acute, fulminant illness. Tuberculous pericarditis results most commonly from extension of tuberculosis in a lymph node contiguous with the pericardial sac. Acute tuberculous pericarditis typically presents with substernal chest pain, often made worse by inspiration and relieved by lying forward. Chronic tuberculous pericarditis typically presents with heart failure. Enlargement of the liver, ascites, and edema of the lower extremities may suggest cirrhosis.

Tuberculous peritonitis is now an uncommon disease in the U.S., but should be considered in cases of unexplained ascites or abdominal pain. It is often secondary to tuberculosis involving an abdominal lymph node, but can also result from miliary tuberculosis, tuberculosis of the gastrointestinal tract (which most commonly involves the ileocecal region), or tuberculous salpingitis (as spill-over from the fimbriated end of the fallopian tube). Classically, the clinical presentation is of one or the other of 2 types: (1) a serous type, presenting as ascites with or without signs of peritonitis such as tenderness with rebound accentuation; and (2) a plastic type, presenting with a "doughy abdomen" that gives the impression of tender masses.

Tuberculous arthritis classically presents as a slowly progressive arthritis involving a single joint, most commonly the knee or hip. A history of previous trauma to the joint is often obtained, as is a predisposing factor. More recently, multiple joint involvement has been emphasized. The joint space is typically preserved, which is explained by an absence of the proteolytic enzymes that characterize acute bacterial arthritis.

Tuberculosis of Solid Organs

Occasional reports document the occurrence of tuberculosis in just about every organ. The disease can, for example, mimic breast cancer, metastatic cancer to the liver, or even acute myocardial infarction. Tuberculosis was formerly the usual cause of Addison's disease (adrenal insufficiency); this presentation is now rare. Tuberculous lymphadenitis is the most common form of tuberculosis outside the chest cavity. Cervical lymphadenitis due to mycobacteria in adults (scrofula) is caused by *M. tuberculosis* in about 90% of patients and non-tuberculous mycobacteria in about 10% of patients (the reverse is true in children; see below). Patients with HIV disease often have generalized lymphadenopathy with fever, weight loss, and evidence of tuberculosis in the lungs or elsewhere.

Genitourinary tuberculosis often becomes expressed clinically when infection of the glomeruli spills over into the renal medulla and then into other parts of the genitourinary tract. In males, the organism can spread to the prostate, causing a form of prostatitis that may be chronic, subtle, and difficult to diagnose. Less commonly, the seminal vesicles, epididymides, and testes become involved. In both males and females, genitourinary infection can also result from direct hematogenous dissemination, for example to the prostate, testes, or fallopian tubes. Tuberculous salpingitis was formerly a major cause of infertility. Sterile pyuria is a classic finding in genitourinary tuberculosis.

Tuberculous osteomyelitis, like other forms of hematogenous osteomyelitis, classically involves the long bones in growing children and the spine in adults. Tuberculosis should be suspected in cases of osteomyelitis in which pyogenic bacteria are not isolated, or in which bacteria usually considered to be non-pathogenic (such as *S. epidermidis*) are isolated. Tuberculosis of the spine (Pott's disease) accounts for about 2% of all cases of tuberculosis and about one-third of all cases of skeletal tuberculosis. The lower thoracic spine is involved most often, followed the lumbar spine and then the cervical spine. Like other forms of hematogenous vertebral osteomyelitis, the disease typically involves 2 adjacent vertebral bodies with anterior wedging (due to disproportionate destruction of the anterior aspects of the vertebral bodies) and narrowing of the intervertebral disk space.

Non-tuberculous Mycobacteria

More than 50 species of the genus *Mycobacterium* are now recognized as potential human pathogens. Species other than *M. tuberculosis* and *M. leprae* have been designated "atypical mycobacteria" or "mycobacteria other than *M. tuberculosis*" in the past and are now called simply "non-tuberculous mycobacteria" (NTM). Many clinical laboratories in the U.S. now find that isolates of NTM outnumber isolates of *M. tuberculosis*, and sometimes by a wide margin. This creates a dilemma for clinicians, since the report of a positive AFB smear or preliminary culture result traditionally calls for prompt initiation of therapy for tuberculosis while awaiting definitive identification of the microorganism. Whereas isolation of *M. tuberculosis* always signifies disease (except for the rare instance of contamination in the laboratory), solation of NTM, especially from pulmonary specimens, is often of little or no clinical importance.

The *M. avium* complex (MAC) is important especially in persons with advanced HIV disease, in whom this organism causes extensive infiltration of tissues including the liver and gastrointestinal tract (with chronic diarrhea); bacteremia, and prolonged fevers. Prior to the introduction of effective drug therapy, *M. avium* bacteremia in patients with HIV disease was associated with an extremely poor prognosis. Pulmonary disease due to non-tuberculous mycobacteria in immunocompetent patients.

Positive sputum cultures for NTM often represent simple colonization but sometimes signify clinically important disease. *M. avium* complex (MAC) is the most common cause and is widely distributed in the inanimate environment. Person-to-person transmission apparently does not occur. *M. kansasii*, the second most common cause, has a more limited geographic distribution. Unlike *M. avium*, *M. kansasii* is not found in soil or natural water; however, it has been isolated from tap water in certain cities. Rapidly-growing mycobacteria such as *M. fortuitum* are the usual causes in persons with aspiration secondary to esophageal disease. The clinical presentation of non-tuberculous mycobacterial infection of the lungs is usually nonspecific and varies to some extent depending on the microorganism and underlying conditions.

M. avium complex (MAC) disease commonly presents with a productive cough, and about 20% of patients have hemoptysis. Fever is present in only about one-fourth of patients. In the past, MAC pulmonary disease was found most commonly in white males with chronic obstructive lung disease. In these patients, MAC tends to cause upper lobe cavitary lesions resembling tuberculous cavities but usually smaller and with thinner walls. The spectrum of illness has shifted, to the extent that a majority of patients are now women, usually nonsmokers. In these patients, MAC tends to cause nodular infiltrates often with bronchiectasis. Disseminated MAC infection in non-immunocompromised persons is extremely well.

M. kansasii also causes cavitary disease resembling tuberculosis but more indolent. *Mycobacterium chelonei* tends to cause pulmonary disease resembling MAC in middle-aged and older women. *Rapidly-growing mycobacteria* such as *M. fortuitum* and various "opportunistic" mycobacteria are frequently associated with indolent pneumonias in patients with esophageal disease and aspiration, which may be occult. Achalasia (a disease of the esophagus) is a classic underlying disease for this presentation.

Cervical Lymphadenitis

Cervical lymphadenitis is the usual expression of non-tuberculous mycobacterial infection in children, who, in contrast to adults, seldom develop pulmonary infection from these microorganisms. *M. avium* is the most common isolate (80% of cases), followed by *M. scrofulaceum*.

Skin, soft tissue and Rheumatologic infections

Non-tuberculous mycobacteria occasionally cause localized dermatologic and rheumatologic infections that require a high index of suspicion for correct diagnosis and appropriate therapy. *Mycobacterium marinum* causes a characteristic skin lesion known as "fish tank granuloma" or "swimming pool granuloma." Numerous nontuberculous mycobacteria sometimes cause infections of skin and subcutaneous tissue. These include

M. fortuitum, *M. chelonei*, *M. ulcerans*, and *M. abscessus*. *M. marinum* and *M. avium* complex organisms are frequently associated with tenosynovitis of the hand. *Mycobacterium terrae* complex has also been associated with hand infections.

REFERENCES

Aly R, Maibach HI. 1978. *Clinical Skin Microbiology*. Charles C Thomas, Springfield, IL.

Crissey JT, Lang H, parish LC. 1995. *Manual of Medical Mycology*, Blackwell Science, Cambridge, MA.

Gorbach SL, Bartlett JG, Blackkow NR (eds.). 1992. *Infectious Diseases*. Philadelphia: W. B. Saunders.

Brillman JC, Quenzer RW, eds. 1992. *Infectious Diseases in Emergency Medicine*. Boston: Little Brown Co.

Niederman MS, Craven DE, Fein AM, Schultz DE. 1990. "Pneumonia in the critically ill hospitalized patient", *Chest* 97:170.

Scheld WM, Mandell GL. 199. "*Nosocomial Pneumonia*: Pathogenesis and recent advances in diagnosis and therapy". *Rev Infect Dis* 13(Supp):S743,

17

Immune Defence against Microbes

From a microbe's point of view, our bodies are excellent places to live and as a result we are covered with microorganisms. Most of these associations are benign or even beneficial, but a few are damaging to the animal host and they would rapidly spread throughout our tissues if it were not for the immune system. This intricate and complex collection of proteins, cells and tissues has the daunting task of keeping the tens of billions of little travelers attached to our bodies from entering areas where they ought not to be. It is nearly unbelievable that the immune system is up to the task, yet it certainly is and our lives depend on it.

Almost all eukaryotic organisms have some way of dealing with the pathogenic microbes in the environment. Even unicellular eukaryotes have mechanisms to defend themselves, although these mechanisms are very simple. In these cases, the cell either keeps the microbes outside or rapidly destroys them once they enter the cell. In many cases, these unicellular eukaryotes (e.g., Paramecium and Amoeba species) actively take up bacteria into digestive vacuoles where they rapidly kill and then degrade the bacteria, releasing the residue as food. Some cells of the immune system (phagocytes) have inherited this capability. As time passed and evolution proceeded, descendants of the single-cell microbes began to group together and form multicellular organisms.

Once an organism becomes multicellular, the game changes a bit. Now, in many cases, there are places inside the organism that must be protected from microbes. Many small worms have a digestive tract and it is undesirable for them to have certain microbes grow there. To this end, the worms secrete a sticky substance that glues the microbes to the sides of the tract, causing them to be pushed along the tube and excreted. Somewhat distantly related systems (e.g., production of antibodies) serve a similar purpose in our bodies. The above systems are crude and not very effective and many individuals of these species fall victim to disease. Primitive multicellular organisms assure the survival of the species by producing many progeny. As long as sufficient numbers survive and reproduce, the species continues to exist.

Further down the time-line of life on this planet, the number of cells in organisms continued to increase and so did the size and complexity of the animal. Being a large animal has advantages such as being better able to defend yourself against predators, having the option of eating smaller folks around you and being better able to forage for food. However, it takes longer for a large organism to reach reproductive age and therefore they need to hold off the hoard of pathogens for a much longer time.

Guarding the body from infection was now reliant on specialised cells. These protective cells used mechanism of their ancestors, but the much slower generation times of animals demanded the immune system be more effective and sophisticated. The most complex multicellular arrangements are found in vertebrates (e.g., mammals, reptiles and birds, all of which are relatively long lived species) and it follows that they will have the most complex immune systems.

Why mammalian immune systems are some of the most complex and how they have evolved over time from simpler defense systems. As each new function evolved, it was added to the capabilities that were already there. In most cases the old functions remain and still serve a useful role, resulting in mammalian systems having many layers to them. This may explain why immunity is a patchwork of different systems, some of which overlap in their duties.

IDENTIFICATION

A central part of the job of the immune system is to differentiate self from non-self. In other words, invading pathogens must first be differentiated from the background of the host. When the immune system fails at this function, really bad things happen, including autoimmune diseases and even death.

At the macroscopic level, the difference between a human and a bacterium is obvious, but at the molecular level, this is less clear. We are made of the same basic building blocks of protein, sugars, DNA and lipids. Yet the host must be able to tell its own macromolecules from those of invading pathogens. Our immune system creates a huge population of cells (lymphocytes) that collectively are capable of responding to many macromolecules, both those that are part of our bodies (self) and those of invading pathogens (non-self).

Self, in these terms, means the proteins, sugars and lipids floating around that are part of our body and are accessible to the immune system. The key step occurs during the maturation of these lymphocytes. Those cells that respond to self-macromolecules are eliminated or suppressed, while those that do not react to self are allowed to mature. In this way, the immune system learns to tolerate the host: the assumption is that any immune cell that survives this process will respond only to non-self macromolecules.

The macromolecule it reacts with may be present in the outside world, potentially part of a pathogen. If that molecule enters the body, the cells that recognise it will activate and alert the immune system, eventually raising an immune response. But how do these cells recognise macromolecules? Before we tackle that question, let us first describe these non-self macromolecules in more detail. Immunologist call these macromolecules antigens.

ANTIGENS

Antigens are defined as any type of molecule that causes an immune response in a host by interacting with antigen-specific receptors on the membrane of host lymphocytes. Proteins by far are the strongest antigens, followed by sugars, lipopolysaccharides and then lipids and DNA. This arrangement makes some sense since it is loosely based on the uniqueness of the molecule in each species. Protein structures tend to have distinct characteristics in each organism.

Carbohydrates are less distinctive, and lipids and DNA are nearly universal in structure. Therefore, it is wise for the immune system to focus on proteins when trying to identify foreign macromolecules. Host T cells and/or B cells recognise these molecules as foreign and then marshal an immune response against them.

B Cells and T Cells

So how do lymphocytes recognise these antigens? By using antibodies and T cell receptors. Antibodies are a class of proteins made by B cells that each contain a unique variable region that recognises just one antigen. Your body is capable of making millions of different types of antibodies and together they are capable of recognising almost any substance that is non-self. You need this tremendous variability to respond to the numerous antigens encountered in a lifetime. B cells use membrane-bound antibodies that act as receptors for foreign antigens. When membrane-bound antibody receptors on B cells recognise their antigens, they bind to them and trigger the synthesis of antibodies.

A second line of defense are T cells. These also recognise antigens through proteins on their surface called T cell receptors. T cell receptors have several antibody-like features and their binding of antigen is somewhat analogous to that seen with antibodies. Once a T cell binds an antigen, it excites the rest of the immune system to attack the source of the antigen. T cells are also the part of the immune system that destroys cancer cells and cells taken over by viral infection. The big picture is that the immune system recognises invaders by the antigens they carry using antibodies and T cell receptors and then enlists a whole collection of cells and proteins to rid the body of the pathogen.

SUSCEPTIBILITY TO PATHOGEN

Before we begin an examination of the immune system it is worthwhile to consider whether an animal is even susceptible to a disease. While not really part of the immune system, this can still determine the outcome of an encounter of an animal with a pathogen.

Differences in Susceptibility between Species

Whether a pathogen can cause disease in a host is dependent not only on the virulence of the pathogen, but also on the genetic background and health of the host. Some species have an innate susceptibility to a pathogen not shared with other related hosts. For example, humans are the only host for the agents of syphilis, gonorrhea, measles and poliomyelitis. In contrast, we have innate resistance to canine distemper virus and feline leukemia virus. These differences in susceptibility may be related to a number of factors.

The resistant host may lack a cellular receptor required by the pathogen for attachment or penetration of the host. The temperature of the host may also preclude the growth of a potential pathogen. For example, *Mycobacterium tuberculosis* does not cause illness in frogs since it cannot grow well at temperature much below 37 °C. Being cold-blooded, frogs do not normally reach these temperatures. Pathogens may also require a nutrient that is not available in a resistant host. Purine-requiring strains of Salmonella typhi cannot cause disease in rats, since rats do not make purines available for the pathogen to grow on.

A final possibility is the lack of a target site for a toxin that the microbe produces. Rats injected with diphtheria toxin show no ill effects because the rats do not contain a receptor on their cell surfaces that allows the toxin to enter the cells. Since the toxin cannot enter the cells, it cannot have its toxic effect.

Differences in Susceptibility within a Species

Individuals within a species can also exhibit different susceptibility or resistance to a pathogen when compared to others. The age of an individual can have an overall effect on disease resistance, with the very young and the very old being more susceptible to infection by a wide variety of pathogens. Stress in the form of extreme exertion, shock, a change in environment, climate change, nervousness or muscle fatigue can have a negative impact on health. Each of these conditions is thought to increase the release of cortisol from the adrenal cortex, causing a suppression of the inflammatory response, thereby facilitating infection. A chronic disease or the treatment of that disease can also weaken the body of an individual and open it up to secondary acute infections. The normal immune defenses can be impaired by serious underlying illnesses, such as AIDS, Hodgkin's disease and diabetes.

Treatment of many cancers involves killing fast-growing cancer cells. This has the unwanted side effect of killing fast-growing non-cancerous cells, including the ones that make up the immune system. Patients undergoing chemotherapy and radiation therapy are therefore more susceptible to infections.

The increasing numbers of transplant patients, who must take immunosuppressive drugs to prevent rejection, are also at higher risk for infection than the general population. Poor diet is another factor that can suppress the immune system. A number of studies link vitamin and protein deficiencies to a higher rate of infection and general immune suppression.

Malnutrition is in part responsible for higher rates of infection and higher infant mortality in developing nations. Being obese is also detrimental to the overall health of an individual. Morbidly obese people are more susceptible to invasive streptococcal infections. Obesity also leads to type 2 diabetes, which in turn makes people more susceptible to infectious diseases. As you can see, eating too much of the wrong kinds of foods and becoming overweight can lead to all kinds of trouble.

There are also sexual differences in the degree of susceptibility to disease. In some cases, anatomical differences cause members of one sex to be more resistant to infection than those of the opposite sex. Obviously, men cannot suffer infections of the uterus and women do not develop prostatitis. More subtly, urinary tract infections (UTI) are 14 times more likely in women, because bacteria more easily travel up the short 4 cm female urethra than the longer 18 cm male urethra to reach the bladder. In addition, because of the anatomical closeness between the opening of female urethra and anus, intestinal bacteria easily gain entry into their urinary tract.

Differences in genetic background between individuals can also have an influence on susceptibility. For example, Eskimos, Native Americans and Asians are more susceptible to tuberculosis than are Caucasians. Also, individuals with sickle-cell anemia are resistant to the protozoan that causes malaria.

Another example comes from the AIDS epidemic. There are rare cases of individuals who have had frequent unprotected sexual encounters with virus-carrying partners yet have not contracted the disease. It turns out these individuals have a mutation in a cellular surface protein that makes it impossible for the AIDS virus to enter their cells. Without the presence of the normal surface protein, the virus cannot infect and cause disease.

IMMUNE SYSTEM

The immune system consists of a complex network of organs and tissues that work together to prevent infection. Many of these systems are unleashed by activation of other parts of the immune system. The extensive interdependence of the players of the immune system can make it difficult for the beginning student to understand. To try to make it a bit easier, we will first describe the anatomy of the immune system, then examine the various cell types involved in immunity, and finish this section by describing how these parts work together in reacting to an invading pathogen.

Many different organs and tissues in the body contribute to the function of the immune system. These include the circulatory system, bone marrow, thymus, spleen, lymphatic system and Mucosal Associated Lymphoid Tissue (MALT). Together these tissues are responsible for the creation, transport and successful operation of mammalian immunity. The tissues of the immune system fall into two groups based upon their role in host defense. Primary (or central) tissues look after immature immune cells, creating and educating them during their differentiation into mature cells.

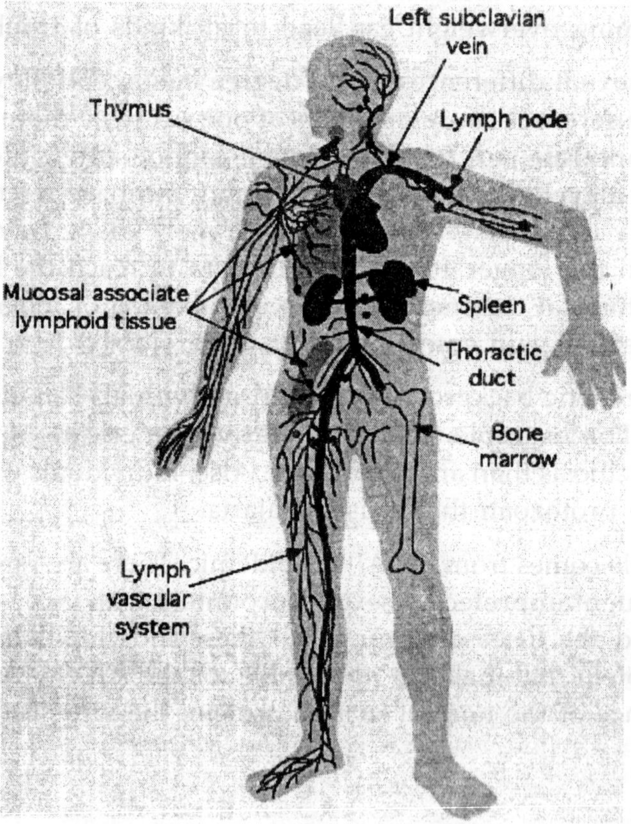

Figure 1. The Tissues of the Immune System

The bone marrow and thymus gland are parts of the primary immune system. Secondary (peripheral) immune organs look after mature cells that are an active part of defense. The secondary system encompasses the rest of the immune tissues: the spleen, the lymphatic system, lymph nodes and MALT. Of course it is not this simple and the spleen and MALT also help in the maturation of immune cells. A large collection of organs and tissues are involved in the immune system of animals. Some of these tissues create and or educate the immune system, bone marrow and thymus, while other parts are involved in fighting infections, lymph systems, lymph nodes, spleen and MALT. The thoracic duct collects liquid from the lymph system and returns it to the circulatory system at the left subclavian vein near the heart.

Role of Circulatory System

The circulatory system is responsible for the transport of blood throughout the body and consists of the heart (a pump), the lungs (a gas exchanger) and the vascular system of arteries, capillaries and veins (plumbing). Blood, which runs through this vascular system, contains both cellular and non-cellular components. The major cell types found in blood are red blood cells, whose role is to transport oxygen and carbon dioxide into and out of the body respectively.

The circulatory system also plays a secondary role as one of the routes immune cells use for transport around the body. Blood contains many types of what are called white blood cells, which are made up of mainly neutrophils, but also monocytes, T-lymphocytes and B-lymphocytes. The non-cellular portion of the blood is a liquid called plasma that contains the protein fibrinogen as a major component. Under the right conditions, fibrinogen participates in a complex series of reactions that eventually result in the formation of a fibrin clot in the blood.

Platelets also participate in this process. Blood clots are important in stopping bleeding and also in inhibiting the invasion of advancing pathogens by entrapping them. Blood removed from the body rapidly clots if anticoagulants such as sodium citrate or heparin are not added. If blood is allowed to clot, the liquid left over is serum. A major constituent of serum is immunoglobulins (antibodies). It also contains various proteins of the complement system, which are equally important in the immune system.

Role of Bone Marrow

The bone marrow is the source of many immune and blood cells in the healthy adult animal. If the bone is split lengthwise, a marked difference in tissue is noticed. Part of the tissue is red, which is the source of red and white blood cells. The other portion is yellow adipose tissue that is inactive.

During an infection, the yellow marrow can be reactivated to become red marrow to help in the production of larger numbers of immune cells. In the adult animal, all immune cells originate from stem cells located in the bone marrow. Stems cells constantly divide and differentiate into various types of immune cells under the influence of cytokines. Cytokines are small signaling proteins that help to regulate the behavior of the cells of the body.

The bone marrow is ultimately responsible for the synthesis of eight types of cells: red blood cells, platelets, neutrophils, basophils, eosinophils, mast cells, monocytes/ macrophages, and T lymphocytes and B lymphocytes. Some of these cell types mature in the bone marrow itself, while others migrate through the circulatory system and undergo final maturation in other tissues. At this point, we will take a short stop and look briefly at some of the features of the cells created in the bone marrow as an introduction to their function. You will learn more details about each immune cell type later when we cover specific immune responses. In humans stem cells differentiate into different types of immune cellsin the bone marrow under the influence of different cytokines.

Stem Cell Research

James Thompson of the University of Wisconsin-Madison was the first to obtain embryonic stem cells by taking them from a growing embryo. These cells are the progenitors of all the cells in the body and are unique in a number of ways.

1. Embryonic stem cells can proliferate indefinitely. Typical mammalian cells are capable of a limited number of divisions. After about 20-50 cell cycles, they are incapable of dividing further. This occurs because at each division a small portion of the ends of the chromosomes, called telomeres, is lost. Initially, this is acceptable because the telomere does not contain vital DNA. However, as subsequent divisions proceed, the cells lose more and more of the telomeres and eventually the chromosomes fail to function. Embryonic stem cells get around this difficulty by activating an enzyme called telomerase. Telomerase adds extra DNA onto the ends of chromosomes during division so that they do not shorten.

2. Because of the role of embryonic stem cells in normal development, they have the capability of differentiating into any cell type. In contrast, mature differentiated cells have their function dictated by a fixed pattern of gene expression. We currently do not know how to revert differentiated cells to a more neutral state. As a consequence stem cells offer the best hope for generating useful cell types for various treatments.

Research on embryonic stem cells may result in a number of promising developments. Having these cells available for research will give us new insight into how humans and other mammals develop. It seems clear that embryonic stem cells develop into other cells types and tissues under the influence of cytokines. By studying the process we will learn what these signals are and how they work. Being able to create an unlimited supply of different cell types may eventually allow the development of new tissues and organs for use in transplants. Presently, there is a great need for organs and tissues for ailing patients. Scientists have already been able to create lungs, bladders and kidneys in animal experiments. These cells may also lead to the creation of cures for brain disorders such as Alzheimer's and Parkinson's diseases.

Embryonic stem cells are also controversial because to obtain them, it has been necessary to destroy a developing embryo. A fertilised egg is allowed to divide for a period of time and then the cells are extracted. Some consider this equivalent to killing a human being, but the key to the controversy is when we decide a human life begins. Does it begin the minute that a mass of cells has the potential to develop into a human? Does it begin when that groups of cells can survive on its own? Medicine is getting better at saving babies born prematurely. What if a time comes when an embryo can be raised completely out of the uterus? Scientists cannot and should not make these decisions by themselves. Human society must decide what is acceptable and the key to making good decisions is educating ourselves about the issues.

Cells made in the bone marrow

The major cell type made by the bone marrow is red blood cells. Platelets also form in the bone marrow and assist in formation of blood clots following any kind of injury. Neither of these cell types plays a role in the immune response, but we mention them here because they also originate from bone marrow stem cells and are essential components of the blood.

Polymorphonuclear granulocytes is the general term given to neutrophils, eosinophils and basophils. The first half of the name describes the appearance of the nucleus that seems to be split into a number of different lobes. In reality, the nucleus is contiguous, but contains many infoldings, which give it a polynuclear appearance. The rest of the name comes from the appearance of the cytoplasm, which looks speckled.

The cytoplasm is full of granules that contain compounds and enzymes important in fulfilling the function of each cell type. Polymorphonuclear granulocytes make up 50-70% of the white blood cells found in blood. They last only about three days and have to be replaced at a rate of 80 million cells per minute.

Neutrophils are the most common type of polymorphonuclear cells, making up 90% of granulocytes in the blood. These cells function as phagocytes in attacking and destroying infectious agents. We will cover their roles in more detail when we discuss phagocytes.

Eosinophils make up 2-5% of granulocytes in the blood, but this number can rise considerably in people with parasitic diseases as well as asthma, eczema or other diseases associated with allergies. They are primarily found in the blood, but also near epithelia that have high bacterial populations (e.g., intestines, vagina, nasal passages).

The granules in these cells bind the red dye eosin, giving the cells their name. Eosinophil granules contain a number of different enzymes including, acid phosphatase, glucuronidase, cathepsins, RNase, and arylsulfatase and peroxidase. They also produce toxic basic proteins. They respond to the chemical signals put out by other immune cells and can then participate in an immune response. The major reactions take three forms.

1. They can down-regulate an immune response by destroying histamine secreted by mast cells using the enzyme histinase. Eosinophils also liberate arylsulphatase that breaks down the slow reactive substance of anaphylaxis (a dangerous form of allergic response) that is released by mast cells.
2. Eosinophils combat antigenic challenges too big to be attacked by phagocytes. Examples of such challenges are parasitic worms or helminths. In battling these infections, the body first covers the worm with antibody. This then activates eosinophils, which bind to the parasite and release the contents of their granules, thus causing external digestion of the worm.
3. As is the case with neutrophils, eosinophils can phagocytize microorganisms, but this is a secondary role.

Basophils are small cells that make up less than 1% of all white blood cells. The granules of these cells contain heparin, histamine, decarboxylase, histidine, dehydrogenase and diaphorase. Heparin is an important anti-clotting compound, and histamine finds its use modulating the immune response. Histidine is converted to histamine by decarboxylase.

The role of basophils in the immune response is not yet clear, but they seem to play a role in the defense against parasitic worms and in severe allergic reactions. They have a very high affinity for IgE antibodies and they are usually found coated with IgE in tissues. Binding of IgE may set in motion a series of events that causes other immune cells to respond to the high concentrations of IgE. Basophils may be cellular alarms that notify the rest of the immune system and help to concentrate the point of attack.

Mast cells are closely related to basophils but are distinct in their reactions to antigens. They are found throughout the body in lymph nodes, spleen, bone marrow, around blood vessels, nerves, glands and in the skin. Mast cells have granules that, like basophils, contain heparin and histamine. They have a high affinity for IgE as well and their activation by antigen triggers histamine release. Until recently, they were mostly thought to trigger unwanted allergic reactions, but it is now becoming clear they participate in immune responses to gram-negative bacteria. Their wide distribution indicates that they are important in many immune responses.

Monocytes and macrophages are long-lived specialised phagocytic cells. Monocytes are migrating phagocytic cells found in the bloodstream and when they enter other tissues, they differentiate into macrophages. Macrophages are found in the brain, lungs, liver, spleen, lymph nodes, joints and peritoneum. The key functions of monocytes and macrophages are to remove our own dead cells when they reach the end of their useful life and also to remove pathogens. Macrophages in the liver, called Kupffer cells, phagocytize old erythrocytes from the blood and remove them. Another one of their functions is the creation of important immune proteins and peptides. They are responsible for synthesizing transferrin (an iron-binding protein), complement proteins and various cytokines necessary for immune function.

B lymphocytes or B cells are antibody-producing cells. They are very important in fighting many different types of infections, especially, bacterial infections. T lymphocytes are involved in regulating the immune system and destroying host cells that are out of control, either due to a breakdown in cell division regulation (cancer) or infection by a virus or even an intracellular parasite. We will discuss the functions of these cells in more detail when we cover the adaptive host response.

The thymus is a fist-sized organ located above the heart that is involved in the maturation of T lymphocytes. T cells produced by the bone marrow are immature and journey to the thymus through the bloodstream. The blood vessels that supply the thymus with oxygen and other nutrients also contain a blood-thymus barrier that only allows immature T cells in and mature T cells out. The thymus is also connected to the lymphatic system through lymph vessels.

lymphatic System

The lymphatic system is a separate vascular system, distinct from the bloodstream, through which the lymph moves. It is a branching system whose vessels get ever smaller as it penetrates tissue. Unlike the blood system, the lymphatic system is not circular, but ends in blind capillaries. The system focuses its attention on areas of the body that are most likely to be entry points for pathogens: skin dermis, respiratory tract, gastrointestinal tract and genitourinary tract. It functions are:

1. To collect excess fluid from surrounding tissue and return it to the bloodstream

2. To absorb fat from the villi of the small intestine.
3. To Participate in the immune response.

Here we will focus on its major role — harboring and transporting many of the cells involved in the immune system. Lymph fluid consists of leukocytes and many components of plasma, but does not contain red blood cells. Liquid enters the lymph system from non-vascular tissue draining into lymph capillaries. The capillaries then drain into lymph nodes that sit at the junction of a number of lymph vessels.

The smallest lymph nodes are about the diameter of the tip of a pen while the largest are the size of an almond. Large volumes of liquid and cells pass through lymph nodes each day and are filtered to detect antigens and remove microbes.

The nodes also interact with phagocytes to begin various immune functions that we will elucidate later. Lymph nodes are dynamic, densely packed structures with the bulk of the cells inside them being mobile. Phagocytes and antigens enter through as many as five vessels, called afferent ducts. The lymph node contains large numbers of B- and T-lymphocytes, but macrophages and plasma cells are also present.

Macrophages carry out the job of filtering the incoming fluid and plasma cells (effector cells that differentiate from antigen-stimulated B cells) secrete antibodies that exit along with other immune cells by the single exit, the efferent duct. The lymphatic system then merges into larger vessels and eventually reconnects to the circulatory system through ducts near the heart.

The flow of liquid is therefore from the lymphoid capillaries, into lymph vessels, then lymph arteries and eventually exiting into the bloodstream at the heart. There are actually two lymphatic drainage systems. The right lymphatic duct drains the upper right side of the body, including the right side of the head, the heart and lungs. The rest of the body drains into the thoracic duct. Both of these systems connect back to the bloodstream near the heart. Liquid entering the lymphatic system from the blood and extravascular tissue filters through typically eight to ten nodes before returning to the bloodstream.

Immune Cells Fight against Infection

Mucosal-Associated Lymphoid Tissue (MALT) is scattered throughout the connective tissues of the body, but especially beneath moist epithelial membranes such as those that line the upper respiratory tract, intestine and urinary tract. MALT is strategically distributed to help the body prevent infection by organisms that have penetrated beyond the mucosal surface. MALT consists of small masses of lymphatic tissue (up to a millimeter in diameter) containing mainly lymphocytes.

These tissues are far less organised than the lymph nodes. Most MALT consists of small groups of cells, but in certain areas it is found in large clusters. For example, large aggregates of MALT occur in the wall of the lower portion (ileum) of the small intestine and are known as Peyer's patches. Tonsils and adenoids are also aggregates of MALT that protect your body from microorganisms present in the upper respiratory tract. Unlike lymph nodes, MALT is not connected to a vascular system.

The spleen is a very important secondary lymphoid organ. Individuals who have had their spleens removed due to rupture caused by a car accident or atrophy from sickle cell anemia can lead nearly normal lives, but they tend to be more susceptible to infection. Some functions of the spleen are similar to those of the lymph nodes; however, it also produces lymphocytes and removes senescent (old) red blood cells from the circulation.

The spleen contains a circulatory and lymphatic system allowing access to it either through the blood or lymph. Cells in the spleen organise around the blood vessels into two tissues: red pulp and white pulp. Red pulp contains mainly red blood cells, and white pulp is made up predominantly of lymphocytes.

The white pulp focuses tightly around the arterioles of the spleen while the red pulp fills the rest of the interstitial space. The location of various lymphocytes in the white pulp further differentiates it. T cells are found near the arterioles and further away are areas of B cells. The spleen is a major area for B cells to congregate in the body, where they wait to be activated by antigens. During an illness, activated lymphocytes are released from the spleen to fight the infection.

The spleen is located on the middle of the abdomen to the left of the stomach. This cartoon of a cross section of the spleen shows both red and white pulp. Both the circulatory and lymphatic system have vessels in the spleen. The magnified section shows the arrangement of cells. The white pulp near the arterioles consists of T cells and B cells and the red pulp is filled with red blood cells.

Innate and Adaptive Immune Systems

Host defenses can be divided into two categories, innate and adaptive immunity. Innate immunity involves general mechanisms in a healthy animal that prevent colonisation by microorganisms and antagonise or kill those that do enter the host. They are always present and the strength of their response does not increase with repeated exposure to the inducing microbe. Adaptive immunity develops through the mechanisms that are turned on in response to a pathogen. This involves activation of the immune system, where the initial response to a pathogen is weak but becomes quite vigorous over a period of a few days. Adaptive defenses also have a memory of encountered pathogens such that a second infection by a pathogen is met with a more rapid and vigorous immune response.

Innate immunity is the first line of defense against a pathogen. The system must be somehow circumvented by the pathogen before it can enter into the host. In most cases, the adaptive immune system is only activated after the innate immune system has been breached. Parts of the immune system that are always present and whose reaction against a pathogen does not increase with exposure to it. Natural killer cell activation does depend on antibodies, but they are not inducible. Therefore we include them here.

Microbial Resistance in Anatomical Structures

The skin is an extremely effective barrier to microorganisms. Besides preventing the cells of our body from escaping, it also prevents microorganisms from getting in. Skin contains several layers of tightly packed, heavily keratinised cells (keratin is a fibrous protein that gives skin, hair and nails its toughness). It is very difficult for most organisms to squeeze in between skin cells. Skin cells are also continually being shed and replaced by new ones, thus removing any microorganisms attached to them.

The skin surface is also very hydrophobic and dry, which prevents the growth of many microorganisms. Sebaceous glands are present throughout the skin and they secrete hydrophobic oils that further repel water and microorganisms. The oil also helps keep the skin supple and flexible, preventing cracking that might allow microbial access to internal layers. Finally, melanin in the skin also helps to reduce the harmful impact of UV light by absorbing it. UV light can be damaging to all cells, including cells of the immune system.

Parts of the body are designed to prevent the passage of unwanted microorganisms. Some examples of structures that inhibit the advance of pathogens include the skin, hair, blinking eyelids, nose hairs, cilia in the lungs and the peristaltic action of the digestive track.

Hair helps to restrict access of airborne pathogens to the skin. It protects the most sensitive or exposed body orifices, including the nasal cavity, the eyes and ears. Hair also serves as a cushion lessening the severity of cuts and grazes, which decreases the depth and number of breaks in the skin. The movement of various body parts can help to rid the body of microorganisms trying to colonise. Blinking the eyelids constantly sweeps microorganisms out of the eye.

The entire respiratory tract, except bronchioles and alveoli, is coated with cilia that beat upward and push microorganisms out. The peristaltic action of the gut not only moves food along our digestive tract, but also flushes microbes out of our system.

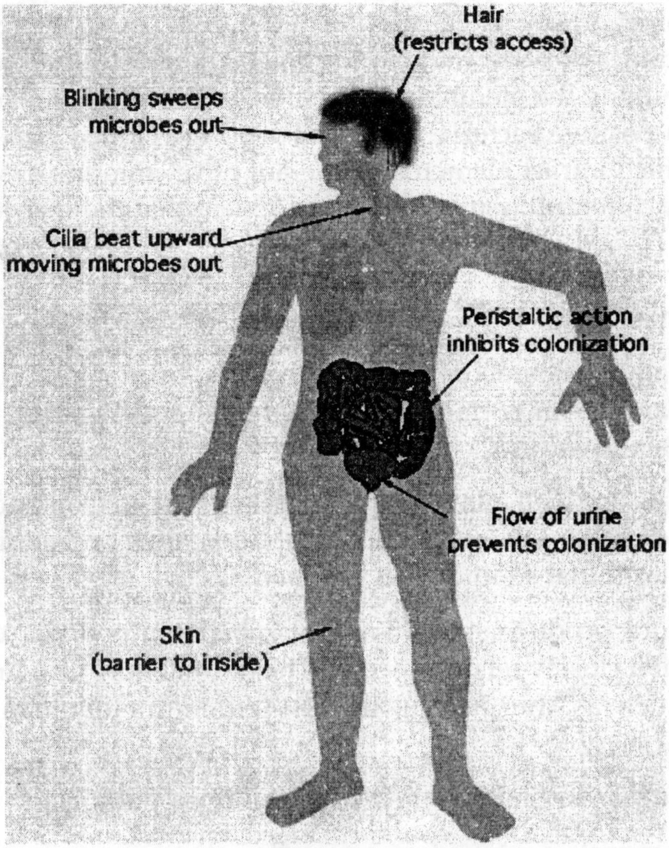

Figure 2. Anatomical Immunity

If microorganisms do not have a method of attachment, they rapidly leave the gastrointestinal tract. The constant flushing of the urethra by urine helps to keep the bladder free of bacteria. All of the above movements continually remove microbes from our bodies.

Antimicrobial Secretions

The body produces a number of antimicrobial substances that inhibit or kill microorganisms. Collectively these substances are known as tissue bactericides. Tear ducts, sebaceous glands, ears, nose, lungs, mouth and digestive tract all secrete antimicrobial substances.

One important group of broad-spectrum antimicrobial peptides is the defensins, which participate in the host defense of mammals, birds, plants and insects. Their presence in a wide variety of species probably indicates that they are an ancient form of antimicrobial antagonism. Defensins form a family of cysteine-rich, cationic and structured polypeptides of 29-42 amino acids that contain three or four disulfide bridges. They work by disrupting the membrane of a wide variety of pathogens, including gram-negative and gram-positive bacteria, fungi and some enveloped viruses. They also serve as chemokines, which attract elements of the adaptive immune system. Defensins are expressed by many different tissues and it is now clear that they are a vital part of innate immunity.

Microbial Antagonism

The normal non-pathogenic microorganisms that live inside and on our bodies as part of normal flora have an important role in defending us against pathogens. This is accomplished in three ways.

1. The normal flora are well adapted to the tissues they live on and they out-compete pathogenic microbes for these sites of colonisation, thus preventing harmful bacteria from gaining an initial foothold on our bodies.
2. The normal flora produce specific antimicrobial compounds (called bacteriocins) that kill or inhibit closely related species. The normal flora also produce non-specific compounds such as the end products of their metabolism that may inhibit pathogenic bacteria.
3. They utilise nutrients, limiting their availability to pathogens.

Activation of the complement system can follow one of two pathways: the classic pathway and the alternative pathway. The classic pathway begins when antibody of type IgG or IgM reacts with antigen. IgG and IgM contain a masked C1 binding site that is exposed upon binding to antigen. These are enzymatic reactions, in which the binding of C1 to antibody results in the formation of a large number of C2a and C4b molecules. These in turn combine to form an enzyme that converts a large number of C3 proteins to C3a and C3b. In this way a single antibody-antigen reaction can be amplified and produce an aggressive host response. Complement activation leads to several ultimate results.

1. The C3a and C5a produced serve as a chemoattractants for phagocytes.
2. C3a and C5a bind to specific receptors on mast cells inducing degranulation and the release of inflammatory substances.
3. C3b remains bound to the Fc portion of the antibody or may bind to the cellular membrane of the trigger. In either case, it serves as a protein that phagocytes can bind to and then envelop and kill the bacteria.

4. Several of the complement proteins (e.g., C5b, C6, C7, C8 and C9) form a pore in the membrane of the initiating microbe that causes leakage of cell contents and results in cell death.
5. When inserted in membranes, C8 and C9 form a phospholipase that can destroy membranes of the pathogen. This can inactivate a virus and remove the outer membrane of gram-negative bacteria, making their peptidoglycan accessible to lysozyme.

The alternative pathway of complement activation does not involve antigen binding to antibody. This pathway involves another serum protein complex, properdin, which reacts with certain insoluble bacterial polysaccharides, including peptidoglycan, LPS and teichoic acids. Note that these macromolecules are unique to bacteria and are not normally found in the host. After binding bacterial polysaccharide, properdin activates C3, forming C3a and C3b, which then triggers the rest of the complement cascade. The alternate pathway allows antibody-independent activation of the complement system, which is important in initial defenses against a new pathogen before antibodies have been synthesized. Individuals who are deficient in complement formation have compromised immune systems. For example, complement deficiency in C3, C5, C6, C7, C8 or C9 leaves the person more susceptible to dangerous bacterial infections.

Phagocytes

An important result of inflammation is the recruitment of phagocytes. These cells function to engulf and attack particles in the host that have been signaled for removal by various mechanisms. In this section we describe the synthesis and properties of phagocytes.

Phagocytes originate in the bone marrow from stem cells that first differentiate into myeloid precursor cells. These are a set of intermediate cells that can take a number of paths. Polymorphonuclear granulocytes (mostly neutrophils) leave the bone marrow and circulate in the bloodstream and body for only a few days before dying. They are the first phagocytic cells that normally encounter an infection and can rise to large numbers during a severe illness.

Monocytes differ from polymorphonuclear granulocytes in that they have an unsegmented nucleus and are much longer lived. This longer existence is critical to their function. After creation in the bone marrow, they circulate in the blood for a period before settling in a tissue and maturing into macrophages.

Finally, macrophages are mature monocytes that are attached to lymph tissues and can be up to ten times the size of monocytes. All phagocytes contain membrane vesicles filled with destructive and degradative compounds (lysosomes) that are available for the annihilation of engulfed microbes.

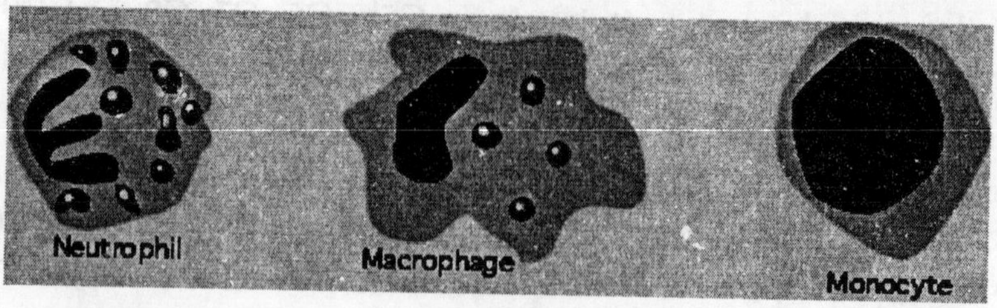

Figure 3: Phagocytes. Neutrophils, Macrophages and Monocytes

Phagocytes are the cellular sentries of the immune system, detecting, engulfing and killing pathogens in our bodies. Theyalso destroy dead or dying cells and cancerous cells.

Phagocytosis

Phagocytes are motile and roam throughout the bloodstream, the lymphatic system and non-vascular tissue in search of particles to assault. When a non-self particle is encountered, it is taken into the phagocyte, combined with destructive compounds and destroyed.

The steps of phagocytosis, detection/chemotaxis, attachment, engulfment, fusion, and killing. Macrophage are also capable of egestion and antigen presentation.

1. Detection of the foreign particle and movement of the phagocyte to the area.
2. Attachment of the foreign particle to the phagocyte.
3. Engulfment or ingestion of the foreign particle.
4. Fusion with lysosome and formation of the *phagolysosome.*
5. Intracellular killing and digestion.
6. In the case of macrophages, egestion and antigen presentation.

For the sake of simplicity, we will focus on the example of the attack of a phagocyte on a microorganism. However, the process is similar when a phagocyte attacks a virus or other foreign particle.

The presence of an infecting microbe sets into motion a number of host responses, one of them being inflammation. This in turn causes the creation of gaps in the blood vessel walls that allow the migration of phagocytes across vascular walls. Once inside an infected tissue, phagocytes must then detect the location of a microbe, which can be signaled in a number of ways.

1. Many microbial proteins contain N-formyl methionine at their amino terminal end, while eukaryotic proteins do not. Certain N-formylated bacterial peptides serve as chemoattractants for phagocytes.

2. The binding of antibodies to antigens triggers the classic complement pathway with subsequent release of C3a and C5a. These proteins are also chemoattractants for phagocytes.
3. The complement cascade also induces an inflammatory response, whose mediators of inflammation also draw phagocytes to the area.

From these three examples, it is obvious that the body creates a number of strong signals in response to microbial invasion, and that these signals quickly draw phagocytes to the area of infection. The rapid removal of microbes is critical in preventing a disease and having multiple attractants insures a swift reaction.

The phagocyte must then attach to the target microorganism. Without assistance, this can be a somewhat inefficient task for the phagocytes and they are not effective killers of pathogens. Phagocytes have over 40 specific types of receptors on their cell surfaces. Some of these receptors are for chemoattractants that draw phagocytes to an area; others recognise molecules that enhance the binding of phagocytes to their targets. This enhancement of binding is termed opsonisation and macromolecules that bind to an antigen and increase the efficiency of phagocytosis are opsonins.

Opsonins provide molecular handles for the phagocyte to grab onto. Several different macromolecules can serve as opsonins for phagocytes. IgG antibodies have a site on their constant region that can react with an antibody binding receptor on phagocytes. This phagocyte-binding region is masked in a free antibody, but becomes accessible to phagocytes when the antibody binds to antigen. Initially, masking this site prevents free antibody from binding to phagocytes.

The complement protein C3b can also serve as an opsonin. Whether triggered by the classic or alternative pathway, C3b eventually binds to the membrane surface of the complement-activating microbe. Phagocytes contain a C3b receptor, which binds to C3b and thus to the complement-triggering microbe. Opsonisation of bacteria greatly increases the rate of attachment and ingestion by phagocytes. Bacteria in the blood are quickly cleared only if they are bound by opsonins.

In the absence of opsonins, phagocytes are still capable of binding to microorganisms by a mechanism that can be thought of as nonspecific attachment. This nonspecific attachment probably involves electrostatic or hydrophobic attraction between the phagocyte and microorganism. It is also possible for the phagocyte to physically trap the microbe against a tissue surface and initiate ingestion—a mechanism called surface phagocytosis. These nonspecific mechanisms of attachment are probably important early in infection to slow the microbe's progress until opsonins are available to speed up phagocytosis.

Attachment of the microbe to the phagocyte results in some sort of signal that triggers ingestion of the microbe. Ingestion involves encircling the target particle with phagocytic membrane so that it is eventually taken inside the cytoplasm of the phagocyte engulfed in a membrane vesicle called a phagosome. This process requires ATP and is triggered by the attachment of the target to the phagocyte's cytoplasmic membrane. Contact between a microbe and a phagocyte also changes the phagocyte's metabolism from an aerobic respiratory to anaerobic fermentative, with lactic acid being the final end product. The increase in lactic acid in the phagocyte lowers the pH of the cytoplasm, including the phagolysosome and this enhances the activity of many of the degradative enzymes present.

Phagolysosome Formation

The phagosome containing the microorganism migrates into the cytoplasm and soon collides with a series of lysosomes forming a phagolysosome. When the membranes of the phagosome and lysosome meet, the contents of the lysosome explosively discharge, releasing a large number of toxic macromolecules and other compounds into the phagosome. The killing processes inside the phagolysosome are confined to the organelle of the phagolysosome, thus protecting the cytoplasm of the phagocyte from these toxic activities.

Intracellular Killing and Digestion

Several minutes after phagolysosome formation, the first detectable effect on the microorganism is the loss of the ability to reproduce. Inhibition of macromolecular synthesis occurs sometime later and many pathogenic and non-pathogenic bacteria are dead 10 to 30 minutes after ingestion. The mechanisms phagocytes use to carry out this killing are diverse and complex, consisting of both metabolic products and lysosomal constituents. Each type of phagocyte (neutrophils, monocytes or macrophages) has a slightly different mix of killing methods. The killing mechanisms that phagocytes use can be organised into two broad groups: oxygen-dependent and oxygen-independent mechanisms.

Oxygen-dependent Mechanisms

Binding of Fc receptors on neutrophils, monocytes and macrophages and mannose receptors on macrophages causes an increase in oxygen uptake by the phagocyte called the respiratory burst. This influx of oxygen is used in a variety of mechanisms to cause damage to microbes inside the phagolysosome, but the common theme is the creation of highly reactive small molecules that damage the biomolecules of the pathogen. Binding of these receptors activates an NADPH oxidase that reduces O_2 to O_2- (superoxide).

Superoxide can further decay to hydroxide radical (OH.) or be converted into hydrogen peroxide (H_2O_2) by the enzyme superoxide dismutase. In neutrophils, these oxygen species can act in concert with the enzyme myeloperoxidase to form hypochlorous acid (HOCl) from H_2O_2 and chloride ion (Cl^-). HOCl then reacts with a second molecule of H_2O_2 to form singlet oxygen ($^1O_2$), another reactive oxygen species. Macrophages in some mammalian species catalyse the production of nitric oxide (NO) by the enzyme nitric oxide synthase. NO is toxic to bacteria and directly inhibits viral replication. It may also combine with other oxygen species to form highly reactive peroxynitrate radicals. All of these toxic oxygen species are potent oxidisers and attack many targets in the pathogen. At high enough levels, reactive oxygen species overwhelm the protective mechanisms of the microbes, leading to their death.

Oxygen-independent Mechanisms

The pH of the phagolysosome can be as low as 4.0 and this alone can inhibit the growth of many types of microorganisms. This low pH also enhances the activity of lysozyme, glycosylases, phospholipases and nucleases present in the phagolysosome that degrade various parts of the microbe. A variety of extremely basic proteins present in lysosomal granules strongly inhibit bacteria, yeast and even some viruses. In fact, a few molecules of any one of these cationic proteins can damage the membranes of a bacterial cell, causing death by an unknown mechanism. The phagolysosome of neutrophils also contains lactoferrin, an extremely powerful iron-chelating agent that sequesters most of the iron present, potentially inhibiting bacterial growth. Monocytes and macrophages also secrete a number of the following substances that may play a role in killing pathogens.

1. Oxygen metabolites such as O_2-, H_2O_2, NO and OH. are also secreted into the local environment when the cells bind an Fc receptor on an antibody or when activated by a-interferon.

2. Prostaglandins are vasodilators and may increase vascular permeability that accompanies infection and inflammation.

3. Monocytes are responsible for the synthesis and secretion of all the complement components, which then circulate in the blood and lymph as potent antibacterial factors.

4. Stimulation of macrophages by endotoxin causes the synthesis and secretion of interleukin-1 (IL-1). IL-1 along with other cytokines, such as IL-6, stimulates T-lymphocytes to proliferate, thus assisting in the activation of the rest of the adaptive immune system. IL-1 also serves as a chemoattractant for neutrophils and keeps them at the site of inflammation by enhancing the adhesiveness of endothelial cells for neutrophils.

Once a microbe is killed, the phagolysosome employs a large collection of digestive enzymes (e.g., lysozyme, proteases, lipases, nucleases, and glycosylases) that break bacterial macromolecules into low molecular weight components. In neutrophils, spent phagolysosomes accumulate in the cytoplasm and the phagocyte eventually lyses and dies. Dead neutrophils make up most of the material in pus.

REFERENCES

Biron, CA: 1994. "Cytokines in the generation of immune responses to, and resolution of, virus infection". *Currrent Opinion in Immunology* 6:530, 1994

Herberman RB, Ortaldo JR. 1981. "Natural killer cells: their role in defenses against disease". *Science* 214:24.

Mogensen LA. 1979. "Role of macrophages in natural resistance to virus infections". *Microbiol Rev* 43:1.

Ogra PL, Leibovitz EE, Zhao RG. 1989. "Oral immunization and secretory immunity to viruses". *Curr Top Microbiol Immunol* 146:73.

Sissons JGP, Oldstone MBA. 1980. "Antibody-mediated destruction of virus-infected cells", *Adv Immunol* 29:209.

Bibliography

Alcamo IE. 2001. *Fundamentals of microbiology.* Boston: Jones and Bartlett.

Almela MJ, González ME, Carrasco L (May 1991). "Inhibitors of poliovirus uncoating efficiently block the early membrane permeabilization induced by virus particles". *J. Virol.* 65 (5): 2572–7.

Atlas RM. 1995. *Principles of microbiology.* St. Louis: Mosby.

Bai J, Rossi J, Akkina R (March 2001). "Multivalent anti-CCR ribozymes for stem cell-based HIV type 1 gene therapy". *AIDS Res. Hum. Retroviruses* 17 (5): 385–99.

Belland R, Ouellette S, Gieffers J, Byrne G (2004). "Chlamydia pneumoniae and atherosclerosis". *Cell Microbiol* 6 (2): 117 – 27. doi:10.1046/j.1462-5822.2003.00352.x.

Beringer, Paul; Troy, David A.; Remington, Joseph P. 2006. *Remington, the science and practice of pharmacy.* Hagerstwon, MD: Lippincott Williams & Wilkins. pp. 1419.

Biron, CA: 1994. "Cytokines in the generation of immune responses to, and resolution of, virus infection". *Current Opinion in Immunology* 6:530, 1994

Blum P (ed). 2008. Archaea: New Models for Prokaryotic Biology. Caister Academic Press.

Bordenave G. 2003. "Louis Pasteur (1822-1895)". *Microbes Infect.* 5 (6): 553–60. PMID 12758285.

Burkovski A (ed). 2008. *Corynebacteria: Genomics and Molecular Biology.* Caister Academic Press.

Burkovski A (editor). 2008. *Corynebacteria: Genomics and Molecular Biology.* Caister Academic Press.

Collier, Leslie; Balows, Albert; Sussman, Max. 1998. *Topley and Wilson's Microbiology and Microbial Infections* (ninth edition), Volume 1, *Virology*, volume editors: Mahy, Brian and Collier, Leslie. Arnold.

Diaz E (ed). 2008. *Microbial Biodegradation: Genomics and Molecular Biology.* Caister Academic Press. ISBN 978-1-904455-17-2.

Dimmock, N.J; Easton, Andrew J; Leppard, Keith. 2007. *Introduction to Modern Virology* (sixth edition), Blackwell Publishing.

Drews G. 1999. "Ferdinand Cohn, a Founder of Modern Microbiology". ASM News 65 (8).

Fish D. "Optimal antimicrobial therapy for sepsis". *Am J Health Syst Pharm* 59 Suppl 1: S13 – 9. PMID 11885408.

Funke BR, Tortora GJ, Case CL. 2004. *Microbiology: an introduction* (8th ed ed.). San Francisco: Benjamin Cummings.

Garcia-Pichel F 2008. "Molecular Ecology and Environmental Genomics of Cyanobacteria". *The Cyanobacteria: Molecular Biology, Genomics and Evolution.* Caister Academic Press.

Heuner K, Swanson M (ed). 2008. *Legionella: Molecular Microbiology*. Caister Academic Press

Holt JC, Bergey DH. 1994. *Bergey's manual of determinative bacteriology* (9th ed. ed.). Baltimore: Williams & Wilkins.

Hugenholtz P, Goebel BM, Pace NR. 1998. "Impact of culture-independent studies on the emerging phylogenetic view of bacterial diversity". *J Bacteriol* 180 (18): 4765–74.

Kitani, Osumu and Carl W. Hall. 1989. *Biomass Handbook*. Taylor & Francis US. p. 256.

Knipe, David M; Howley, Peter M; Griffin, Diane E; Lamb, Robert A; Martin, Malcolm A; Roizman, Bernard; Straus Stephen E. 2007 *Fields Virology*. Lippincott Williams & Wilkins.

Langford, Roland E. 2004. *Introduction to Weapons of Mass Destruction: Radiological, Chemical, and Biological*. Wiley-IEEE. p. 140

Lerner, Brenda Wilmoth & K. Lee Lerner (eds) 2006. *Medicine, health, and bioethics : essential primary sources* (1st ed.). Thomson Gale.

Ljungh A, Wadstrom T (ed). 2009. *Lactobacillus Molecular Biology: From Genomics to Probiotics*. Caister Academic Press.

Madigan M, Martinko J (ed.) 2006. Brock Biology of Microorganisms (11th ed.). Prentice Hall.

Martinko JM, Madigan MT.2005. *Brock Biology of Microorganisms* (11th ed. ed.). Englewood Cliffs, N.J: Prentice Hall.

Mogensen LA. 1979. "Role of macrophages in natural resistance to virus infections". *Microbiol Rev* 43:1.

Moses, Vivian et al. 1999. *Biotechnology: The Science and the Business*. CRC Press. p. 563

Ogra PL, Leibovitz EE, Zhao RG. 1989. "Oral immunization and secretory immunity to viruses". *Curr Top Microbiol Immunol* 146:73.

Rehm BHA (ed). 2008. *Microbial Production of Biopolymers and Polymer Precursors: Applications and Perspectives*. Caister Academic Press.

Ryan KJ, Ray CG (eds). 2004. *Sherris Medical Microbiology* (4th ed.). McGraw Hill.

Saiman, L. "Microbiology of early CF lung disease". *Paediatr Respir Rev*. Volume = 5 Suppl A: S367–369. PMID 14980298

Schmidt A, Ragazzi E, Coppellotti O, Roghi G. 2006. "A microworld in Triassic amber". *Nature* 444 (7121): 835.

Shively, Jessup M. 2006. *Complex Intracellular Structures in Prokaryotes* (Microbiology Monographs). Berlin: Springer.

Shors, Teri . 2008. *Understanding Viruses*. Jones and Bartlett Publishers.

Sissons JGP, Oldstone MBA. 1980. "Antibody-mediated destruction of virus-infected cells", *Adv Immunol* 29:209.

Sodeik B, Griffiths G, Ericsson M, Moss B, Dom. 1994. "Assembly of vaccinia virus: effects of rifampin on the intracellular distribution of viral protein p65". *J. Virol.* 68 (2): 1103–14.

Soni, S.K. 2007. *Microbes: A Source of Energy for 21st Century*. New India Publishing.

Verstraete W. 2007. "Microbial ecology and environmental biotechnology". *ISME J.* **1** (1): 4–8.

Witzany, Guenther. 2008. *Bio-Communication of Bacteria and its Evolutionary Interrelations to Natural Genome Editing Competences of Viruses.*. Nature Precedings. hdl:10101/npre.2008.1738.1.

Index